OURANOS THEOREMA

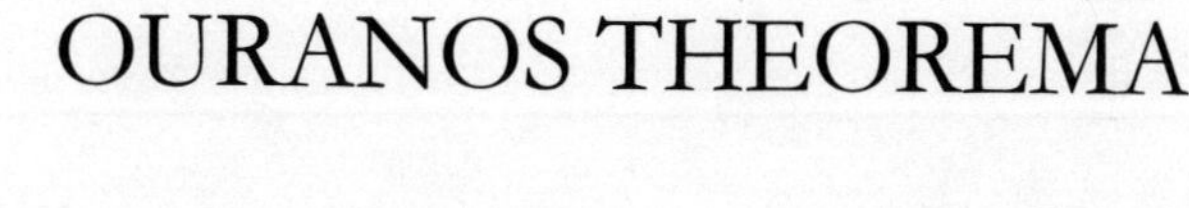

OURANOS THEOREMA

BY

Alberto Buffo

BOVOLO PRESS

Cambridge, Massachusetts

2000

Published by Bovolo Press

Cambridge, Massachusetts

www.bovolo.com

1st Edition

Text laid out with TₑX and typeset in Monotype Garamond

Manufacture in the care of Sheridan Books

Library of Congress Card Number 99-75570

ISBN 0-9675538-0-6

Cataloging under   QB41   or   523.1

Printed in the United States of America

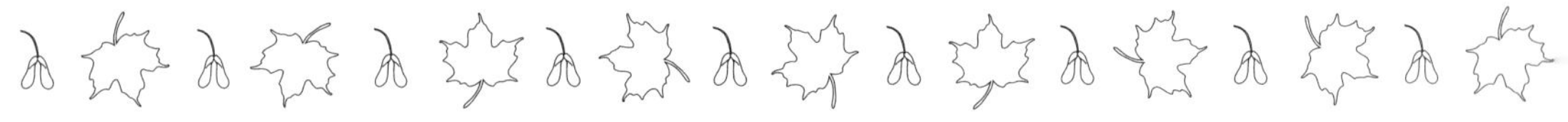

In the beginning Gaea, Mother Earth, was awakened by the birth of her son Uranus, who covered her entirely. Later he would proceed to seed the world. But Ouranos (following the original Greek spelling $O\upsilon\rho\alpha\nu\acute{o}\varsigma$) is also said to have been Gaea's husband and their progeny included the Cyclopes and the Titans. Kronos, the youngest of the latter, is presumed to have been the one who mutilated his father. His own children from his sister Rhea included Hera and Poseidon and later Zeus himself. It is probable that Ouranos was not a Greek invention, but is instead an older Arian god dressed in Greek garb. In any event, the fact that he covered the Earth entirely remains and in Greek literature his name was later used to denote the sky, the firmament, or what we would call today the Universe. Theorema, the Latin word that English adopted as theorem, is an adaptation of the Greek word $\Theta\epsilon\acute{\omega}\rho\eta\mu\alpha$. In its original form it meant a sight, a spectacle, an object of contemplation, an object of speculation. Subsequently the Greek themselves used the word to denote a mathematical statement that can be proven from first principles.

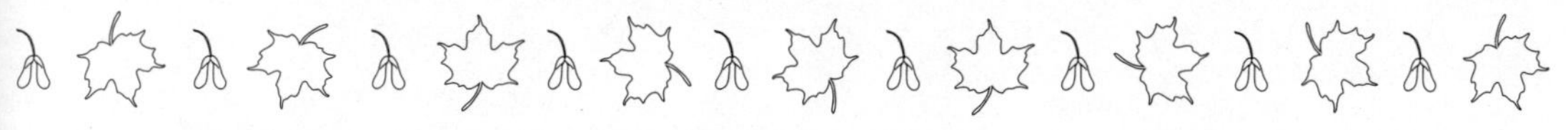

OURANOS THEOREMA

A Dialogue

on the subject of how the distances to the farthest reaches of the Universe have been measured and on the many attempts since Antiquity to understand the architecture of the Cosmos, with a digression or two on a few related matters.

Plebeius : Good day, Albertus! I hope I have not kept you waiting for too long. I did not realize it was getting late. As I was approaching the park I decided to take a detour and enter it through the southern gate where there is a grove of magnolias. I thought that they would already be in bloom at this time of year. Indeed they are and it is a magnificent sight.

Albertus : Greetings, Plebeius! I had barely arrived myself and did not have to wait for you. I am very glad you suggested that we should meet in this park. We have been very fortunate in choosing such a splendid day. You will have to guide me to this grove of magnolias sometime; I have walked through the park many times but I have never noticed it.

Plebeius: It is not really hidden. You will find it midway along the path that leads to the pond. During this past week I have been awaiting our encounter today with some anticipation and I am eager to hear all about the investigations you have been pursuing. It was very kind of you to send me a copy of your book fresh out of the press; I began browsing through it immediately, reading a few fragments here and there, and I already have a few questions I shall be asking you.

Albertus: Whenever a new book comes to my hands I also begin by browsing through it, reading bits and pieces, before I decide to read it from the beginning, which seems to me quite reasonable and fair; yet, from the point of view of the author, one must suppress a slight annoyance at having to concede the same liberty to the reader, as if the reader did not allow the author to make his case cogently by following his line of thought in an orderly fashion.

4

Plebeius: During my readings, as I was preparing for our meeting today and think-ing of the grandiose questions that you address in your book concerning the nature and fate of our Universe, I was reminded of the Timaeus, Plato's celebrated dia-logue. I believe it is Socrates and Critias who are assembled for the discussion when Timaeus, after an invocation, launches on a lengthy exposition recounting the creation of the world, explaining how it came into being through a mixture of intelligence and necessity, how the Creator made mind persuade necessity to bring things to perfection and why there can only be one world, for the Creator, in all his goodness, free of jealousy, made it in the image of himself and, like a living creature, gave to its body the most perfect shape, which is spherical, without extremities, with the motion of a circle upon itself. At its center he placed the soul, says Timaeus, which is only a manner of speech, for the soul is not part of the body; although it is one, it is diffused through every part of the body, self-sufficient, able to converse with itself, without the need of other friendship or acquaintance. And then, before describing how our individual souls came into be-ing and all the sensations and passions that give them life, or how a star is assigned to each individual soul, Timaeus states that the Creator had wished to make the world eternal, since the ideal being was everlasting, but this was impossible for a created creature, so he fashioned a moving image of eternity and thus, setting the Heavens in motion according to number, created the image of time, while only eternity, outside of time, could rest in unity. Until then, days and nights, or months and years, had not existed, so that the Moon and planets were created precisely for this purpose, as well as the Sun, a fire lit at the center of the world to make evident the passage of the hours. As I indulged in reminiscing about these things, I fancied that, in our meeting today in this park, we might be doing just as Plato had done with his friends in the Academy, gathering to discuss some of the deeper riddles concerning the origin of this world.

Albertus: You remind me very vividly of that lengthy account given by Timaeus, but I must confess that it has never aroused my interest or sympathy in any degree. It is too whimsical, too fanciful and at times too solemn and almost portentous in its tone. It is very difficult to tell what parts were meant in earnest and what in it was giving free rein to the imagination. I would very much prefer if we keep our feet to the ground in our discussion; we may not arrive at explaining the meaning of the Creation but, taking only a few cautious and sure steps, we shall not be led astray so much and, in the end, we may succeed in being more persuasive.

Plebeius: It is difficult indeed to discern what Plato presents as an account of true events and what is merely fiction. Yet, I would not have expected you to be so plainly dismissive of what he wrote in his dialogue. After all, to make this lengthy exposition he summons Timaeus, who is presumed to be familiar with the work of the mathematicians of his day and must represent their viewpoint to some extent. Moreover, being yourself a mathematician and considering all the praise that Plato lavished on your science, are you not inclined to be more agreeable and sympathetic toward his points of view?

Albertus: Perhaps I am being somewhat unfair, especially considering that the Timaeus is not fully representative of Plato's thinking. It is only disappointing that the one time that he addressed a subject directly related to the sciences he did it in the manner he did. My friends tell me that I should apply myself to improve the smattering of Greek I learnt, so that I will be able to appreciate the wit and grace of his prose, but my curiosity has not yet moved me into action. I am more suspicious of his enthusiasm for mathematics. At the Academy he certainly welcomed and encouraged the work of a few excellent mathematicians, but it does not appear that he was involved personally in any kind of mathematical activity. He was inclined to extol the virtues of our science but, as far as I can tell, it was more the pleasure of flirting with its vocabulary than with the thing itself; by no means was it the praise of an artisan familiar with his craft. It is unfortunate that for so many centuries the Timaeus remained the only known work of Plato in the Latin West and that, having acquired such a great reputation, it served as a model for many inquiries of this kind in the future, perhaps even to our own days, for now we are confronted with a tale about the origin and evolution of our Universe which is no less intrepid and fanciful. Nonetheless, it is said to carry the full weight and authority of the sciences and thus, under this mantle, it has gained great respectability and almost unanimous approval. It is not presented as merely suggestive or as an approximate representation of the events that have shaped the Cosmos but, instead, it presumes to be an objective account of all things as they have happened, the chronicle of an eye-witness, as it were, the bare facts without any tendentious interpretations. In the beginning there was nothing. This is not new. But we are told that this nothing was unstable, it underwent fluctuations, not quite real yet, virtual fluctuations they are called. Hence, it had to fluctuate between nothing and something else that was not nothing; therefore it had the quality of something. I do not wish to deride this theory of virtual fluctuations because our resourceful physicists today have made perfect sense of them, if only in quite a different context, but here it simply begs the question. In any event, out of these fluctuations a big explosion occurred. It was not a virtual explosion but a real one and out of it came the world. How and why this took place is still a murky mystery and the details and ornaments that are constantly added to this theory to make it more plausible only add to the mystery. But the explosion remains and, as it blew apart, space and time and matter came into being. Space itself blew apart, expanding out of itself although always within itself. Time, unlike space, was passive and merely held its pace in check. Matter, out of a myriad particles, bouncing against each other by the force of the explosion, began to coalesce into atoms and molecules and yielding to the laws of physics, which for much of the time amounted to not much more than the force of gravitation, they shaped themselves into everything that there is, everything indeed, the stars and the cooler planets, the birds and the sunrises, the sense of smell and the desire to dream. This could be a pleasant myth, but it has become instead a rather insidious one, in that it claims it is nothing of the sort. It claims to be an absolute truth and those who advocate it with zeal and conviction present it as a final and complete theory of the Universe. Even if it was a faithful account and we had finally singled out the correct ingredients of a true theory, they refuse to acknowledge that we may have only done like a child who dismembers a toy and examines accurately each part, without having the least understanding of what it means as a whole. It is not just the fact that science ought to be seen as more fragile and ambivalent, subject to reinterpretations and self-corrections; beyond this, what is most obfuscating is the inability to see that a theory, any theory, is always partial, it offers only a perspective, it is a way of ordering our experience and it is, by necessity, limited in its view. A theory that claims to have arrived at final and conclusive answers or, even more, that claims to have arrived at the final questions, merely shows its lack of imagination to entertain new ones.

Plebeius: You are quite vehement in your opposition. Your colleagues of today may claim that they are, at last, in possession of a complete and objective account of our Universe, but they certainly would not be the first to lay such a claim. As I listen to your reaction against their arguments, I imagine that you might be doing battle against the Manichaeans. They are mostly forgotten nowadays, but for many centuries they were a force to contend with; they too were persuaded of having arrived at a thorough and conclusive explanation of the world and its meaning. The Church in Rome thought of them as the most dangerous of heretics and went after them relentlessly. Their account of how the Universe came into being is one of the more elaborate and fanciful stories ever concocted. It all began, so they said, when the Prince of Darkness, living in the South, wandered into the North and discovered the abode of the Father of Greatness. Until that time they had been separated, but now the Prince of Darkness wished to conquer the North. The Father of Greatness, to protect his dwellings, creates the Mother of the Living, who gives birth to Primordial Man along with his five children. These are all devoured by the demons under the command of the dark Prince, causing that a portion of light remains now trapped by darkness. The good God then sends forth a so-called Living Spirit, who rescues the Primordial Man, returning him to the Heavens. As the Living Spirit vanquishes the demons, he fashions with a portion of the liberated light the Sun and the Moon and, lastly, with the rest of the demons' bodies, he creates the Earth. But, as it turns out, not all the demons have been dealt with and the Father sends forth a third Messanger who, attempting to liberate the light that remains trapped, evokes twelve Virgins of Light. For the demons, these have the appearance of beauty, so that they are tricked into spreading their semen through the Earth giving birth to all the vegetation. The story continues with further implausible twists and contortions until Adam and Eve are created from the union of two demons. As a result, the remaining light still imprisoned in the material world also resides in them, where it must remain, through their descendants, locked in perpetuity. But just as it had been the case with the Primordial Man, a messiah is summoned to descend to the world, to awaken the spirit of Adam and Eve, remind them of their true origin and lead them to salvation. It is in this manner that the Manichaeans construct the entire history of the world. In all its details it is a vivid, colourful and intricate story, to some quite grotesque with its crude cannibalism amongst demons or its brutal sexual encounters but, as it is in all mythic accounts, these are signs or symbols pointing to a deeper meaning that is never made explicit: the confinement of the seeds of light to a unending cycle of deaths and rebirths in the material world, the purification and regeneration by, in turn, devouring and being devoured. It was in part the Zoroastrian religion, dominant in Persia during the Seleucid era, at the time their doctrine was shaped, that gave the Manichaean view its dualistic color. The two principles of Light and Darkness do battle for control of the Heavens. The history of the world is divided then into three stages, an anterior time when the two principles coexisted separate from one another, the present time of intermingling and strife, which will culminate in the last stage, when Light will triumph and Darkness will be thrown in its own bottomless pit for the rest of time. Now, you may think that all this is a good deal more fantastic and arbitrary than the tale that your scientific colleagues have devised by following the laws of physics. The only similarity might be perhaps that the origin of the world is caused by a chance event that remains unexplained, in this case, the accidental wandering of the Prince of Darkness into the realm of the North, which arouses his desire of conquest. However, from the Manichaean point of view, their account was meant to be highly rationalistic, not merely a moralistic tale, but a truthful chronicle of the cosmic drama, in which everything that comes to pass is explained from cause to effect. In this regard, you may liken their attempt with the one elaborated by the ambitious scientists of today, since they too were striving for a universal science that could claim to be final and would terminate all further inquiry. They were influenced not only by Zoroastrian beliefs but also by Jewish and Christian ideas and by Buddhism as well, with which their prophet and founder Mani came into contact during his pilgrimage to India. Their intent was to incorporate and supersede all previous religions. But what brings them closest to what I would call a modern scientific attitude is the fact that theirs was based not so much on ritual and devotion, or even on faith, but on the conviction that knowledge is the only worthy objective, that the soul must come to understand its true condition, led astray in the world of matter and, reawakened to its divine origin, must find its path to deliverance. For this reason, perhaps, it is said that Mani and his followers put down their doctrines in writing very laboriously and with great attention to detail. This also explains in part their asceticism, their emphasis on teaching and their missionary zeal. Of course, their underlying view of life was utterly somber and gloomy, for they saw the material world as deceitful, as darkness and evil, calling it often an abyss of waste. Our present existence had its origin in nothing but a supreme, divine humiliation and there was little to rejoice about. It is not surprising that their self-righteous attitude antagonized the followers of other religions and that, for a long time, they were vilified as dangerous heretics. Mani himself had successfully gained the support of the king of Persia and was allowed to proselytize but fell out of favour in the end and died in the gallows. It is said that his head was severed and exposed in public, while his body was thrown to the dogs. In this regard we must have advanced somewhat, for I do not think that we would treat the truth-seekers of today in that manner, however adamant they might be about the validity of their theories, or however much they may have strayed into error. The Manichaeans and their doctrines are ignored today, but now that their force is spent, it has been found that many of their hymns and prayers, unearthed by archaeologists in recent times, are quite beautiful and of some literary value.

Albertus: I will not venture to say how apt or pertinent is the comparison you have just established, but your evocation of this fantastic doctrine of the Manicheaens is a good reminder to those of us who may think that at last we are closing in on the truth. I did not intend to say that those who peddle the prevailing scientific theory of the day do so with the eagerness of missionaries wishing to win converts. Firstly, instead of presenting their view in as clear a language as possible, they have a tendency to hide behind the complexity of their own technical language, arguing always that this complexity is due more to the difficulty of the subject matter than to the uncertainty or confusion in their own minds. Moreover, they often seem not to have any particular attachment to the specific ideas they advocate, for they are very ready to change them as soon as they do not serve their goals. They lack all memory of the evolution of their science; they easily make yesterday's convictions today's ridicule, so that they cannot be accused really of adhering to a strict dogma. This would be a healthy attitude if it were not for the fact that it shows them more interested in winning the argument than in the things the argument is supposed to explain.

Plebeius: I admit that sometimes I have been disenchanted myself for I have seen some new scientific theory presented with confidence and excitement and, carried over by the sense of discovery, every so often I have taken the trouble to try to understand their convoluted reasoning and the intricacies of their theories, to which I have little access. But soon thereafter, as the rest of us begin to take them seriously, they decide that the theory was not quite right and move on lightheartedly to explore new ideas. It is all the more bewildering since those of us who are outsiders, only mildly interested, seem to be more disappointed at their failure than they were, despite all the energy and effort that they had invested in their work. But let us now turn to the ideas that you discuss in your book. There is, for example, this notion of the expansion of space that after all these many years seems to have been accepted almost unanimously and without hesitation. Without attempting to understand the process of creation in the original explosion, I have often wondered what is the concrete meaning of this expansion, which is said to continue to this day. We learn that it is not quite appropriate to say that things are moving apart; instead, it is the distance scale which changes as the Universe gets older. We see the distant galaxies receding from us, but neither they nor we are moving; it is the intervening space that grows out of itself. Space expands everywhere and all distances increase in relation to what they were in the past. If this expansion is all pervasive, why is it no possible to seek some measurable effect nearby, that we have to rely on the light arriving from distant galaxies to verify its existence? Furthermore, if all distances keep increasing, in relation to what standard are they increasing? Doesn't the spacing between the markings of my measuring rod also expand and in equal proportion? What tool do I use to compare the scales of today with those of yesterday?

Albertus: Those questions are most pertinent and I do not think that you will be able to find satisfactory answers. You will certainly be told that the apparent velocity with which two objects are seen to move apart is proportional to their distance and is therefore readily noticeable between two distant galaxies. You will also be told that the expansion of your measuring rod in the laboratory is so insignificant that no experiment would be able to detect it, but that this difficulty does not make it less real. As to the existence of an absolute standard of length, fixed throughout the ages of the Universe, the theory is silent. You might say that the way it formulates the laws of physics does not address the matter and that, consequently, this theory is only half baked.

Plebeius: But your point of view is, in fact, that all my questions are meaningless, that the expansion of space is a concoction of the human imagination and that this notion of the birth and evolution of the Universe is as fictitious as the Prince of Darkness. Are you arguing that this whole theory should be abandoned, discarded, thrown somehow onto a pile of refuse?

Albertus: Not entirely, Plebeius. Besides, old theories have a hard time dying. I would say that, in due course, and sooner rather than later, you will be able to find that theory offered by some antiquarian at bargain prices. There will always be an alert and ingenious buyer, who will refashion it and make bits and pieces of it come alive again. But, as a whole, the theory is beyond rescue. However, that is not the main thrust of my argument. Let me first state what is the subject of my inquiry. When I speak of the history of the Universe, I have in mind the manner in which we order all the events that take place under the Sun and beyond, across the immensity of space and from the distant past to the distant future. We seldom give much thought to how this ordering takes place. After all, we are confident, as Plato said, that the Moon and the Sun will continue to circle through the skies with unfailing regularity, giving us days and nights and the sequence of months and years. Likewise, in countless stars throughout the Galaxy, similar clocks will be ticking away in synchrony. And all across space we see millions of other galaxies, signposts and beacons of the Universe, go about their business very much in the same manner as we do. Thus, altogether apace, throughout the entire Cosmos, we march in unison, toward our ultimate fate, bearing witness to this majestic procession that we call the history of the Universe. However, we have now come to learn, through the power of mathematics, that not only the idea of an expanding Universe is a delusion but the very notion of the history of the Universe is also a product of our fancy. Despite its being rooted in common sense, it does not correspond to anything real and I am afraid we will have to abandon it if we wish to understand the true nature of things. We have now known for a long time that light, space and time play tricks on our senses and behave in ways that are contrary to our intuition. However, we never felt the need to relinquish this very elementary notion of history, the assumption that there is a hidden clock in the Universe, which assigns to each event the epoch of its occurance. Now we will have to do so, the Universe we inhabit precludes such a possibility. This is not a complete theory, a full account of where things have come from and where they are headed. It is at most a viewpoint, some sort of inventory that allows us to put every thing in its proper place in relation to others in space and time and that enables us to understand our own place in the Cosmos. It is a viewpoint dressed in geometric language and, as such, it is surprisingly simple and succinct, but I think that in its simplicity one will find its strength as well.

Plebeius: It is this economy of expression, the succinctness of mathematics, that some of us, who are not specialists, fear most. Out of curiosity and taking advantage of our friendship, I am interested in learning at least the rudiments of these investigations of yours, but I must confess that I share this reticence in the face of the difficulty of mathematics, of its elusiveness. Even when I can follow its logical steps, one at a time, the meaning to be inferred from its conclusions remains often inaccessible. In truth, I have given very little thought to mathematics since my school days. I suspect I could recite Pythagoras' theorem and a few of the rudiments in Euclid, but that would be the extent of my recollections.

Albertus: The reputation of mathematics as being difficult and too abstract, too remote, is not easy to explain. Is it not one of the first things we teach children in school and a subject in which they easily excel? It is only when it comes to the teaching of history or ethics, subjects in which the weighing of complex arguments is required and where the experience of life is most helpful to form a judgment, that we notice the immaturity of children. Mathematics, on the contrary, is of such simplicity that they understand it fully and without limitations. In any event, if you can recall Pythagoras' theorem, that will be a good starting point for us.

Plebeius: I certainly could recite that the square of the hypothenuse in a right triangle is equal to the sum of the squares of the legs, but if you asked me to put this statement to use or to draw some practical conclusion from it, I would find myself in some difficulty. Neither could I prove it from first principles. For some reason, I vaguely associate this theorem with the figure of two squares, one placed askew inside the other, with its vertices touching the sides of the outer one.

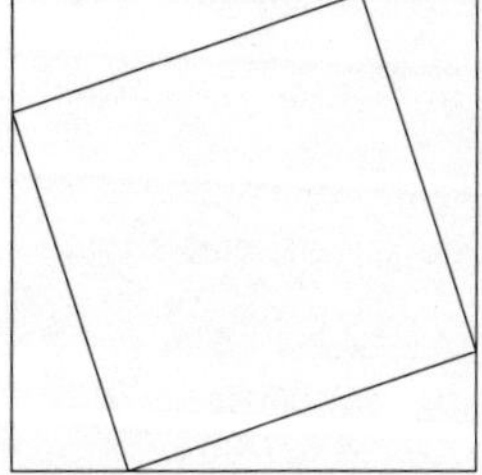

Even if I could find my way through a proof of this theorem, I could not explain why it has to be true, that is, I could not think of anything that should inspire me to suspect that such a relation amongst the sides of a triangle should hold.

Albertus: It would take you only a brief moment to realize that the figure of the two squares allows you to prove the theorem quite easily. If we label c the length of the side of the smaller square, then a and b the two segments composing the side of the larger square,

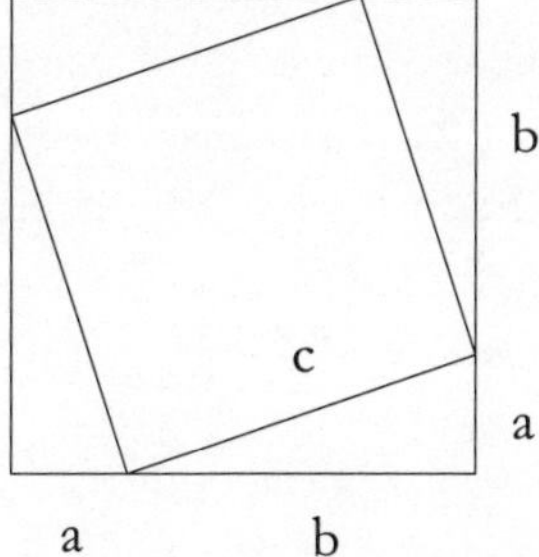

12

it is clear that the area of the larger square is equal to $(a+b)^2$, which can also be written as the sum of the area of the small square c^2 plus the areas of four equal triangles $4.(a.b/2)$, that is,

$$(a+b)(a+b) = c^2 + 4.(a.b/2).$$

Once you simplify these expressions, you notice that the term $2.a.b$ appears on both sides of the equality, which must remain an equality after we subtract them from both sides, to obtain

$$a^2 + b^2 = c^2.$$

Why should we have expected to have a premonition of the validity of this theorem? It is satisfactory enough that one might have stumbled upon it, especially considering that there are many paths that could lead to it. For example, it is clear that a right triangle is always contained within a square, whose side is the length c of the hypothenuse, so that its area, smaller than c^2, must be a fraction thereof.

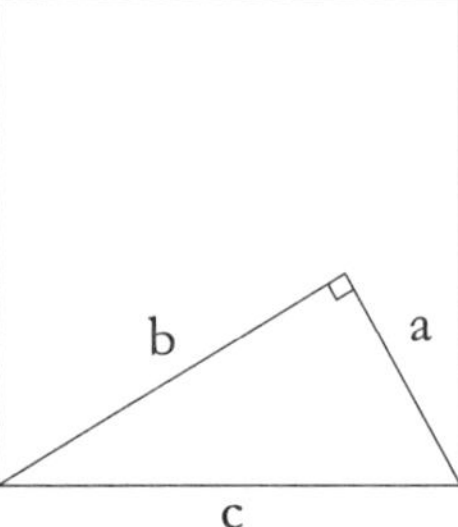

It is also true that the length of the hypothenuse and one angle completely determine the right triangle, given that the missing leg must meet the other one at a right angle.

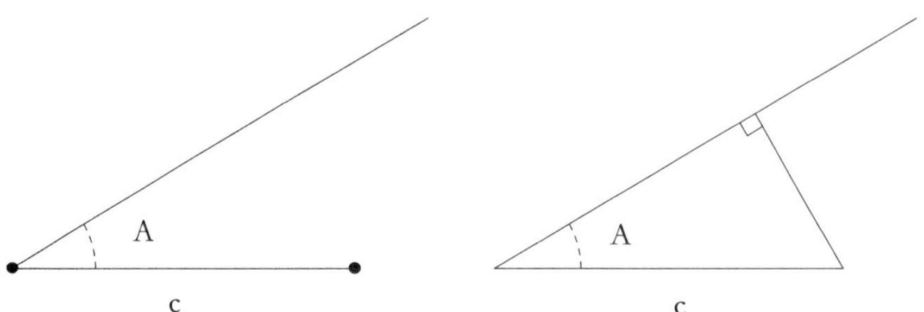

Therefore, the area of the triangle is a fraction of c^2 and the fraction ought to be completely determined by the choice of one angle A. To put it in a concise way, I shall write the area as

$$f_A c^2,$$

where f_A, a number, merely denotes a fraction, smaller than 1, that is known once I have chosen the angle A. Now, drawing the height from the hypothenuse to the opposite vertex we decompose our original triangle into two right triangles which are similar to the original one, each containing the angle A as well,

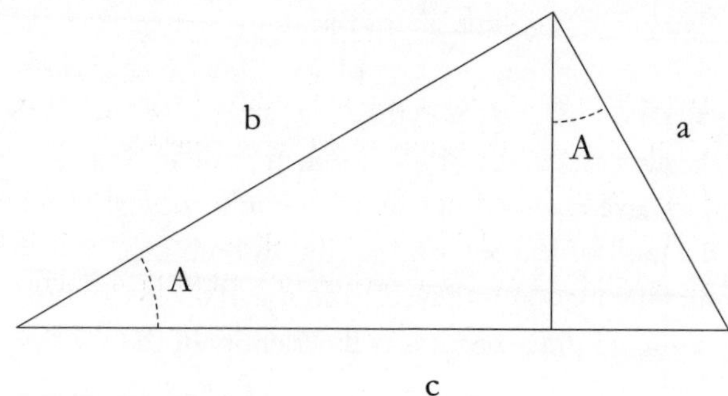

so that their areas, using exactly the same principle, are given by $f_A a^2$ and $f_A b^2$, since a and b are their respective hypothenuses. Finally, combining the two we obtain

$$f_A a^2 + f_A b^2 = f_A c^2,$$

from which we can strike the common factor appearing in each term to write

$$a^2 + b^2 = c^2.$$

Plebeius: This latter proof has the air of being more deliberate and rational; nevertheless, it retains an element of magic. You invoke this fractional number that is never made explicit, only to make it disappear later without leaving a trace.

Albertus: Every mathematical proof ought to have that element of magic. Perhaps for that reason it is said that Pythagoras had an ox gored after he had proved his theorem. In fact, according to some reports he had many oxen gored to celebrate the occasion, but this seems to be only one of the many apocryphal stories ascribed to him many decades after his lifetime. The truth is that we do not know with certainty whether he was the first to prove his famous theorem. Indeed, we know very little about Pythagoras' life. Some say that he traveled extensively for many years, with long sojourns in Egypt and Babylon. He was a native of the prosperous island of Samos and it is possible that he may have remained there well into his adult life until he was driven into exile by the tyrant Polychrates. Eventually he

14

settled in Crotona which, along with Sibaris and Tarentum, was one of the more developed Greek settlements in southern Italy. Pythagoras is credited with having discovered incommensurate quantities, what we call irrational numbers, and he must have been the first to experiment with the vibrations of strings and of air in pipes and to associate the ratios of their frequencies with the harmonic intervals of music. Any of these accomplishments would suffice to put him amongst the most prominent early mathematicians. Pythagoras may also have been the first to speak of the sphericity of the Earth, although we do not known what might have led him to this opinion. We know very little about the study of astronomy before his time. Both Thales and Anaximander, the two philosophers from nearby Miletus who belonged to a generation older than Pythagoras, are said to have been interested in astronomy. Thales was credited with the prediction of an eclipse, whereas Anaximander was said to have invented the gnomon, the vertical shaft whose shadow is used to follow the path of the Sun on the sky. In both cases it is quite likely that they were relaying information retrieved from much older Chaldean sources. Another discovery assigned to Pythagoras is the identity of the evening and morning stars. On other occasions the identification of Venus is assigned to Parmenides, who belonged to the Eleatic school, although he followed Pythagoras' opinions in many other respects; he is also said to have believed in the sphericity of the Earth. We may speculate that if Pythagoras himself was the first to hold this opinion, he may not have done so on account of any direct observations but could have arrived at this conclusion arguing from general principles. Pythagoras died at quite an old age around 500 B.C. In the century that followed, with the exception perhaps of Anaxagoras and Democritus, the scientific study of the Heavens cannot be said to have advanced a great deal. Anaxagoras seems to have been the first to state that the Moon does not shine with its own light but reflects the light of the Sun. As the Moon followed the Sun's path, he could explain its phases as well as the occurence of the lunar eclipses, caused by the interposition of the Earth on the path of the Sun's rays illuminating the Moon. Anaxagoras was also from Ionia, more precisely from Clazomene, near Smyrna, but he was one of the first philosopher to emigrate to Athens where he became acquainted with Pericles. According to some stories, he was imprisoned for his heretical ideas and only saved his skin thanks to the intercession of Pericles. One of the ideas he proposed was that the annual cycle of the Sun moving from the summer to the winter solstice and back was caused by the repulsion the northern cold exerted on the Sun. That also explained why the Moon, being much less powerful than the Sun, went through its cycle faster and much more often. Anaxagoras also expressed some original views regarding the Milky Way. Being aware that a bright light may cause fainter ones to appear dimmer than they are, he speculated that the majority of the stars are overpowered by sunlight, even at night, and that the higher density of stars seen along the Milky Way was caused by the shadow that the Earth casts on the Heavens. By this means he tried to infer the true shape of the Earth which he assumed to be flat and somewhat elongated. These conclusions appear quite strange and in conflict with his own interpretation of the phases of the Moon, as was later pointed out in a simple manner by Aristotle. Pythagoras had a great influence, not only because of his individual contributions or because of the large following he attracted, but particularly because he started a school of thought or, as we might say, he struck a light that in one form or another has been burning ever since. In its most synthetic form, this school of thought states that the Universe is ruled by number and harmony. Perhaps it is nothing more than a frenzied dream or, at best, a manner of reasoning. It is difficult to elaborate on it, for every path that we can follow seems to have its shortcomings and limitations, but the accumulated effects of all such attempts in the past has certainly made us more knowledgeable and wise. The Pythagoreans themselves may have understood the constraints imposed by their point of view, for it was said that they were taciturn and that they gained more respect by saying little than did others by saying much. Aristotle, however, was a little dismissive of them, thinking that they had been led astray by assuming that everything was ruled by numbers. For a while the Pythagoreans led a communitarian life and isolated themselves from the rest of society. They may well have fallen into the trap of believing that they were in possession of some esoteric knowledge revealed only to them. Some say that there were two classes of Pythagoreans, those who were only followers or sympathizers and the ones who had been initiated into the secret learning. Others contend that the true division was between those who believed that the Universe should be studied according to numbers and those who had lost their way into numerology. The latter thought that numbers had all sorts of meanings and had to be interpreted; the former believed that the powerful meaning of mathematics was within itself. It would be tedious and fruitless to follow in detail all the ideas put forth during and after Pythagoras' time concerning the nature of the Universe, whether the Sun was a different Sun every day, whether it was an incandescent rock or a crystal that reflected some other light and many like discussions. Even those who belonged to the Pythagorean school were inclined to engage in this type of speculation. They propagated the very unusual notion that the Earth was not at rest; it circulated on a circle around a central fire that occupied the center of the Universe. This we learn from some fragments by Philolaus, who lived approximately a century after Pythagoras. There was also an anti-Earth, which occupied a middle position between the central fire and the Earth we inhabit. Because the anti-Earth might at times also prevent the Moon from being illuminated, this seemed to explain in part why there were more lunar eclipses than solar eclipses, the latter occurring only when the Moon comes into the path of the Sun. The notion that the Earth is spherical gradually became more acceptable and widespread. Those who had traveled to Egypt or to northern latitudes into the continent pointed out that new stars were seen to rise above the horizon revealing the curvature of the Earth. Similarly, the shape of the Earth's shadow on the surface of the Moon during lunar eclipses was visibly round. As a further demonstration, Herodotus makes the remarkable assertion that, according to Phoenician seamen who had traveled along the African coast beyond the columns of Hercules, they had seen the Sun on their right while traveling west. As far as the Pythagoreans is concerned, it is ironic that, despite their belief in the sphericity of the Earth and their being willing to accept its motion through space, it never seems to have occurred to them to explain the daily rotation of the stars by allowing for the rotation of the Earth. We may speculate that the reason for this is rather fortuitous; the only object in the Heavens for which a rotation on its axis could have been observed was the Moon and the Moon did not seem to be rotating at all. But the Pythagoreans were also the originators of the theory according to which the Sun, the Moon, the planets and the stars, each rotated around the heavens in a sphere of its own, producing a sound in the process. They conjectured that the velocities of rotation, or their periods, were in appropriate ratios, so that the music produced from their movements would be harmonious. They proposed, we could well say, that the Heavens sang; of course, our ears are deaf to this music, but the possibility existed that having been listening to it continuously since birth, we have grown accustomed to it and would be unable to conceive what the world would be like without it. In retrospect, when we look at what they were trying to explain and the notions they were employing, it appears to us that it was all a rather mad idea; nevertheless, we have to acknowledge that the method, the general principle that animates it has endured and countless generations of scientists have since attempted to explain various natural phenomena on account of harmonious ratios and proportions. One idea that occurred to these Pythagorean philosophers and their contemporaries was the notion that the rotation of these spheres in the heavens, one encased into another, would affect the rotation of those inside. It was therefore important to understand the correct sequence of spheres, a problem that had also preoccupied the astronomers of ages past. We find now in surviving fragments of Greek works references to the Egyptian and Chaldean traditions. The former was inclined to place the Sun between the Moon and Mercury, whereas the Chaldeans had a preference to place Mercury and Venus below the Sun. Let me now say a word about this idea of having each sphere dragging somehow those that were inside it. From a modern point of view, we imagine the planets rotating on one and the same plane around the Sun. Although the axis of rotation of the the Earth is not exactly perpendicular to this plane, we can still speak of the northern and southern hemispheres relative to the equator or to the plane of the ecliptic, the plane along which all planets circulate. Now, if we remove ourselves above this plane, in the direction of the northern pole, we would then see all the planets rotating counterclockwise around the Sun. From that vantage point, the Earth would also be rotating counterclockwise around its axis, and the

Moon as well would rotate counterclockwise around the Earth. Imagine now seeing an observer on the Earth looking away from the Sun precisely at midnight and seeing the Moon directly at the zenith.

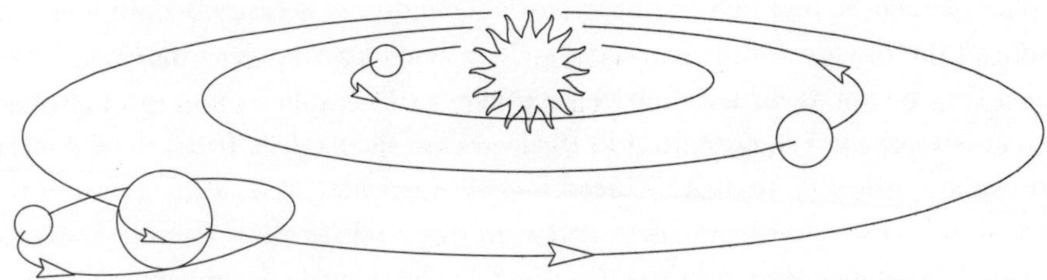

After twenty-four hours the Earth has made a full rotation; meanwhile, the Moon has not been idle but has advanced a few steps counterclockwise as well. Thus, our observer would have to wait a little longer until he catches the Moon at the zenith; in other words, from his point of view the Moon was still in the East, lagging behind the motion of the stars, or, he might also say, the Moon has a motion of its own that pushes it eastward contrary to the general daily motion of the Heavens from east to west. The same argument can be applied to the planets, but their daily delays, or their proper motions eastward against the stars are much less pronounced than that of the Moon. Therefore, coming back to the relative motion of the spheres, it was thought at one time, as pointed out by Democritus, that the sphere carrying the stars is more efficient dragging behind it the spheres that carry the planets, whereas the sphere that carries the Moon, encased farther inside, is more successful in pursuing its own motion in a contrary direction. The first satisfactory account of the motions of the planets by the method of concentric spheres was provided by Eudoxus. A native of Cnidus, he was probably one of the last great scientists to emerge from the cultured cities of Ionia, before the upheaval caused by the Persian invasions. He traveled quite extensively during his not very long life. In his twenties he was in Athens, then traveled to Egypt, established a school in Cyzicus, returned then to Athens and finally back to Cnidus, where he became a legislator until he died at age fifty-three. Even before his trip to Athens he is said to have studied mathematics with the great Archytas, the mathematician of Tarentum, in southern Italy. Both Archytas and Eudoxus brought Greek geometry to a high level of perfection. Unfortunately, none of their writings have survived and we know of their work only through the reports given by others. Nevertheless, these reports are sufficient for us to appreciate the sophistication of their techniques and the subtleties of their methods, which put them on a par with any generation of mathematicians that has come ever since. To Eudoxus we owe the theory of proportions that Euclid compiled in Book V of the Elements and also the theory of exhaustion, which enabled him to evaluate volumes of cones and other solids. Now we should concern ourselves only with his work in astronomy and more particularly with his theoretical account of the motion of the planets by means of concentric spheres. Eudoxus is also said to have acquired some expertise in carrying out observations when he was in Egypt and it is possible that he founded an observatory in Cnidus. It is easiest to begin the account of the motions of the Heavens by describing the motion of the Sun, for we know that the Sun is carried daily by the rotation of the sphere of the stars, but we also know that the Sun does not follow always the same circle. From the northern hemisphere it is seen to move instead toward a smaller circle in the south in the winter and back toward the north and a greater circle in the summer. Eudoxus accounts for this motion by affixing a second sphere inside the sphere that carries the stars, concentric with it, but with its axis slightly tilted. The Sun is now placed on the equator of this second sphere. It is immediately apparent then that, as the great sphere completes a revolution every twenty-four hours, it carries with it the second sphere and the Sun with it, but the second sphere has a slight rotation of its own that takes a year to complete and, as it does so, the Sun, so to speak, rotates daily on different circles of the big sphere.

You may notice that the equator on which the Sun is fixed wobbles inside the larger sphere within narrow bands that depend on the inclination of the axis of the inner sphere, so that the Sun would be allowed to travel south and north only within certain bounds. To describe the motion of the Moon we encounter a difficulty that also manifests itself in the case of the planets. The Moon is carried daily like the Sun by the sphere of the stars and, in addition, its circle of rotation is also seen to vary on a cycle that lasts, not a year, but only twenty-eight days. These two motions could also be accounted for by means of two spheres. However, Eudoxus knew also that the positions in the Zodiac where the Moon reaches its most northern circle to begin then its trip south is not always the same; it moves slowly backwards along the Zodiac completing an entire revolution after 223 lunations, a long cycle that had been noticed centuries earlier and was used to predict the occurence of eclipses. It is clear then that the second sphere carrying the Moon on its equator is in fact affixed inside an intermediate sphere, whose axis is tilted again and whose period of rotation, close to eighteen years, causes then the rotation of the innermost sphere, producer of the lunar month, to go through a cycle that takes this long period to be completed. Today we would say that the Moon circulates around a disc that wobbles slightly or, if you wish, that the axis of the plane on which the Moon rotates around the Earth does not remain fixed in space, it sweeps a narrow cone which stands perpendicular to the ecliptic, the plane on which the Earth and the Moon travel around the Sun.

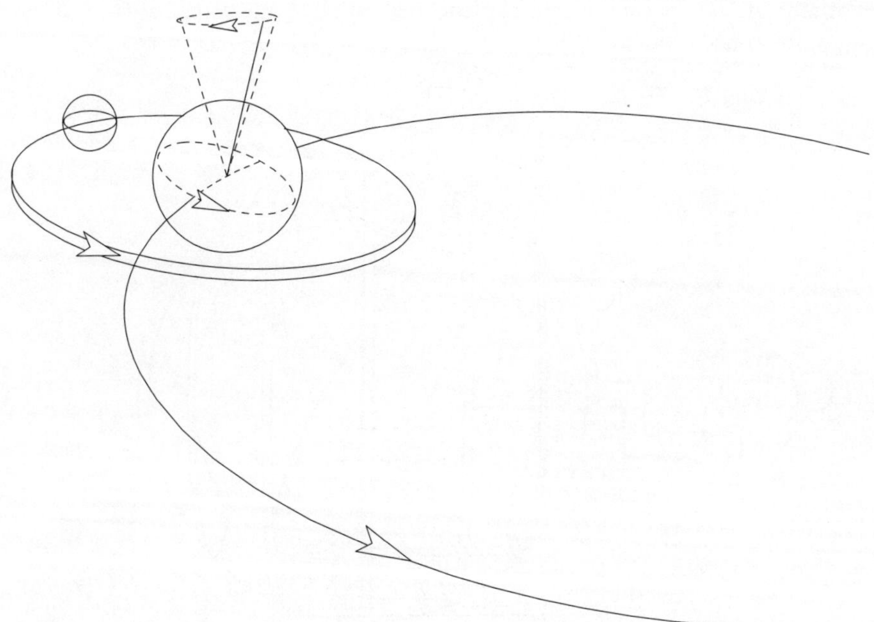

Interestingly, Eudoxus thought that this anomaly in the motion of the Moon, which is also present in the case of the planets, had to be present in the motions of the Sun as well, so that he prescribed three spheres to account for its motion too. The case of the planets is readily seen to be more complicated. Here we should distinguish the cases of Mercury and Venus on one hand, both of which are seen to accompany the Sun in its yearly trip, shifting periodically from one side to the other on the Zodiac and, on the other hand, the exterior planets, Mars, Jupiter and Saturn, all of which perform great circles of their own with increasing periods. Consider one of the exterior planets for a moment and let us adopt once again a modern point of view. The planet's rotation around the Sun proceeds more slowly than the Earth's rotation. When the Earth, moving swiftly along its orbit, catches up with the planet and we see it directly in opposition to the Sun, the planet appears briefly to be moving backward along the Zodiac whereas, at all other times, its motion will proceed forward. This leads to a peculiar pattern of advances, stations and retrogressions, as they are called, which the ancients had difficulty in explaining. Eudoxus resorted here to the following procedure. Let us consider once more two concentric spheres, one encased into the other and with its axis of rotation slightly tilted relative to the axis of the larger sphere. For simplicity, concentrate on the uppermost point fixed on the equator of the inner sphere. Imagine now that the larger sphere begins to rotate in one direction, while the inner sphere rotates in the opposite direction, so as to keep our equatorial point always at the highest position possible. It will not take you long to convince yourself that this point remains always on the central plane, oscillating along a small arc as it moves right and left while the two spheres complete each rotation.

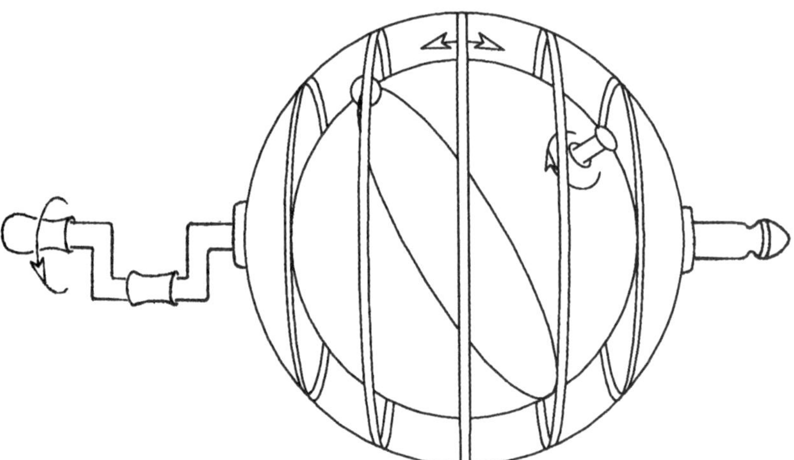

Eudoxus thus proceeded to employ five spheres to account for the motions of each superior planet, the first three are similar to those used for the Moon, but inside these three spheres he placed two more in order to produce this periodic movement of the planet back and forth along its orbit. I am simplifying matters somewhat, this is not quite exactly what Eudoxus proposed. The rotation of the two innermost spheres, one against the other, were not arranged so that they would keep the equatorial point at its highest position as I stated. He proposed instead that the velocities of rotation of the two spheres, one in one sense, the other in the opposite sense, should be uniform and equal. If this is the case, then our equatorial point will not remain on the same plane; as it oscillates right and left, it will describe the figure eight. One might think that this could be made to correspond with the departure of the planets from the plane of the Zodiac but this is only true in a very rough and imprecise way. All this Eudoxus presented in a work he titled On Speeds, now lost, and we do not know what numerical values he assigned to the relative inclinations of all the spheres and their velocities of rotation, so that we cannot say how accurate he could claim these arrangements to be in explaining the motions of the Heavens. He was certainly aware of some difficulties that could not be resolved within the context of his model. These were addressed in part by his successors, amongst them Callipus, an able astronomer from Cyzicus, on the coast of the sea of Marmara, who may have studied with Eudoxus and certainly with the mathematician Polemarchus, a disciple of Eudoxus, with whom he traveled to Athens at the time Aristotle was also in the city. It had been known for a very long time that the seasons, the period of time between solstices and equinoxes, are not all of the same length and this could not be accommodated by the uniformly rotating spheres of Eudoxus. Callipus performed the most accurate observations of his time to determine the length of each season and attempted to modify Eudoxus scheme, increasing the number of spheres he assigned to the Sun by one and by two spheres for the planets, in order to account for some of the discrepancies. Aristotle also tried his hand at elaborating on Eudoxus' theory, although his concerns were not astronomical in themselves, but more of a mechanical nature. He was interested in explaining the nature and causes of motion, so he devised a scheme in which all spheres were material, encased one into the other, in such a way that the exterior ones would drag in their rotation all that were inside, so that for each successive planet he was forced to increase the number of spheres in order to undo the motions of those that were external and that were carrying the other planets around. Clearly, in this manner he did not introduce any new principle or advantage in the description of the orbits. Another contemporary who is said to have devoted some attention to these problems was Heracleides, a native of Heraclea, in Pontus, and who was active in Athens during Aristotle's time. We do not possess any of his writings, but he is said to have written profusely and on numerous subjects, from music and poetry to history and astronomy, in very animated dialogues. His renown in astronomy derives from the fact that he appears to have been the first to suggest that the daily rotation of the Heavens could be accounted for by the rotation of the Earth around its axis. This is quite intriguing because his suggestion was not adopted by his successors until much later times. Heracleides also appears to have suggested that Mercury and Venus might in fact rotate around the Sun. One of the difficulties in Eudoxus' arrangements of concentric spheres was the fact that, in the case of Venus, it was known that the planet takes much longer to travel from its farthest eastern point to the extreme western point than it does on his trip back and this could easily have been explained by assuming that the planet rotates around the Sun.

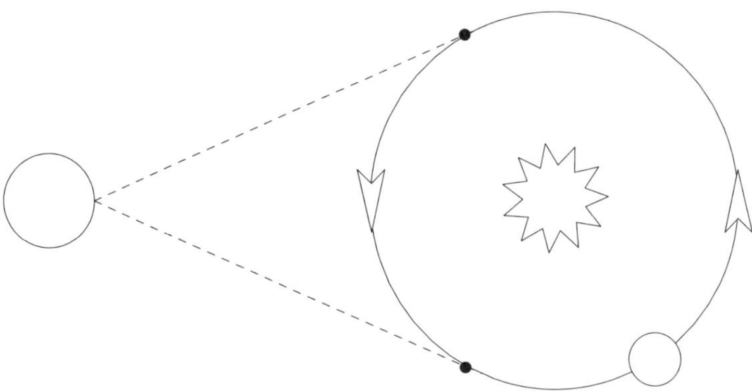

There was one more difficulty to be faced in the attempt to describe the motions of the planets, something that was well known to almost all astronomers and that Aristotle commented upon, but it seems that they all disregarded it in order to simplify the task at hand. This was the fact that the planets are seen to vary in brightness, often interpreted correctly as implying that their distances to the Earth did not remain constant. Some attempted to do away with this difficulty by observing that we never perform our observations from the center of the Earth and that it was the distance from the planet to the observer that varied but not its distance to the center of the Heavens. However, the same difficulty manifested itself during solar eclipses when the observer might be aligned with the center of the Earth. On different eclipses the Moon was seen to hide a different fraction of the Sun's surface and this could only be explained by assuming that their relative distances are not constant. Considerations of this type slowly led to the abandonment of Eudoxus' method to account for the orbits of the planets. One simple but unpleasant idea that soon suggests itself is to place the Earth off the center of their circular orbits. This had the advantage of immediately accounting for the planets' variations in brightness. It also led rather simply and directly to the next stage of complication that one had to ponder. Consider for example the case of Mars. It was well known that the planet is at its brightest when it is in opposition to the Sun or, as we would say nowadays, during the brief period when both the Earth and Mars are marching in phase on the same sector of their orbits. As a consequence, if the center of Mars' orbit is somewhere away from the center of the Earth, it must nonetheless be on the line that joins the Earth with the Sun. Now, it was also known at the time from direct observations that Mars is not always in opposition to the Sun on the same sector of the Zodiac, so that a conclusion followed inescapably: the center of the orbit of Mars was itself in motion and describing an orbit of its own around the sky. This is what came to be known as the theory of movable eccentrics and was later applied to all the planets and the Moon as well.

Plebeius: It never ceases to astonish us that in a relatively short span of time and when communications were slow and hazardous, there were so many men of great learning coming from the most distant corners where Greek colonies had been established, from Syracuse to the Hellespont and all around the Anatolian coast. If we were to search for the causes of these advances, I think there are two that seem almost evident. Firstly, these men flourished in communities that were enjoying relative prosperity and well-being and could afford a good deal of leisure. But, more importantly, they were all raised in a society that had come to place great value in education and this, in my eyes, is what gave to their achievements a distinctive brilliance, the fact that education had been made the foundational rock of their civilization, and hardly any society since has committed as many resources or given as much thought to its advancement. It is not so much the content of their education that matters but its general orientation and more than the knowledge they imparted the desire to acquire it they instilled. Above all, they never seemed to lose sight of the more distant goals of education, which remain forever so elusive, how to lead one's life with equanimity and good judgment, to enlarge the scope of one's soul, to become wise, if you will. By thinking that the human being was malleable, as if made of clay, that by his own efforts he can fashion himself into something more perfect, they certainly gave a superb expression to this aspiration, of expanding one's capacities and of conceiving new goals. It was an aspiration that was not circumscribed to the achievements of the individual within himself or, at least, the fulfillment of the individual was attained only by acquiring qualities that made him worthy of emulation. Without doubt, education was an aristocratic privilege reserved for those who wielded power. In addition, the fruits of education were not always put to dignified ends, but in this regard there is little new under the Sun. I have little patience with those who, in our times, think we should not so much sing the praises of an age past, but value the goods of our own culture and search for the roots of our identity within ourselves. They maintain this when this very attitude belonged to those ancient Greeks they prefer to ignore, for the paramount importance they assigned to education was an attempt to come to terms with their own identity and to affirm a culture, a language and traditions that were still in the making. For all the abstraction, even aridity, that we find in the thinking of not few Greek philosophers, education was a field in which they did not restrict themselves to mere theoretical speculations, but put their ideas into practice, founding schools or gathering disciples. This is certainly where Athens excelled above other cities and, whatever one may think of a practitioner such as Isocrates, with his rather confined provincialism, or the strained and exaggerated detachment of Plato, their ambitious experiments were unprecedented and still deserve our admiration. In the case of Plato, and to some extent in Isocrates's case as well, the emphasis on the relevance and power of mathematics toward a good education must have contributed to promote the study of the sciences.

Albertus: Of the generation that succeeded Eudoxus, it was most notably Aristotle who, with the foundation of the Lyceum, kept alive in Athens still for some years an interest in the study of the sciences. After his time, although the city continued to have excellent schools, it no longer made any significant scientific contributions. Very soon, other centers, such as Alexandria in Egypt, rose to greater prominence.

Plebeius: Clearly the advent of the period of Macedonian influence and dominance brought about many changes throughout Greece, not the least of which was the loss of independence for the city states and, to some extent, the termination of what until that time had been an unfinished experiment in political and social organization. The reign of Alexander was as brief as it was tumultuous. He coerced many, in part seduced them, to participate in a great adventure of expansion and conquest that took him to Egypt, then to Syria, Persia and finally to India and Bactria. Along the way, he founded as many cities as he destroyed. At least a dozen Alexandrias were established, an ambitious project of settlements, most of which did not become anything more than fragile military outposts. In many Greek cities there had been a natural decline in population which was exacerbated when so many were enticed to emigrate. Then there were those many thousands who, having moved to the distant East and yearning for a return to their Greek way of life, started on their journey home and were brutally put down and forced to stay. After Alexander's sudden death, a struggle for the spoils of the empire naturally ensued, for control of the land and for possession of the enormous booties acquired. The shifting alliances, the wealth of those who seized on the opportunity, the misery of the many left behind, all this caused tremendous upheaval. Nevertheless, as part of the order that emerged from these convulsions, it turned out now that a group of dominant Greek immigrants, over an unprecedented and vast territory, mixed sometimes with the local populations, began to give a distinctive Greek character to the cities they controlled. These were very different from the Greek cities of old; they were paying dues and levies to rulers or kings and were disengaged from their own defense. In other respects they achieved greater success, in education and in the provision for the population's health care, for example. It was during this time that the primary schools and gymnasia became widespread and began to teach a fairly standard curriculum. These institutions contributed perhaps more than any other to the dissemination and the longevity of Greek culture throughout the ancient world. As for the advancement of the sciences and for those who were pursuing higher studies, I suspect that the landscape had changed in some measure from decades past. Whereas previously they could find their home in the free associations of educated citizens devoting their leisure time to disinterested studies and counting on the support of the city's authorities, now they were dependent on the patronage and employment of chieftains or kings, who sought them with an utilitarian goal in mind, as architects or engineers, or merely as ornaments to their kingdom, as was the case with poets and sculptors. The court of the Attalid kings in Pergamum, on the Mysian coast across from Lesbos, was a case in point and so was, of course, the court of the Ptolemeys in Egypt. The famous Museum, that sanctuary of the Muses that was to attract so many scholars to the city of Alexandria, was in fact appended to the king's palace. The kingdom of Egypt, controlled by the Ptolemeys for three centuries until the territory became a Roman province, stands out in particular for the degree of prosperity they achieved. This is true most notably during the first one hundred years, under the rule of the founder of the dynasty, Ptolemy the Saviour, followed by his son and grandson. They managed to amass a true Empire, extending their jurisdiction to Palestine, parts of Syria and southern Anatolia, Cyrenaica to the west and Ethiopia to the south. Their rule cannot be called anything but efficient and enlightened. They improved irrigation, reopened a canal to the Red Sea, introduced the camel for transportation and generally expanded both industry and commerce. Of course, Egypt had remained during this time the granary of the world and the extraordinary revenues derived from foreign trade were sufficient to satiate the powerful and even trickle down to the less fortunate. Although the administration was in the hands of a Macedonian and Greek elite, in local matters considerable autonomy was granted to the native Egyptian population. The different ethnic groups never assimilated completely although, after a time, when Alexandria became a gigantic urban center, it incorporated inhabitants from the most diverse origins, Arabs, Jews, Indians, all living in close proximity more or less harmoniously. The rise of Alexandria was spectacular and its endurance has been matched by few other cities of the world. The Ptolemaic kings provided lavishly to its embellishment, building a stadium, gymnasia, a hippodrome and, of course, the colossal lighthouse on the island of Pharos that protected the sea harbour, a tower of some 120 meters and an architectural wonder of the time, proclaiming to the newcomer the grandeur of the city long before he had set foot ashore. On the mainland, facing the island, there was the elegant district of Brucheion, where the royal palaces and main temples were located and where the Museum and the great Library were also established. Of all the scientists that came to work within its walls, I suspect that Euclid must have been one of the earliest luminaries. The scholars who lived in the Museum and worked there at the discretion of the king must have formed a rather isolated community; a contemporary poet, as I recall, already referred to the Museum as a bird cage devoted to fatten a lot of contentious and argumentative pen-pushers.

Albertus: As far as I know, it is possible that Euclid had no direct association with the Museum and may have been a private teacher in Alexandria, although it is unlikely that he would have remained there without having any contacts with the members of the Museum or without being a frequent visitor to the Library. Let me now turn to Eratosthenes, who belonged to one of the first generations that grew out of the Greek diaspora during Alexander's time. He was born in Libya and we know that he spent several years studying in Athens. Later he was invited by the king of Egypt to move to Alexandria, where he became the second or third head of the great Library, a post he retained for several decades until his death at very old age. Eratosthenes was probably a man of great learning, but most of his works are now lost, in particular his poetry and his comments on the literature of his times. He was also an accomplished mathematician, a skill that he applied to the work that would give him lasting fame, namely, the drawing of a map of the known world from Gibraltar to the Himalayas, which accompanied his extensive geographical survey of regions, their climates and the peoples that inhabited them. Obviously, the great resources of the Library at Alexandria must have served him well in this encyclopaedic work. Although his maps have been lost, we know of his efforts to make them accurate because of the references in a later and disgruntled commentator, who objected to his use of mathematical arguments in the discussion of his maps, while Hipparchus, the great astronomer of a century later, complained that Eratosthenes' maps were still inaccurate and could not be relied upon. Here I wish to single out a minor outgrowth of this great work of Eratosthenes, the determination of the size of the Earth, an accomplishment that represents the first documented instance in which a sound method was applied to the task and carried out with a certain degree of precision. Eratosthenes knew that Syene, in the upper Nile near Aswan, lay approximately on the tropic, since the Sun casts no shadow there at noon on the summer solstice. At the same time in Alexandria, which Eratosthenes assumed to be at the same longitude as Syene, he observed the inclination of the Sun by measuring the shadow cast by a vertical stab inserted at the bottom of a bowl that had the shape of a half sphere and whose radius coincided with the length of the stab. The shadow amounted to 1/25 of the total length of an arc of longitude on the hemisphere. Thus, the Sun's rays were falling directly vertically at Syene but with an inclination slightly greater than seven degrees at Alexandria. This is tantamount to saying that the distance from Syene to Alexandria covers seven degrees of the full circumference around the Earth.

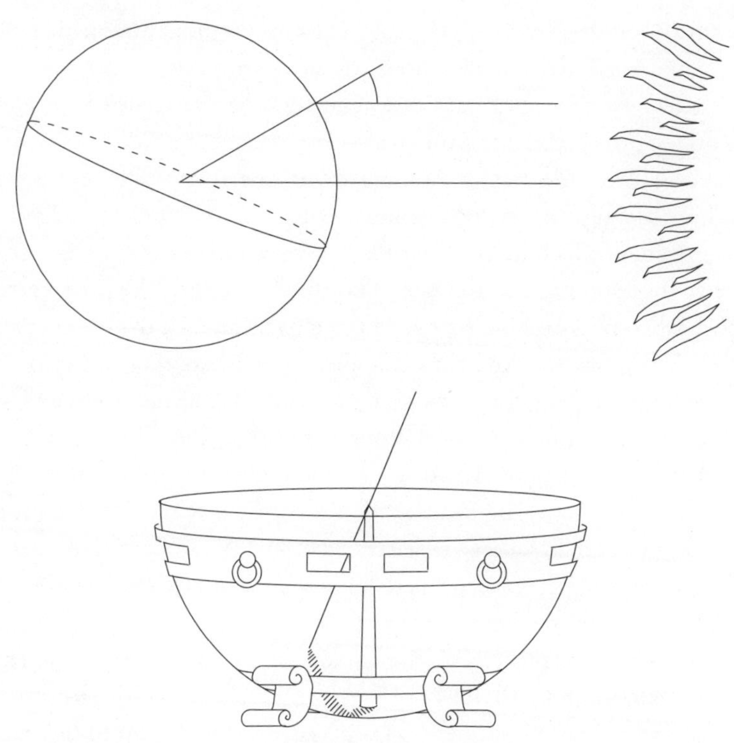

It must be said that for this argument to be valid one must assume that the Sun's rays are falling at both places parallel to each other, that is, that the distance to the Sun is so great that the difference in direction can safely be neglected. Eratosthenes could now rely on an estimate of the distance between Syene and Alexandria provided to him probably by the king's pacers, a distance that he set at 5000 stades. This resulted in 250,000 stades for the circumference of the Earth. There have been some arguments as to the true length of the stade that Eratosthenes used and, therefore, on the accuracy of his measurement, but in one case or the other, he certainly did not err by more than ten percent, which is quite remarkable indeed, considering the rather imprecise way in which distances were measured at the time, as well as the probable inaccuracy in his measurement of the shadow at Alexandria or the fact that the places are not exactly on the same longitude nor was Syene on the tropic. We know of a few previous occasions when the dimensions of the Earth were estimated. Aristotle mentioned a circumference of 400,000 stades but does not explain how he arrived at that number. Archimedes, who was a contemporary of Eratosthenes, also provided an estimate of 300,000 stades without any explanation. Archimedes was in Alexandria for a time, before returning to Syracuse, and we know of his acquaintance with and appreciation of Eratosthenes because he dedicated one of his works to him. The method of Eratosthenes was later modified and the inclinations of certain stars were used to perform the measurements. To really validate the assumption that the large distance to the Sun made the directions toward it from Syene and Alexandria indistinguishable, we must turn back one generation before Eratosthenes and refer to the work of Aristarchus of Samos, also known in his time as Aristarchus the mathematician. We know next to nothing about Aristarchus' life, only a few references to him, amongst others by Archimedes himself. Fortunately, a little tract of his has survived, in which he elaborates on a geometrical method to determine the distances to the Moon and to the Sun. This work, as well as the praise he receives from those who speak of him, suffice to place him amongst the greatest scientists of his time. We know that Aristarchus studied with Strato, the man who had been responsible for organizing the Museum of Alexandria at its inception and who was also an accomplished philosopher and physicist. Strato was originally from Lampsacus, in Mysia, or northern Anatolia, until he moved to Alexandria and later to Athens, where he became the head of the Lyceum, the second successor to Aristotle. We are not certain where Aristarchus became his student. Even later in his life, we ignore entirely the places where he resided. Aristarchus' claim to posterity comes above all from the fact that he was the first to propose that the Sun was at rest at the center of the planetary system and that the Earth and the other planets rotated on orbits around it. In his book on distances, however, there is no mention of this heliocentric hypothesis, perhaps because it was irrelevant to the purpose of his arguments or, possibly, because this is a much earlier work. In any event, let us now sketch the methods Aristarchus proposed to estimate the distances to the Moon and to the Sun.

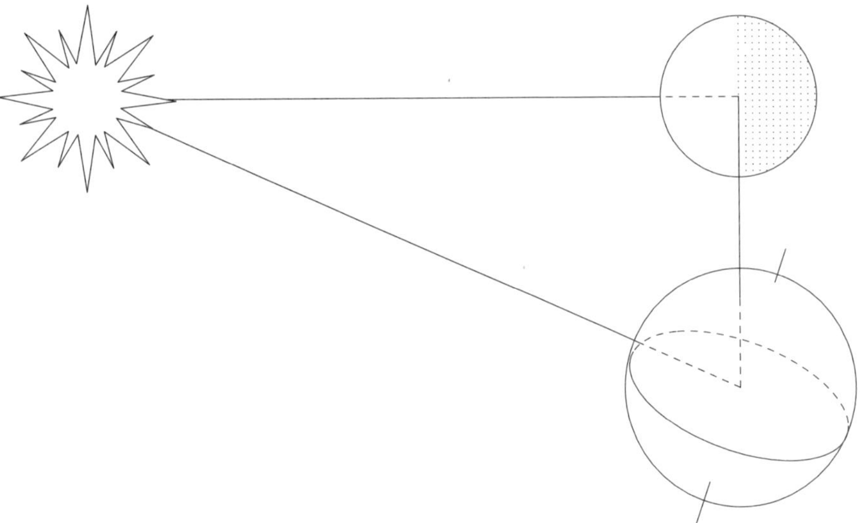

The first method has received the name of lunar dichotomy. Since the Moon receives its light from the Sun, he proposed to observe it at the instant when the illuminated part of the Moon, as seen from the Earth, is exactly one half of its full circle and to measure then the angle between the Moon and the Sun. Considering then the triangle formed by the Sun, the Moon and the Earth, we know that, at that moment, the angle at the vertex occupied by the Moon is a right angle, so that we can determine all angles and the proportions amongst its sides. Aristarchus' measurements resulted in an angle of 87 degrees between the Sun and the Moon at the moment of dichotomy, or just three degrees for the separation between the Earth and the Moon as seen from the Sun. As it turns out, this is much too big by a factor of almost twenty, so that Aristarchus underestimated by that factor the distance to the Sun in relation to the distance to the Moon. However, it is difficult to fault him for the inaccuracy of his measurements. The method, although sound in principle, remained forever very difficult to perform, not only because of the intrinsic difficulty of measuring the angle between the Moon and the Sun at any given time, but also because it is almost impossible to determine with any precision the true moment at which the Moon is exactly one half illuminated. The method was attempted with renewed vigour many centuries later, after telescopes had come into use, but not with greater success. As it was, Aristarchus concluded that the Sun was some 20 times farther removed than the Moon, not 400 times as it really is the case. Next, in order to determine absolute distances, he devised an ingenious method that involved the observation of lunar eclipses. Firstly, he determined that the angle subtended by the Sun as seen by an observer on the Earth amounted to half a degree, which is quite satisfactory. Moreover, he assumed, to a first approximation at least, that the Moon subtends almost exactly the same angle, as it is quite apparent in solar eclipses. He also made use of the assumption that the distance to the Sun was twenty times the distance to the Moon. One last ingredient was now needed to determine absolute distances. Observing the progress of lunar eclipses Aristarchus estimated that the width of the shadow cast by the Earth at the distance where the Moon is located was twice the diameter of the Moon.

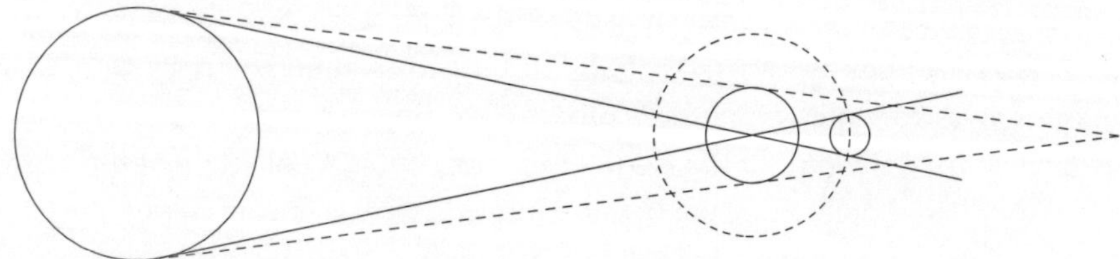

It is now easy to convince oneself that if one places the Moon too close to the Earth, it ought to be small and it fails to fill any sizeable portion of the cone comprising the Earth's shadow. As we slowly remove the Moon farther away, the distance to the Sun is equally expanded, but the width of the Earth's shadow can never exceed the Earth's diameter, so that the Moon will come to fill greater and greater portions of the cone until it satisfies the condition we have imposed on it, namely, of occupying exactly one half the width of the cone. Interestingly enough, Aristarchus carried out this geometric argument to its conclusion, but did not give numerical values to his results. Clearly, the only known length scale in this diagram is provided by the diameter of the Earth. We can now determine that his estimates give a distance to the Moon equal to 80 times the Earth's radius, which puts the Sun at a distance of 1600 Earth radii. Thus, he overestimated slightly the distance to the Moon, but underestimated the distance to the Sun by a factor of 14 or a little more. Returning now to Aristarchus' proposed heliocentric hypothesis, there is a very enigmatic reference to it by Archimedes, amongst the very few that have survived. Archimedes quotes Aristarchus to the effect that the sphere occupied by the stars stands in such a ratio to the sphere that contains the orbit of the Earth around the Sun that is comparable to the ratio of a sphere in relation to its center. This appears to be a figure of speech since the ratio of a sphere to its center should be considered infinite, and he may only have meant to indicate that the sphere containing the stars is enormous compared to the sphere containing the Earth's orbit. This would account for the fact that we do not detect any perceptible motion within the great vault of the stars throughout the year as the Earth moves from one location to another. Nevertheless, Archimedes says that Aristarchus obviously could not have meant literally what he said and he reinterprets the statement with rather undue liberty to mean that the ratio of the sphere of the stars to the Earth's orbit is comparable to the ratio between the diameter of the Earth's orbit and the diameter of the Earth. Clearly, despite all his brilliance, Archimedes had no basis to arrive at this conclusion. All the same, given that Aristarchus had put the distance to the Sun at 1600 times the Earth's radius, Archimedes considered now plausible that the stars might be more than a thousand times farther than the Sun. In retrospect, it is rather mysterious why Aristarchus' hypothesis of making the Earth orbit around the Sun did not prosper and was not explored further. We know of an astronomer by the name of Seleucus of Babylon, who advocated Aristarchus' model at a later time, but for the most part these ideas were abandoned. We also know that Cleanthes of Assos, the leader of the Stoa and undoubtedly without an ounce of an astronomer within him, proposed to bring charges of impiety against Aristarchus for daring to set the Earth in motion, but it is unlikely that this represented the shared opinion of the better educated, especially considering the great reputation that Aristarchus enjoyed as a geometer. Perhaps we should not be so surprised that Aristarchus' successors did not take up his suggestion or that they left it to languish. From our point of view, by not considering it in greater detail, they refrained from making a great discovery, but we may be assigning too much importance to something that, in their eyes, did not carry so much weight. They had quite a clear conception of what were the tasks of the astronomers and mathematicians and how they differed from those of the physicists. The latter were supposed to analyze the nature of things, their substances and the origins and causes of change in the phenomena we perceive. As far as the heavenly bodies were concerned, this was a matter of far-fetched speculation. The task of the astronomer and mathematician was more concrete and specific. He had to account for the paths followed by the planets, the velocities with which they traveled, their stations and retrogressions. For this purpose he was not bound to anything but the requirement to describe what had been seen in the past and anticipate what would be seen in the future. He was only constrained by the power of his imagination to contrive the appropriate hypotheses and models that could save the phenomenon, as they used to say. From this point of view, it is not that adopting a heliocentric hypothesis was a radical idea, but that it was unnecessary or superfluous. Let us examine in broad generality a few of the tools they employed to describe the motions of the planets.

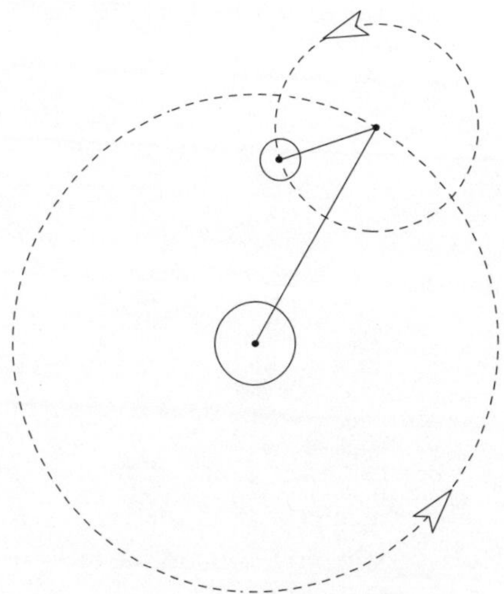

More interesting than the detailed construction of the planetary orbits is the simple originality of the geometric methods they devised. For the outer planets we saw that an eccentric circular orbit might account for their variations in distance and luminosity, but the center of that eccentric circle was also suspected of describing an orbit around the Earth. For Mercury and Venus, it had seemed reasonable to postulate that they accompanied the Sun in its yearly orbit, while revolving at the same time around the Sun in circular orbits. This latter configuration of a large circle carrying the center of a small circle, on whose edge the planet is revolving, came to be known as the method of epicycles and was later applied successfully to the outer planets as well. When the main circle and the epicycle both rotate in the same direction, and provided the rotation of the epicycle is sufficiently swift, we easily see that the planet may appear to be moving backwards precisely when it is nearest the observer located at the center of the main circle. It soon became clear that the method of epicycles and the method of movable eccentrics were two ways of describing the same thing. To verify this it is sufficient to move back to the center of the main circle and draw from it a replica of the radius of the epicycle. The endpoint of this new radius becomes now the center of the movable eccentric and the segment joining this endpoint to the planet becomes precisely the radius of the eccentric circle. Likewise, we can convert the case of a movable eccentric back to an epicycle; the process, once again, does not amount to more than completing a parallelogram.

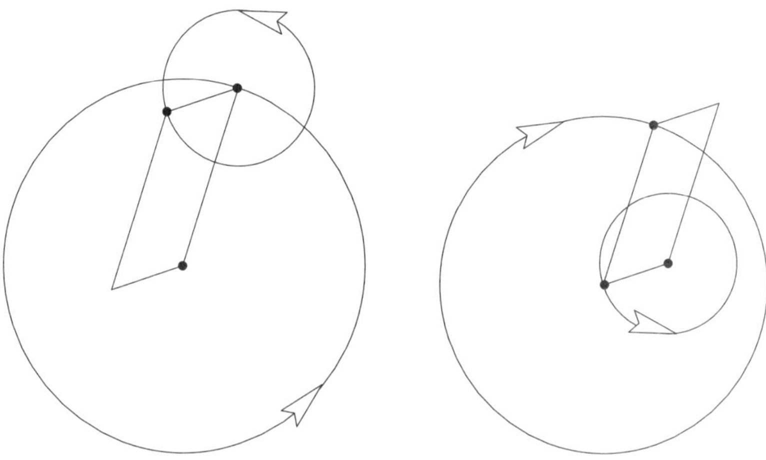

A word should be added regarding the sense of rotation of each circle. In the instance of an epicycle both circles rotate in the same sense, as in the cases of Mercury and Venus. However, when referring to the eccentrics, the Greek mathematicians were in the habit of thinking that the eccentric circle is carried rigidly by the rotation of the circle whose radius rotates around the Earth, in the same manner that a person would turn holding a round tray with both hands, so that they were forced to conclude, such was the case of Mars, that the planet, pacing along the edge of the outer circle, at the edge of the tray in my image, rotated contrary to the direction in which its center was rotating around the Earth. As for the Sun, its annual motion around the Zodiac was accounted for simply by an eccentric orbit whose center remained fixed; this was sufficient to explain the differences in the lengths of the seasons. However, it was also found that it could be converted into an orbit on an epicycle, similar to the procedure used for the orbits of the planets. This is easily seen by making the Sun rotate on the edge of the epicycle with the same speed but in the opposite sense to the one its center follows rotating around the Earth. The combination of the two motions simply produces an eccentric orbit around the Earth.

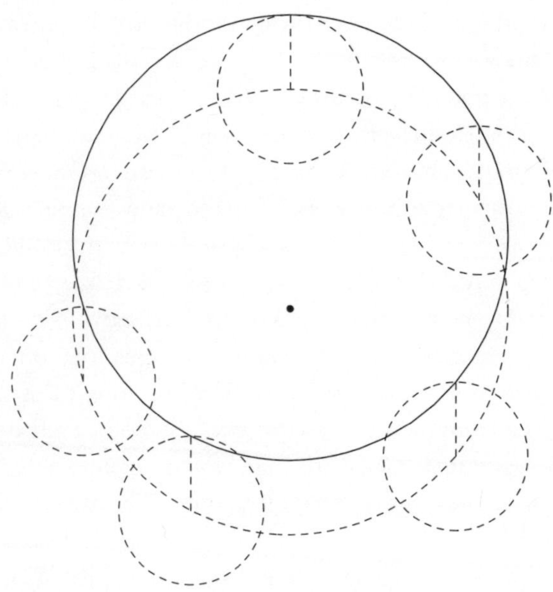

These ideas were well known to the mathematicians of the time, in particular to Apollonius of Perga, one of the greatest of antiquity. We know of his prowess because we still possess his book on conics, a treatise perhaps short in clarity and accessibility but long in sophistication and the reach of its results, so much so that few of his successors dared to follow in his wake. His study of the sections of cones, the curves for which he introduced the names of ellipse, parabola and hyperbola, would become all important in the theory of planets after Kepler's time, but this, of course, was very far from Apollonius' mind. He did concern himself with astronomical matters, although we are not aware that he wrote anything on this subject. Nevertheless, he is repeatedly mentioned in relation to a theory of the motion of the Moon and he is also believed to have used his geometrical skills to perfect the theory of epicycles. I have only described the construction of the planetary orbits in a rather qualitative manner. Apollonius may have found the appropriate relations amongst the radii of the various circles and the velocities with which the circles rotate, so that the planets' movements with their retrogressions were approximately reproduced. It is uncertain where Apollonious carried out most of his work. We know that he spent some time in Alexandria, when the Museum had already been in operation for a hundred years, and he also visited Pergamum and Ephesus but we know little else. It was not until half a century later, at the time of the destruction of Carthage by the Romans, that another scientist of comparable stature came to the scene. Hipparchus was a native of Nicaeia, in the region of Pontus, in northwest Anatolia, where he is most likely to have started his astronomical work, although he later took residence in Rhodes and it is here where he probably spent the greater part of his life. A skillful geometer, he was primarily an astronomer and not a great innovator in mathematics like Apollonious. But if we can rank Apollonius amongst the best mathematicians of any later epoch, it is safe to say that Hipparchus was the first astronomer in a modern sense and he shines amongst all those that were to come after him. I shall first mention a relatively minor work of his but the one that concerns us most, namely, his determination of the distance to the Moon. For this purpose Hipparchus made use of some previous observations of a solar eclipse that could later be reconstructed to have taken place in March 189 B.C. He sensibly made the assumption that, at mean distance, the Moon subtends the same angle as the Sun for an observer on Earth; this implies that a typical solar eclipse is seen as total on an insignificant portion of the Earth at any given moment. Furthermore, he estimated that the Sun is sufficiently far from the Earth so that two observers, even at opposite poles of the Earth, when pointing toward the Sun, would be pointing very much in the same direction. Of course, this is strictly true only if the Sun were at infinite distance. He then used two observations of this particular eclipse which had been recorded as being total in the Hellespont but only four fifths in Alexandria. Hipparchus knew the latitudes of both places and the dimensions of the Earth, so he could evaluate the distance between the two observation posts. Picture in your mind now these two widely separated observers pointing two imaginary cones in the direction of the Sun, one parallel to the other, cones whose aperture coincide with the diameter of the Sun. It is easy to convince yourself that there will be only one possible distance where the Moon could cover the entire aperture of one cone and only four fifths of the other.

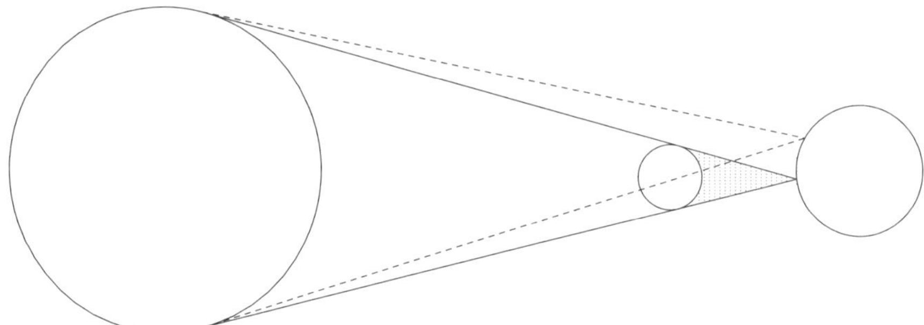

From his computations, which were a good deal more subtle than what I have just made them to be, Hipparchus arrived at the conclusion that the distance to the Moon was greater than 59 and smaller than 67 Earth radii. Since the mean value is close to 60 Earth radii, Hipparchus must be credited with the first true measurement of the distance to the Moon. In fact, he went much farther in studying the geometry of the motion of the Moon's shadow on the curved surface of the Earth in order to better predict the location and extent of solar eclipses. As it was probably the case with Apollonius, the problem of understanding the Moon's orbit occupied him for a long time and apparently never arrived at a solution that satisfied him entirely. In this task he followed the tradition of using eccentric and epicycle models. The motion of the Moon is quite complex and its study has filled the lives of many astronomers since the time of Hipparchus. For example, the Moon does not travel along its orbit with the same constant speed, nor are the phases of the Moon of equal length. This is similar to the difference in length of the seasons and can be accounted for, as in the case of the Sun's motion, by an eccentric orbit or, equivalently, as we saw before, by an appropriately chosen epicycle. Hipparchus knew also that the line joining the farthest point in the Moon's orbit to the point of closest approach to the Earth does not remain fixed in the sky but displaces itself slightly every time around, thus aligning itself with different diameters of the underlying orbit, traveling backward along it, so to speak, completing the full circle in some nine years. Hipparchus could account for this anomaly by breaking the match between the period of rotation of the Moon on the epicycle and the period of the epicycle in going round the Earth. One further complication in the Moon's orbit is the fact that its plane is tilted relative to the ecliptic, the plane swept by the Earth in its motion around the Sun; otherwise, every full Moon and new Moon would cause an eclipse. Each month the Moon spends half its time above the ecliptic and the other half below it. Those nodal points, where the Moon dips below the plane and where it resurfaces, form an axis that also displaces itself, making a full circle after some 223 lunations. Although the inclination of the Moon's orbit is only of 5 degrees, the combined effect of all these different variations made it quite impossible for a simple epicycle model to trace the Moon's motion with any accuracy. Nevertheless, Hipparchus had considerable success with his model, thanks in part to his own observations of the Moon as well as to the access he gained to Babylonian records compiled over several centuries before him. We should not part company with Hipparchus without mentioning two more amongst his great accomplishments. Firstly, we owe to him the first catalogue of stellar positions. Although this work of his is now lost, like most of his other writings, the catalogue did survive in part thanks to Ptolemy, who used it when compiling his own catalogue some two hundred years later. Finally, we must mention Hipparchus' greatest discovery, the precession of the equinoxes. This has to do with the motion of the Sun in relation to the Earth, from north to south and back in the course of a year, and its simultaneous motion in relation to the stars along a great circle in the sky. In relation to the Earth, the Sun spends half the year below the equator and the other half above it. Hipparchus discovered that the length of the year as measured by two alternate passages of the Sun across the equator did not coincide with its return to the same position in relation to the stars, but occurred some twenty minutes earlier. More precisely, thanks to his meticulous determinations of stellar positions and making use of old observations performed some 200 years earlier at the Museum in Alexandria, Hipparchus was able to conclude that the position of the Sun at the equinoxes had been drifting persistently along the Zodiac, in such a way that it would complete a revolution in some 26,000 years.

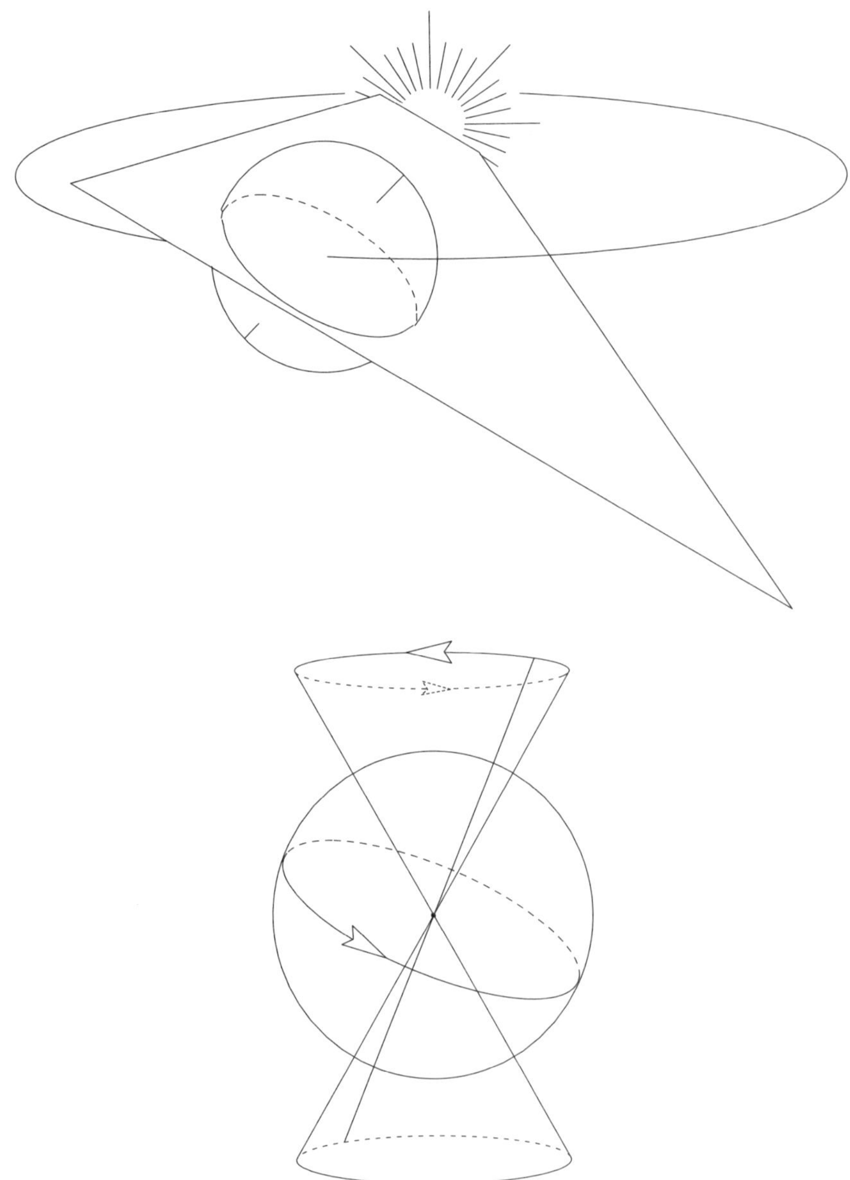

Today we express the same things differently; we ascribe the discrepancy to the drift of the axis of rotation of the Earth, which precesses around the plane of the Ecliptic, just like the axis of a top precesses above the ground when it is about to tumble. Thus, after the occurence of the equinox, when the Earth returns to the same position on its orbit in relation to the stars one year later, the Sun is not found on the equator's plane but has already crossed it, an effect that is caused by the drift in the orientation of the Earth's axis, which precesses around the ecliptic in a sense opposite the orbital motion of the Earth around the Sun. Hipparchus' legacy includes not only the detailed results of his investigations but also the methods he employed to acquire them, his emphasis on the accuracy of the observations, the supremacy of observations over theory and the need to keep records of them over long periods of time. We do not know what kind of recognition he was awarded during his lifetime but posterity has been kind to him and his work has been celebrated according to its greatness. Even in antiquity there were coins minted in his honour and carrying his image. But it is also true that his example was not followed by his immediate successors. In fact, for a hundred years after Hipparchus there are essentially no records of any attention given to astronomy and, throughout the period of Roman expansion, the sciences appear to have been largely neglected. The names of a few mathematicians stand out, such as Menelaus and Heron of Alexandria, and some astronomical observations must have been carried out, without doubt. Cleomedes, who wrote on astronomy, reports the astonishing sighting of a lunar eclipse in full view of the Sun, something that could only be accounted for by the effect of refraction in the atmosphere near the horizon, but Cleomedes must be counted almost as a contemporary of Ptolemy. Judging from their work, one could easily be deceived into believing that Ptolemy was a direct disciple of Hipparchus, when in fact more than two centuries elapsed between them.

Plebeius: The Romans had little inclination toward abstract thinking and perhaps the closest they came to any scientific interest was through their accomplishments in engineering. For the rest, they had little interest in abstract philosophy and I imagine that their efforts in astronomy must have been oriented more in the direction of fortune telling than in the direction of mathematics. By the time Hadrian became emperor Rome had long incorporated most of the Hellenic world. It is sometimes said that the Romans never managed to produce a culture entirely their own but this seems rather unfair and a bit of an exaggeration. It is true that Latin never replaced the Greek language, particularly in the East. But more than betraying a certain hesitancy on their part, a sense of inferiority or an inability to impose their own language, it speaks perhaps of their ready disposition to incorporate a language that was richer and with a longer history, capable of greater nuance and subtlety than their own. They deferred to the Greeks on many matters and imitated them, for example in the arts. Otherwise, they incorporated only the things they found valuable and congenial and rejected the rest. One of the things they surely adopted as theirs was the emphasis on education and the respect for learning. In their schools, however, mathematics was never emphasized as were grammar and rhetoric. This reveals their character and originality for, even when they followed the Greeks in emphasizing education, they fashioned it to their own needs and idiosyncrasy. They never attached much importance to gymnastics, which was so important to a Greek education. Jurisprudence, instead, was a discipline they created out of their own genius and they took great pains to inculcate a sense of the law, not as an abstract concept of justice, but as a practical method of apportionment and of resolution of disputes. Practical, frugal, little inclined to idleness, they were not so much interested in the why but in the how of things and they probably sensed a certain leniency or softness in the way the Greeks had nurtured their own culture. The times of Emperor Hadrian were probably the more prosperous for the Empire. The coins of the period carried the pompous inscription saeculum aureum but it must be admitted that this was truly a golden age. This was the time when Aristides Aelius delivered his famous panegyric that moved the Emperor to tears. A well known orator, he later acquired further fame battling a mysterious disease for some thirteen years, something he accomplished by becoming an almost fanatical follower of the cult of Asclepius, with whom he communicated in his dreams. The Romans had in general a penchant for believing in oracles and for interpreting their dreams. In his famous panegyric this eccentric man lavished great praise on the accomplishments of the Empire, heaping platitudes on commonplaces, saying amongst other things that the entire Earth had now become a garden. Nevertheless, despite his extravagance, Aristides probably expressed something that was felt in general by the population at large. Hadrian had retreated from some of the more adventurous conquests of Trajan and the Empire was for the most part at peace. Perhaps the Romans no longer had the ascendancy of an aggressive power or their former ambitious resolve, but neither did they feel the dissipation of a lost cause or of a world that was coming to an end. Quite to the contrary, they exuded confidence and optimism and, if only by the way in which they launched public works, they expected that their prosperity would continue for ages to come. One gets the impression of a society solid on its foundations and firmly rooted in its customs, in force not necessarily through power or coercion, but working thanks to a healthy and efficient contract of civic life. Throughout the Empire, the cities had been made largely autonomous, they held responsibility for the collection of taxes and for ensuring the supply of corn to the population, as well as providing for the public schools that were by now ever present, even in outlying regions. Roman society always remained a hierarchical one; the fruits of wealth were enjoyed by a minority and many of their accomplishments were obtained through the toil of their many slaves, but even if their conditions were harsh it does not seem that they were treated any worse than the marginal members of society have been treated at any time, even in our own days. Besides, the Romans managed to strike an acceptable balance between their notion that each citizen had few rights of his own, being at the service of the larger community and, on the other hand, the notion that the community was at the mercy of the magnanimity and good will of its citizens, so that an individual derived his pride from his own contributions to the common good. Thus, it could be said that it was an elitist society but public spirited at the same time. An individual could buy public office with money, thereby making perhaps a contribution to some public work, and his action did not have then the sinister meaning it has later acquired. In no other way could they have financed their great and ambitious works, nor built their magnificent cities with their paved roads, their solid houses, their public monuments and fountains. Their aqueducts and water works were nothing short of remarkable. Hygiene seems to have been one of their obsessions and they were even successful in providing private houses with running water. These achievements were attained in particular during the reign of Hadrian, who was very much interested in urban matters; he founded quite a few cities during his extensive travels and stimulated many construction projects. He visited Egypt, incidentally, for most of the year 129 and he spent that entire summer in Alexandria. Having an intellectual inclination himself, it is said that he spent many a day at the Museum exchanging ideas with the resident scholars, so it is quite possible that he met Ptolemy, who at the time must have been a young investigator. It was during this trip that he launched several public works to embellish the city of Alexandria, specifically the quarters that had been damaged during the uprisings of the Jews. It is so very sad that barely a trace has survived of this great city and its past splendour. It all appears so far removed from us now. When I contemplate those surviving paintings from nearby Al Fayyum, the Roman citizens depicted in those portraits seem to speak to us in such an eloquent manner, with their self-assurance, their restrained elegance, that we wish we could learn more from and about them. It must have been most stimulating to visit Alexandria in those times, with its two harbors bursting with activity and its imposing Pharos looking out to the sea. The city was rich with monuments, the Gymnasium, the temple Cleopatra had dedicated to Marcus Antonius and, above all, the Serapeum with its Museum that was said then to rival the Roman Capitol. It must have seemed somewhat of an oddity, a largely Greek city, long under Roman control, in a land that had not yet lost any of its exotic charm, with a majestic river coming from the desert and laden with ancient history. At the time, with half a million inhabitants, it was undisputably the second city of the Empire after Rome, with an incessant maritime traffic linking it to the capital. After a two-week trip, long lines of ships would form at Ostia, the harbor near Rome, bringing African corn to satisfy the needs of the city. As I mentioned earlier, from the very beginning Alexandria had prospered exporting corn to Greek cities, but by this time there was considerable trade with India and with

China too, mostly in luxury goods, silk, ivory, precious stones, as well as spices and aromatics. This contributed to a truly cosmopolitan atmosphere in the city. The population was not just composed of the traditional Egyptian, Greek and Jewish segments but included Ethiopians, Persians, Indians as well as the Roman traders and soldiers. The city had a burgeoning industry too, mostly in paper, glass, carpets and jewelry. It is difficult for us to imagine nowadays the amount of trade there was at the time, as well as the traffic in people. Linen from Alexandria was being sold in Britain while horses from Spain could be bought in Syria. Romans traveled not only on government business or to join the army, but doctors also traveled far and wide advertising their treatments, teachers were constantly traveling on lecturing circuits and students traveled seeking the best teachers. In this respect Alexandria was a great magnet, certainly above Rhodes, Antioch or Carthage but also above Athens. Romans traveled also to attend festivals or for health reasons, to change climates. The network of roads was quite remarkable, their ships were sturdy and reliable, but considering the length of the trips, the precarious accommodations along the roads and also the perennial danger of brigands along the way, it is still astonishing the amount of traveling people were willing to endure. Many of these things we know today thanks to the works of Pausanias, who left a marvelous chronicle of life in the Empire during his times.

Albertus: And we can only rely on the accounts provided by such historians to learn about the general atmosphere in which Ptolemy worked, for he was most reluctant in his works to comment on the society in which he lived or to write about the conditions under which he was working, much less to refer to events of his own private life. Indeed, we know almost nothing about Ptolemy himself. If our aim, in this brief survey of the development of astronomy, was strictly to follow the efforts and achievements directed at understanding the scale of distances in the Cosmos, we might mention Ptolemy in passing and race ahead to an age more fruitful in its result. But the work of Ptolemy is so central to the evolution of our science that we could not possibly ignore him; his contribution laid the foundations and defined the problems of theoretical astronomy for more than a thousand years after his time. His great mathematical compilation, or Almagest, as we know it today through a deformation of the name given to it by the Arabs, is one of the great scientific documents of history. Visiting once the Pantheon in Rome, I realized that it had been built at the same time that Ptolemy had been writing his treatise and I came to think that these were two monuments that had stood gallantly through the centuries and are today as eloquent as ever expressing the genius that created them. At the risk of irritating you with some details, let me attempt to describe the contents of Ptolemy's imposing volume. The Almagest is divided into thirteen books; it is very thorough, didactic and clear in style. It begins at the beginning, describing the construction of the Heavens with a spherical and stationary Earth located at its center. It then presents the mathematical tools that will be employed, some elements of geometry and trigonometry. This occupies the first two books. Then he launches on the theory of the Sun's yearly motion, for which he adopts the simple model of Hipparchus, with the Sun orbiting on a circle and the Earth slightly off its center. The motion is uniform when seen from the center, not from the Earth. The eccentric orbit accounts for the differences in the length of the seasons. Thus far, the theory is exceedingly simple. Today we know that the line joining the point of closest approach to the most remote point on the orbit, known as the line of apsides, does not remain fixed in relation to the stars; it slowly rotates around on the plane of the orbit. Unfortunately, Ptolemy used some observations, perhaps his own, that were inaccurate enough so that he was unable to notice the change that had taken place since Hipparchus' time. In addition, some mistakes in the interpretation of old data led him to miscalculate the length of the tropical year, erring by some eight minutes. Compounded with the fact that he did not improve on Hipparchus' estimate of the rate of precession of the equinoxes, which is slightly faster than he thought, all this put his account of the solar motion in a rather unsatisfactory state. The simple theoretical model was capable of performing well enough, but its quantitative details, its relation to the observed Sun were out of joint. One subject that Ptolemy managed to elucidate adequately was the so-called equation of time. During the day, a sundial tracks the motion of the Sun and measures the length of the hours. The inclination of the Earth's axis of rotation in relation to the Sun makes the days vary in length throughout the year, thus producing unequal hours if we were to insist on always having twelve hours of day and twelve hours of night. But there is another effect which is independent of the Earth's inclination and is due to the eccentricity of the Sun's orbit and the fact that the Sun does not travel at the same speed each season. Imagine for simplicity that the Earth rotates daily around a vertical axis, perpendicular to the plane of its orbital motion. An observer sees the Sun at the zenith; he awaits a full rotation of the Earth, but the Sun is now not yet at the zenith. There is a slight delay due to the fact that the Earth has advanced in its orbit and it is this delay that Ptolemy also took into account, studying how it varies throughout the year. The following books of the Almagest are devoted to the study of the Moon. Here again Hipparchus' model of an eccentric circular orbit is presented first, in the form of an epicycle carried around by a large circle, the so called deferent circle, with the Earth at its center. We saw the reasons for this arrangement, namely, the epicycle rotates backward as the deferent carries forward and they are only slightly out of phase, so that the line of apsides, joining nearest and farthest points of the orbit, is progressively displaced as it slowly rotates around along the orbit. This is similar to what we just saw should be done for the Sun, except that in the case of the Moon the phenomenon is much more evident and it was already known by the Babylonian astronomers, who not only knew that the Moon moves with varying speed along its orbit, but that the fast and slow segments of its orbit do not always occupy the same regions of the Zodiac. Let us now concentrate on one of the elementary ingredients of the model, at least for the observer on Earth, namely, the lack of uniformity in the Moon's motion caused by the presence of the epicycle, the epicycle anomaly, as it is referred to. Ptolemy observed that Hipparchus' model was quite accurate in predicting the passage of the Moon through the positions of conjunction and opposition to the Sun, the syzygies, as they are called or, in other words, the moments of new Moon and full Moon. This was not unexpected given that the details of the model, the size of the epicycle, the rate of rotation and such things had all been fashioned from data taken during eclipses, moments when the position of the Moon in relation to the path of the Sun are obviously known with greatest precision. However, the model was inaccurate at quadratures, at half Moon, to the extent that the predicted positions in the sky erred sometimes by several Moon diameters. Furthermore, this discrepancy did not manifest itself always in the same manner but depended on the orientation of the orbit; that is, if the eccentricity of the orbit was aligned with the Sun, the discrepancy was minimal, but reached a maximum as the eccentricity pointed directly perpendicular to the Sun's direction. This latter arrangement corresponded precisely with the case when the observer on Earth notices the epicycle anomaly most vividly and Ptolemy realized that the discrepancy he was trying to solve had the effect of enlarging the epicycle at quadratures. Clearly, in the opposite situation, the radius of the epicycle was aligned with the observer and a change in its length would not have changed the position of the Moon on the surface of the sky. Nevertheless, it seemed illogical to have an epicycle that would inflate and deflate as it was carried around. Ptolemy devised instead an ingenious trick to bring the epicycle closer to the Earth precisely at quadratures, so that it would appear bigger to the observer. In essence he broke the stiff arm of the deferent circle giving it an elbow and the ability to fold on itself. In other words, the center of the deferent, instead of being fixed at the Earth, is asked to rotate around it in a small circle, in such a way that the direction toward it and the direction toward the center of the epicycle, as seen from the Earth, are bisected by the direction toward the Sun. In this way you can easily see after a moment's thought that the arm is stretched at full length during new Moon and full Moon, but is completely folded precisely at quadratures. With this device Ptolemy achieved greater success than Hipparchus had with his simpler model. The scheme, however, presented a glaring difficulty. It predicted too great a variation in the distance to the Moon, by as much as a factor of two, so that its diameter should have been seen to double in length in the course of one month.

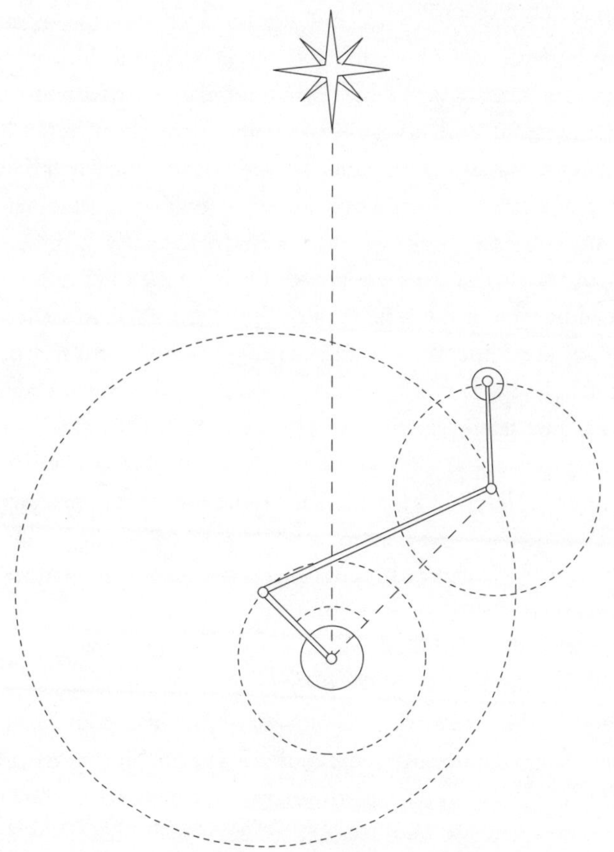

Of course, this is not the case; the variations in the Moon's size are quite marginal. Ptolemy does not discuss this matter. He turns next to the determination of the actual distances to the Moon and Sun, accepting implicitly the validity of his model, which he uses in conjunction with specific observations to arrive first at a value for the distance to the Moon. For these very reasons the procedure is vitiated and Ptolemy concludes that the Moon's distance oscillates between 38 and 64 Earth radii when, in fact, only the latter value is close to its mean distance. Then he turned to Hipparchus' eclipse diagram I discussed earlier, but instead of using it to determine the lunar distance, which was the sound thing to do, Ptolemy, having already obtained it by other means, used it to determine the distance to the Sun. In this case the method is totally unreliable. Nevertheless, he obtained a value of 1210 Earth radii for this distance, which is too small by a factor of twenty. It is difficult to say whether he was in any way influenced by the estimates that had been obtained before his time, such as the one by Aristarchus; that is, we cannot be certain that Ptolemy actually obtained the measurement that he wished to obtain. The last five books of the Almagest are devoted to the planets, but before turning to these there are a couple of books devoted to the fixed stars, which are used later to chart the paths of the planets. Here Ptolemy compiled a catalogue of approximately 1000 stars with fairly accurate coordinates, a great accomplishment that remained a fundamental reference work for more than a millennium. I will not annoy you describing the theory of Ptolemy on the planets in any detail, although this work was a personal achievement of his, since no previous models had been developed to any considerable extent. I will limit myself to two brief observations. Firstly, Ptolemy accounted for the orbits of the planets by means of a deferent circle and an epicycle, both rotating in the same sense. This was the old method of obtaining retrogressions along the orbits. The radii of the epicycles were assumed to remain always parallel to the direction pointing to the Sun from the Earth. From our modern point of view, this was a way of transferring the Earth's rotation around the Sun to the planet itself. Furthermore, in the case of the superior planets, the deferent circle was eccentric, that is, the Earth was removed from the center and the motion along the deferent he prescribed to be uniform as seen from the point opposite the Earth across the center. This privileged position was the famous equant, introduced by Ptolemy. I mention this because, just as in the case of the Moon, it represented a rather daring innovation on his part, breaking with a tradition that considered uniform circular rotation as a law not to be violated. Of course, the orbits of the Earth and the planets are all eccentric and Ptolemy had to account for the fact that the retrogressions are not always of equal length or duration and things of this nature. My second observation relates to the motion of the planets in latitude. Their orbits are also inclined relative to the Ecliptic; each of them lies on its own plane going through the Sun. But Ptolemy was describing the orbits as centered at the Earth; as a consequence, he encountered insurmountable difficulties in describing the motion of the planets away from the Ecliptic. Nevertheless, he remained undaunted and devised a way of letting the epicycles vary their inclinations periodically in a complicated way, although the models were too cumbersome and did not prove to be very successful. The Almagest is a massive, comprehensive work and one of the earliest that Ptolemy wrote. In it he does mention the ordering of the planets in relation to their distance from the Earth and corresponding to the lengths of their periods of revolution, although for the most part, throughout the discussions, Ptolemy is not so much concerned with a description of the physical constitution or the dimensions of the Cosmos, but instead with the task of finding geometric models to describe the orbits, merely as luminaries crossing the vault of the sky. During his long life, Ptolemy composed a few other astronomical works and still many other books on a number of subjects. Amongst these there is a book on astrology, which appears to be, particularly from the introduction, very cautious and sober. In the beginning he discusses the prediction of astronomical phenomena, then turns to the subject of how, through careful observation, these phenomena can be associated to weather prediction and climate changes or other cataclysmic events, such as earthquakes or plagues. However, in the end he comes to the question of how the configuration of the planets at the moment of birth affects the character of an individual and, of course, how the stars influence one's fortune and life; here Ptolemy simply reflects the opinions of his time, without exercising any scientific restraint or judgment. Nevertheless, he abstains from the more extravagant claims of fortune-tellers or of those dedicated to cast horoscopes. A work that brought Ptolemy considerable fame was his Geography, a compilation of the latitude and longitude of many cities, along with maps comprising the entire known world. This must have entailed an extensive work of gathering data from places far afield, relying on information provided by men of learning and many travelers. The maps accompanying this work are now lost but Ptolemy's text is purposefully very detailed, with the intention that his readers should be able to recreate them. The regions farthest removed were naturally the least accurately reproduced in these maps, the western Mediterranean, Southern Africa and Persia. Centuries later, the Arabs, who learned their mapping techniques from Ptolemy and whom they held in high regard, did not hesitate to make corrections in the places where he had erred. Ptolemy also wrote a book on Mechanics which has been lost too, and even a book attempting to prove the fifth postulate of Euclid, which is also lost. Perhaps I should mention only one more book, the Optics, which survives only in an Arabic translation and in a rather fragmentary form. As is the case in other instances, it is difficult to assess what was Ptolemy's original contribution to the subject. We know, for instance, that Archimedes had worked on these matters but nothing of his has survived. In any event, Ptolemy is very thorough treating a number of questions in geometrical optics, the formation of images in mirrors, the nature of binocular vision and finally he attempts to derive a law of refraction, referring to experiments done by passing light through air and glass or air and water and so forth. He was obviously aware of the role refraction plays in astronomy. At one time he had said that the increase in size of the Moon as it approaches the horizon was due to the refraction caused by the moist air close to the surface of the Earth, but later he recanted, recognizing that the phenomenon was only an optical illusion. Some have accused Ptolemy of being a poor observer, other times he has been accused of not being an original thinker, having borrowed most of his material from others and limiting himself merely to the role of a compiler. Even if a fraction of these accusations is true, his work is so vast, his method so painstaking and his dedication and patience so admirable that he would still retain a very honourable place amongst the most distinguished scientists. Furthermore, without any contemporaries or followers of a comparable rank, he stands alone at the end of a long scientific tradition, a tradition that was then neglected and completely abandoned at the same time the Empire was losing much of its vigour.

Plebeius: The decline of the Roman Empire after the Antonines has been attributed to the most diverse causes. Most likely, it is not that any one of them proved to be decisive but that all in conjunction created an atmosphere that made it impossible to maintain the previous level of organization and prosperity. To the cruelty of the emperor Commodus one must add the plague and famine that broke out during his reign, a famine that, in turn, was provoked to some extent by a less than honest administration of the food supply. The dissipation and abuses of the Praetorian Guard along with the many conspiracies that followed and the lack of any orderly process of imperial succession proved to have a much more corrosive effect than the external aggression the Empire suffered, either from the first wave of invasions of the Germanic barbarians from the north or as a consequence of the war against the Persian armies in the Orient, leading to the ignominious capture of the emperor Valerian at the hand of the Sassanid king Shapur I. These circumstances all combined to give a greater voice to the army in administering the affairs of the Empire, until the emperors themselves came to be chosen amongst the army generals that were then protecting its borders. All in all, the barbarian raids as well as the internal disputes contributed to disrupt the material aspects of daily life, at a time when excessive consumption and an unfavourable balance of trade with the Orient helped make the Empire a good deal poorer. It soon became evident that the aristocratic class that had previously prided itself in providing for public works and for the general well-being was now more concerned with its own interests and privileges and with preserving the comforts it enjoyed in its own private life. The changes in the Empire may also have been apparent in the overall climate of opinion and the expectations people had for the future. This was made manifest by the influx of Oriental cults and mysteries that attracted a great number of followers and also by the severe persecutions against the Christians that erupted from time to time. The city of Alexandria did not fare well throughout this period of decline and its attempts at controlling its own affairs were dealt with rather severely. First it had to endure the wrath of Caracalla who, during one of his many trips and without much provocation, punished the city with a widespread massacre, building then a guarded wall to divide it in two and prevent further disturbances. Later, during the time of Valerian, taking advantage of his much weakened rule, the city was up in arms again and a protracted civil war ensued that lasted for more than a decade. It ended only during the reign of Aurelian with a cruel repression, much devastation and a further partition of the city. It is said that the district of Bruchium with its palaces and community of scholars never recovered from this onslaught. But this was not all, for Diocletian, toward the end of the century, initiated his own campaign to the East by landing first in Egypt and at once laying siege to the city of Alexandria, a siege that lasted eight months. The city's water supply was cut and, after further damage by fire and pillaging, Alexandria surrendered, although this did not spare it from more punishment and reprisals. During this time and in the midst of all the upheavals, we can say that Alexandria had become the center of the Christian world and the home of some of its most influential theologians, Clement in the first place, who had come to the city from Athens and who could well have met Ptolemy in his old age, and later Origen, whose extensive writings became so influential in shaping the language and the doctrine of the Christian Church. I have long been fascinated by the history of that relatively short period of approximately two hundred years, during which Christianity rose from being an obscure Jewish sect to becoming an ecumenical force and an arm of the Empire. It is not so much how the Church organized itself into a powerful institution that has interested me or even how the Gospels came to be conceived and written but, more specifically, how Christianity shaped itself out of the Jewish apocalyptic tradition, the oriental mysteries that circulated at the time, as well as the hellenistic culture in which it grew. It has never been short of mysterious how this rather small sect, initially confined to Palestine, one amongst many, would become so dominant in such a vertiginous manner, especially considering the hostility and persecutions it initially faced. The courage and determination of those who did not brake their faith and were martyred must have been a most compelling force in gaining new converts. The message that Christianity came to proclaim was undoubtedly quite extraordinary in itself and amounted to a new conception of life, a life that was almost out of this world for, on one hand, to grant salvation, it seemed to require one's transformation and surrender and yet, on the other, it demanded nothing but purity of heart and the generosity of love. It was, in a way, a most optimistic message, it announced that the true and only path had at last been found, it was within reach and promised eternal life. But it had, of course, a darker side too. It spoke of a curse, an unsuspect guilt and the need for atonement and renouncement, to abandon this world, stained and beset by evil. It may very well be that by disregarding the relevance of one's social standing in relation to one's fate, Christianity found a warm reception with the marginalized population, the downtrodden and disinherited. But also the fact that its message was addressed directly to each and every individual soul, regardless of race and tradition, may explain how it could soon move out of its Jewish confines and become an ecumenical religion. I wish that sometime we discuss in more detail the history of this period and its intellectual character, in particular, how the traditions that we would now call scientific and religious interacted with one another. Amongst the scholars who concerned themselves with mathematical questions there may have been those who were only interested in solving practical problems of engineering but, in astronomy, the painstaking and minute study of the planets' orbits must have always had a different goal, never too far removed from the goal of solving the riddle of the Universe. You mentioned the book Ptolemy wrote on astrology. It is unlikely that he did so to satisfy a superstitious ruler or benefactor. Instead, we should consider that the vagaries of the human soul and the mysteries of our existence were as much part of the riddle of the Universe as the wanderings of the planets and it was legitimate to investigate how all these things were related to one another. Why should the cosmic world be divorced from the human world? The advent of Christianity, however much it later came to be seen as opposed to the scientific spirit, ushered in a view that we would recognize as more modern and compatible with the scientific spirit, a view that separates us from the contemporaries of Ptolemy, namely, it introduced the belief that one's fate was within oneself. Man had been created in God's image and the road to salvation was not written in the heavens but in the soul. I think that sometimes it has been stressed too strongly that the emergence and success of Christianity offered a new world view that lay opposite the hellenistic culture in which the rational study of the sciences prospered. In fact, Christianity found its home in the very same community that was immersed in that culture and only became what it was by intermingling with it. The Church historian Eusebius writes about the members of the Christian sect that followed a preacher by the name of Artemon. They were accused of altering the holy Scriptures and abandoning the rules of faith to form their opinions according to the subtle precepts of logic. The science of the Church was neglected for the study of geometry, writes Eusebius, and these Christians had lost sight of Heaven while they were employed in measuring the Earth. Euclid was always in their hands, Arsitotle was object of their admiration and they revered Galen. They were corrupting the simplicity of the Gospel by the refinements of human reason. In the end, as we know, the force of the institutional Church would come to smother this spirit of quest and exploration and would find it more expedient to declare itself the depository of a unique truth that it would later impose dogmatically and with intolerance. But the ambiguities and tensions of the early days speak of a more complex relationship and it is the evolution of that relationship that I think should be interesting to consider in more detail at some time. As the new faith attempted to interpret its doctrine in the context of the old tradition, it was reshaped and dressed in new clothes. In this way, what could have remained a marginal and esoteric cult, like dozens of others against which it had originally competed, became in the end respectable and mainstream, whereas the old hellenistic culture in which it had grown lost its original form and emerged in a new language. The early apologists, writing against the practices of the pagans, attempted to explain the new faith in a rationalistic way and inserting it in the context of natural religion, while maintaining that all knowledge that predated the coming of the messiah and the revelation of the true faith had to be suspicious or provisory. The ancient Logos of the Greek, the order of the Cosmos, the divine Reason, were only a shadow of the true Logos, revealed and made flesh in Christ. In the words of Justin, who had come from Palestine to preach in Rome and who was martyred under Marcus Aurelius, it was not a matter of ignorance versus revelation; the pagans had been able to participate in the true Logos, but they had known it obscurely. Tertullian, a generation later, was less sympathetic toward those inclined to mix their faith with the old philosophy. Seek and you shall find, was their doctrine, Tertullian complained, but that was true when that which was not yet known had still to be sought. The task was therefore to search until one has found and to believe when one has found, then do no more than hold on to what one has believed. All these questions, however, were not easily resolved. What role did reason play in faith? Could it be that only through argument and the conquest of knowledge the soul could really find its path to deliverance, or was revelation to be understood, perhaps not necessarily as blind faith, but as acceptance and surrender to an inscrutable mystery, lying above reason? Clement of Alexandria devoted much attention to these questions in his writings. Unlike Tertullian, who considered that God was inaccessible to knowledge, Clement, to the contrary, thought that faith was the stock the soul then refashions into knowldege. For him the soul, through piety and devotions, ascends to the divine in stages and reason is one of the tools making the way possible. Knowledge was for him the activity of faith seeking to understand its object. Philosophy, he would say, is not where you want to dwell, that is faith, buth faith stands out better in the background of philosophy. In fact, it is often apparent, reading Clement, that he is quite eager to show his command of the classic works of philosophy, offering extensive and unnecessary quotes. His intention was to demonstrate that his theology could not only withstand the attacks from the philosophers' arguments but indeed be supported by them. Yet, his works are not of an apologetic nature, that is, he was not writing against the pagans. Instead, he aimed at composing a comprehensive system of theology to be used within the Church. He did not neglect to write on morality and on the nature of a virtuous life, subjects in which he was influenced by the stoics, but these matters are subordinate to the more speculative ones dealing with the work of the soul in its path toward divine knowledge. In his system the entire history of the world bears the mark of a deliberate, progressive revelation of God's truth and the life of the individual soul replicated this path of progressive revelation, conquering one stage at a time. If Clement wrote against any group, it was in particular against the gnostics, with whom he shared some general views but who were considered by the emerging Church as outright heretics. The gnostics can also be counted amongst the first theologians and they too were influenced by Greek philosophy and the oriental myths. However, being fond of occult mysteries and by employing allegories, they attempted to transform the Christian doctrine into a far more ambitious and fanciful system of theology. They held the opinion that the world in which we live is a corrupt version of a superior divine world, that our being is divine in origin, but that it has since fallen into the evil world of matter. Salvation can only be attained by becoming acquainted with this secret knowledge and is reached when the soul returns to its spiritual home. This knowledge, they insisted, is obtained, not through reason but by revelation, not by argumentation but through initiatory practices. The fall of man, as recounted in the Old Testament, was in itself an allegory of a greater drama of divine history that entailed the fall of the entire material world. Unlike the orthodox doctrine of the Church, for which the history of the world is pregnant with meaning and celebrates the glory of God, according to Gnostic belief everything pertaining to this world is entirely deceitful and evil. In effect, they conceived of the existence of two gods, so that the god of the Old Testament, the creator of this world, turns out not to be the true God. It is only the arrival of the messiah, Jesus, who proclaims the existence of the yet unknown divinity. Of course, by holding this view they created a problem for themselves, for they had to explain how the nature of the true God could enter the material world, become a man and be part of the deception. In one of their explanations, it was only a spirit that had descended to this world, was captured by the evil god and forced into matter, thereby forgetting its own identity and only later, through the intervention of the true God that awakens him, does he become a messanger and reveals the secret history, before being lifted to the spiritual world once again. In another variant of this story, Jesus is the true messanger of the superior God, but is flesh only in appearance. Then, when Yahveh learns from Jesus of the existence of a world above his own, he becomes furious and takes revenge by apprehending the body of Jesus and delivering it to his henchmen. Both Clement and Origen opposed strenuously the teachings of the gnotics. During the persecutions launched by Septimus Severus, Clement left Alexandria and Origen succeeded him as the foremost theologian of the city. Origen's father, who had already converted to Christianity, was also a victim of these persecutions and Origen, at age seventeen, was left to care for his mother and siblings. It is said that because of his distinction and dedication he received some help that enable him to finish his studies and become a teacher of literature. Meanwhile, after being brought to the attention of the local bishop, who was in need of teachers of catechesis, he was encouraged to give up his secular studies and to devote himself to instruct the incoming cathecumens. Origen threw himself wholeheartedly into the new task and, realizing that he was dealing with students who had received a classical Greek education, began himself a thorough study of Greek philosophy that greatly influenced his religious views and his writings, which he continued throughout his life on a great number of subjects. His greatest accomplishment is to have been the first to build an articulate theological edifice, argued with all the tools of philosophy, but springing nonetheless from the text and message of the Scriptures. Origen was primarily an exegete, devoted to elucidating the text of the Bible. However, given that for him the Bible contained, on one hand, both a mythic and a literal meaning and, on the other, a learned, allegorical one, this task acquired for him an enormous scope and impelled him to construct this vast and complex system of theology that had an enormous influence on his followers as well as on his detractors. Neither Clement nor Origen were entirely innocent of the sins and transgressions they imputed on the gnostics. Both of them, once embarked in their prodigious theological adventures, participated in the belief that when it came to the teachings of the faith, there were certain truths that should only be accessible to the few and initiated. Even Clement had hinted of a tacit and somewhat secret and allegorical meaning of the Scriptures, something that was later much developed by Origen. His impact was such that much of the theological writings after his time bear his stamp, particularly in the Eastern Church, which developed further its theoretical and speculative inclinations. In the West men such as Tertullian and Cyprian gave Christianity a different orientation, more introspective, more interested in morality and piety and with greater emphasis on living the life of the faith. This western school is perhaps best represented two centuries later by Augustine who, in part due to his transparent intelligence, his eloquence and his joyous demeanor, came to have an influence just as considerable, although his attitude was predominantly pastoral, interested in guiding the believer and much less inclined toward debating questions of dogma. Nevertheless, it must be said that even if the West remained more fastened to the direct teachings of the Gospels as norms of conduct and belief for the faithful, it progressively put a greater emphasis on the Church as an institution, as home and refuge, it is true, but also as arbiter and as an ever expanding authority, a tendency which, in the end, had stifling and deleterious consequences. Christianity in the East, despite its stress on the cult and its inclination toward idealized, abstract theories, remained, it seems, more thoughtful, more personal and more spiritual. The writings of Origen are very imaginative and fanciful. Perhaps I should pass on to you some fragments to read, although it is difficult to get a sense of the scope of his ideas because he often rambles and does not proceed in a methodical manner. His effort to constrain himself to the text of the Scriptures and then the attempt to search for the hidden, secret meaning in it often lead him far afield. One of his better known ideas is the notion of a world of spirits, created before the beginning of time and before the existence of the Universe; they represent the unfolding of the self-existent, supreme being. By a number of accidents they fall to various degrees, becoming the spirits of angels, of which there is an entire hierarchy, while the most corrupt ones descend to be assigned to demons. In between there are the spirits who animate the human soul. Thus, the soul is the battleground and the prize disputed between the body that attempts to bring it down to its material existence and the corresponding spirit who struggles to raise it and bring it back to its true abode. In this way, some of Origen's writings may be said to be of a mystical nature or oriented toward spirituality, but his interest was always theological or even philosophical. In his reading of the Holy Scriptures, he saw in them a road toward cosmological ideas, to which the soul ascends through contemplation. He would argue that as the soul battles the evils that tie it to this world, educating itself in the various layers of the Scriptures, it rises above its own history and, coming to its final resting place, it finds in the knowledge acquired the substitute to the life it lived, almost as if the life of this world dissipated into a forgotten dream. Thus, Origen's system turns out to be overly intellectual and formal. On astronomical matters, when it comes to the motions of the planets and stars, he states that these are for the angels the equivalent of what the Bible is for the faithful, in the sense that their various layers of meanings are inaccessible to man and only become intelligible and could be read by a superior class of spirits. Perhaps you will find his writing too convoluted and capricious but his ingenuity is often appealing, as much as the ambiguity of his language. After some twenty years of teaching in Alexandria, Origen began to travel. He first visited Palestine where, at the invitation of the bishop of Jerusalem, he delivered some of his first homilies. In fact, it was only during his stay in Caesarea, in Palestine, that he was ordained, causing a rift with the bishop in Alexandria and leading to his condemnation. Because of his presence Caesarea became an important center of theological studies. But Origen continued to travel, to Rome, Athens, Arabia, engaging at each stop in numerous debates. Unfortunately, most of his correspondence has been lost. His older years were not happy ones; during Decius' persecution he was arrested and tortured and, his health broken, he died soon afterwards. His writings preceeded the great controversies and heresies that were soon to engulf and shake the Church and that led to so much rigidity in the interpretation of the dogma, not to speak of the interminable debates over subtleties and minutiae of very dubious significance. I suppose this is what invariably happens in human history when we let ideals decay into orthodoxies or the thoughts that were once focused and purposeful seem to indulge in their own contortions, getting lost in peripheral detours that never arrive at that which had made the road worth undertaking. All the more so when an authority, the Church in this case, establishes itself as custodian of a unique, indisputable truth that it later sets out to impart by indoctrination. Perhaps for this reason I find the early history and evolution of Christianity of much greater appeal, when a renewed awakening to the spiritual needs of man manifested itself in unprejudiced questioning and a restless curiosity that was not yet dulled by a magisterial, elaborate and rigid doctrine.

Albertus: All things considered, Plebeius, I am afraid one would have to conclude that, from the point of view of the evolution of science, the preoccupations and concerns of the Christian theologians stole most of the intellectual vigour of the times and its vast literature drowned the little that was produced directed toward the study of the natural world in the manner that had been promoted by the hellenistic scientists. You mentioned the followers of a preacher by the name of Artemon, who combined their faith with their studies of Euclid and Aristotle. I do not know how widespread their attitude may have been but we have so few records of any sustained interest in any of the sciences during this period that we must assume that they did not arouse much curiosity and ceased to be considered an indispensable part of a good education. The names of only a handful of truly great scientists during this period come to mind, such as the mathematician Diophantus. Even the pagan philosophers who were never seduced by Christianity, were they not speculating on ideas very similar to the ones professed by Origen that you have just described? I am thinking of Plotinus, for instance, who acquired such a large following during his lifetime bringing back to life the theories of Plato. Was he not invoking also a First Principle from which he wished to explain all of existence, a supreme mind, a divine intellect, from whose emanations all the sublunar world unfolded? What could the human soul wish, seeing itself bound to this foreign world of matter, but to return to that original, ethereal world of the spirit? It is probably fair to say that it is not that the soul would find its home in this lofty realm, but that the supernatural world itself finds its home in the soul, being a product of its own inflamed imagination. Perhaps you will be able to teach me to appreciate some of the writings of Origen, but I have always found these philosophic vagaries rather stale and uninspiring and it is puzzling to me why they have never gone entirely out of fashion. Amongst those who attempted to mix philosophy with science in those days, there was a certain Iamblichus, of whom we have several writings to assess his frame of mind. Despite his high praise of mathematics, he saw it through the lense of philosophy, to the detriment of both. He also spoke of the higher life of the soul that aspires to enter another realm and rejoin some sort of supreme mind, but he had a taste for dressing things in a mathematical language, so he spoke of the limited and the unlimited, being at the origin of everything, the limited giving rise to the One, which is good and beautiful, the unlimited is the father of divisibility and plurality and therefore it is unordered and evil. His abstruse system could only be called number mysticism. In it there were various hierarchies of numbers, intelligible numbers as well as physical numbers and then, of course, mathematical numbers. Some say that Iamblichus was involved in practices that were half magical and half religious, including sacrifices and abnormal things like levitation, the purpose of which was to invite the visitation of the higher spirits to one's presence. All this was a great deal of nonsense and it is easy for us to ridicule, but in his time Iamblichus commanded some attention and respect. We must also remember Theon of Alexandria and his famed and beautiful daughter Hypatia, about whom so many fantastic stories have been told. Theon may have been interested in divination and the occult powers of astronomy but he is also known to have written intelligent commentaries on Euclid and Ptolemy. Two centuries after Ptolemy, they were the remnants and last links of a scientific tradition that was on the verge of disappearing altogether. According to some sources, Hypatia excelled above her father in mathematics, although nothing of her work has survived and we have only a few scattered references from her disciples. She was also interested in philosophy and surrounded herself with a group of followers, amongst whom there were Christians as well as pagans. All accounts speak of a very remarkable woman, but the subjects that were debated in her circle or the mastery they had of the scientific achievements of ages past is almost completely unknown to us. Alexandria would soon cease to be the center of studies it had once been and would not leave too many records for us to examine. The brutal murder of Hypatia at the hands of a Christian mob is often portrayed as the final blow to a tradition of science in the city that never resumed again.

Plebeius: The countless religious conflicts no doubt helped to bring it to its ruin. Its decline had not stopped after Diocletian's time but continued steadily for the next century and beyond. During the reign of Constantine there was the great theological controversy about the Trinity stirred up by Arius, who was himself a preacher from Alexandria and Athanasius, the local bishop, or the Patriarch of Alexandria as he came to be known, took the banner to do battle against him. It is not just that these theologians stole the intellectual vigour of the times, as you said, but they even went to war repeatedly to fight one heresy or another. Although Arius and his heresy were eventually condemned at the Council of Nicaea, for a time he seemed to be on the ascendancy and Athanasius was sent into exile. Even after his restitution, many years later, when Constantine's son became emperor, he favoured Arius ideas once again and sent his troops against the resolute bishop in Alexandria, still unwilling to surrender. In the end he was forced to flee and to seek the protection of the monks that lived in the desert, south of the city. This did not prevent the troops from laying waste to Alexandria, sacking and pillaging it for several months. The end of the city's misfortunes was not in sight yet. After the reign of Julian, who had favoured a revival of pagan cults, many Christians felt the urge to retaliate and there was a vigorous campaign to destroy ancient temples. In Alexandria, Theophilus, the new patriarch, incited the Christian populace to tear down the Serapium, considered at the time to be one of the wonders of the world. No matter, the operation was carried out and the patriarch received the commendations of the emperor Theodosius. Upon his retirement, Theophilus left his nephew Cyril as bishop of the city. By this time, this office had acquired sufficient power to be able to compete against the civil authorities. Cyril was probably no less sinister than his uncle in his machinations and ambition, but in those magnanimous times the Church made him a saint all the same. He got embroiled in a struggle with the prefect of Egypt, to the point that a mob of hermit monks, who were sympathetic to Cyril, ambushed the prefect one day and it was only thanks to their ineptness that the prefect escaped with his life. This was not a struggle between Christian and pagan forces, since the prefect was himself a Christian, but probably more of an insurgent class rebelling against a traditional aristocracy that still retained most of the power in its hands. It was at this time that rumours started to circulate about Hypatia, already an elderly and highly respected woman, who supposedly had been conspiring with the unlucky prefect. It did not take much time until she was assaulted in the street and murdered. Many legends have been created surrounding her life, as you said, but by all accounts her murder appears to have been a very brutal and nasty scene. Of course, Cyril did not acquired his sainthood because of this action, regardless of how responsible he was, but because he was also an indefatigable and quarrelsome theologian, always ready to defend the orthodoxy of the Church against any heresy. He was later involved in one more controversy of the many that arose during this era concerning the dogma. In this instance the subject of the Incarnation was in question and his opponent was Nestorius, the patriarch of Constatinople. How could Jesus have had within him a divine nature and, at the same time, been made to suffer. Nestorius had favoured the opinion that the two natures of Jesus, the human and the divine, had united themselves in one person but only in appearance and not in substance. This is the opinion that was later condemned as heretical in various synods, culminating in the council of Chalcedon. Ciryl held the prevailing view that God had descended in substance to dwell amongst men, that he had taken humanity into his being without losing any of his attributes. In any event, this was the contention that exercised the Church, mobilized armies and consumed so much energy and time. We can make little sense of these disputes unless we think that they were also about power and influence. Nevertheless, one cannot be entirely cynic and not consider them also as battles about ideas, in which men were willing to pay with their lives defending what they believed mattered a great deal. Needless to say, the religious controversies of those times became ever more arcane and, to our eyes, almost meaningless. The bishop of Poitiers, not a fellow with a great intellectual inclination, complained at the time that every year, truly, every moon, he said, we make creeds to describe invisible mysteries. We condemn either the doctrine of others in ourselves, or our own in that of others and, mutually tearing each other to pieces, we have been the cause of each other's ruin. The controversy aroused by Nestorius had some other interesting consequences. Nestorius himself was deposed and sent into exile in Libya, where he died some years later. But his disciples were numerous, particularly in Antioch, in Syria, where he had begun his career and also in Edessa, in upper Mesopotamia, from where they spread far and wide. Despite many vicissitudes, the sect survived and occasionally became quite influential. At first they found favour in Persia. Later, either because of some hostility or because of their zeal for proselytizing, many continued traveling as missionaries in all directions and to more distant lands. They preached to the Bactrians and to the Huns upon their migration eastward. They reached Yemen in the south and established relations with the Arabs long before the Arabs began their expansion toward the West. They established missions in India and in the island of Ceylon and they went as far as the southern coast of China. Being more literate than the average population, during all these migrations they carried with them many works that they had translated into Syriac, their native language, and it is through these versions that many people in the East came into contact for the first time with the state of learning in the West. Another factor that many years later would prove to have a great impact in the diffusion and preservation of learning, however unlikely this seemed in the beginning, was the emergence of monasticism. It was in Egypt where, during the time of Diocletian, Saint Antony first became an hermit and Pachomios founded the first monastery. The many monks that soon followed their example led an extremely ascetic and disciplined life. The movement, at least in the East, took two disparate roads. There were those individuals who withdrew entirely from all the cares of the world, sometimes leading a vagrant and miserable life, or else spending their days and nights atop a tall pillar in prayer and contemplation, a practice that became particularly popular in Syria and more common than one might suspect. On the other hand, there were the monks who felt less individualistic and who, organized in monasteries, participated sometimes in the life of the community at large and not always to good purpose. Such were the monks in Egypt to whom I referred earlier and who responded with too much zeal to the incitements of Cyril, the bishop in Alexandria. Monasticism arrived in the West somewhat later and took yet a different form. In the East the monks had emphasized the need for a deprived life and a return to the simplicity of the Gospels, disparaging any kind of intellectual sophistication. The monks in the West were recruited from a better educated segment of society and, as they paid more attention to the study of Scripture, they were also inclined to maintain libraries and schools in their monasteries. It is ironic that in the East the society at large always maintained a higher degree of education without any intervention on the part of the Church, while in the West, after the many invasions, most of the once solid institutions of the Empire had collapsed, including the schools, and it was the task of the monks in their monasteries to preserve a semblance of education, assuming increasingly the major role in the organization of schools, first for the training of new priests and later to educate the children of the entire population. With the breakup of the Empire, life followed very different courses in the East and in the West. The East was better able to withstand the barbarian invasions, a cause of so much disruption in the West. Even in the West, different regions were affected differently; some never really suffered the devastation and plundering of succeeding armies of ruthless men, but were only forced to absorb entire new populations of Germanic people, who themselves had been pillaged and pushed westward. In one case or the other, the central Roman authority soon became merely a facade for the power asserted and often disputed by local chieftains. The old senatorial class of landowners, who had always kept itself one step removed from exercising power directly and had maintained a semblance of tradition and classical culture, saw now its privileges severely curtailed and it was this class, already Christian for several generations, that often found a refuge in the cloister and amongst whom the Church sought to select its bishops. It was not rare that with the crumbling of all civic organization, the bishops would take the role of secular authorities and even became soldiers in a few instances. Amidst so many new immigrants, the Church found that its main mission, besides preserving its own identity, was to spread the Gospel. This is very much unlike the situation prevailing in the East, for in Constantinople the institutions of the Empire had remained intact and there was no opposition between the clergy and the civil authorities. The Eastern Empire was a Christian Empire and the Church, through its patriarchates, wielded a good deal of power. At the same time, amongst the educated classes, an interest in the classical literature of the past was never lost nor was the excitement about philosophical speculation which led, within the Church, to lengthy elaborations of the meaning of the faith and, ultimately, to further disputes on dogma and more heresies. In the West, by contrast, the Church was concentrating on the task of evangelization and on consolidating its own organization as an institution. Broadly speaking, the scale of life in general was very different in East and West. The West splintered into regional powers very loosely held. Trade declined and the old network of Roman roads was left unattended. Communications became slow and difficult. The bishops often found themselves ruling over the towns but even the Church could hardly claim to be an ecumenical force. Power in the East was very much centralized. Throughout the Empire a burdensome tax system was enforced in order to support the lavish court and the many officials in the bureaucracy. Later, during Justinian's reign, enormous funds were also required to finance his overambitious military campaigns. It is not surprising that when his troops landed in Italy seeking to restore the glory of the old Roman Empire, those living in the West were less than enthusiastic about having to send heavy tax contributions to Constantinople. Justinian had some successes and accomplishments, but his absolutist style of government had few beneficial effects. A generation later the great Pope Saint Gregory, when addressing Justinian's successors in Constantinople in his correspondence, was extremely respectful and deferential, while at the same time he would admonish the members of the Merovingean court in Gaul in a rather patronizing way on how to lead a Christian life. Beyond the excesses of the court and the general pomposity of life, the absolutism of the Empire was responsible for a certain amount of intolerance as well. Pagans and Jews were occasionally persecuted and there was an attempt at controlling all forms of expression, leading to the infamous closure of the Academy in Athens by Justinian, at the same time that theological schools were being opened elsewhere. It was during this period that the customs and proclivities that we have come to know as medieval began to take shape, with the consolidation of an elaborate liturgy in the Church, a simpler and more confessional literary taste of a moralizing character, exemplified by the accounts of the lives of the saints that gained enormous popularity, and the emphasis on new forms of expression, such as the art of the icon or the new music of chants brought in by Syrian monks.

58

Albertus: From the point of view of the evolution of scientific ideas, this era was a most sterile one. If we consider the long period of roughly 250 years, during which the city of Alexandria was subject to so much turmoil, from the time of Ptolemy to the disbandment of Hypatia's circle, we witness a complete lack in originality and a progressive indifference toward scientific matters. Thereafter, the scientific tradition that had flourished on the shores of the eastern Mediterranean appears so feeble that one might say it disappeared altogether. In the ensuing 250 years Alexandria gradually ceased to be a city of great learning and no other city acquired a comparable stature until the emergence of Baghdad as a great capital. Nevertheless, before we race ahead to discuss a more fruitful era, I wish to say a few words about the astrolabe, one of the oldest astronomical instruments and probably the one that has been most widely used. Although its history is very uncertain, the earliest references to it date from this particular period. A letter by Synesius of Ptolemais, the scholar who had belonged to Hypatia's circle, has been preserved and in it he briefly describes the uses of the instrument, a version of which, made in silver, he was enclosing with his letter as a gift to a friend in Constantinople. The theory underlying the construction of the astrolabe was surely known to Hipparchus and was explicitly discussed by Ptolemy, as Synesius acknowledges. Ptolemy's text has survived indeed in an Arabic translation. Its basic premise is the use of a stereographic projection to map the surface of a sphere onto a plane. Let us assume that a sphere rests on a plane, making contact with it at one pole. If we extend a line from the opposite pole to any other point of the sphere, piercing through it until the line touches the plane, it is easy to see that we establish a correspondence between each and every point on the sphere, one pole excluded, to each and every point on the plane.

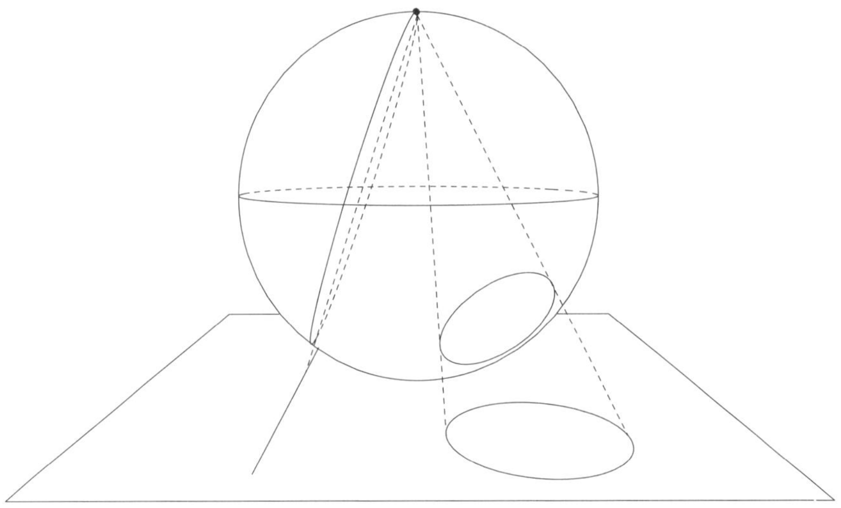

This mapping of the points on the sphere onto the plane has the peculiarity that any circle drawn on the surface of the sphere is made to correspond to a circle or a line on the plane and, conversely, any circle or line on the plane is the projection of a circle on the sphere. Clearly, the lines on the plane, extending to infinity, correspond to those circles on the sphere that pass through the pole from which the projections are drawn. For this reason that pole is sometimes assigned figuratively to a point at infinite distance on the plane. Now, an observer on the Earth can imagine himself at the center of the celestial sphere, on whose surface all the stars are affixed, and employ such a stereographic projection to map the Heavens onto a plane. For an observer on the northern hemisphere it is convenient to use the plane that is parallel to the Earth's equator and that makes contact with the celestial sphere at its northern pole, since the stars are seen to rotate daily around this point which remains forever fixed. The projection is now done from the southern celestial pole onto the plane or, alternatively, from the center of the sphere.

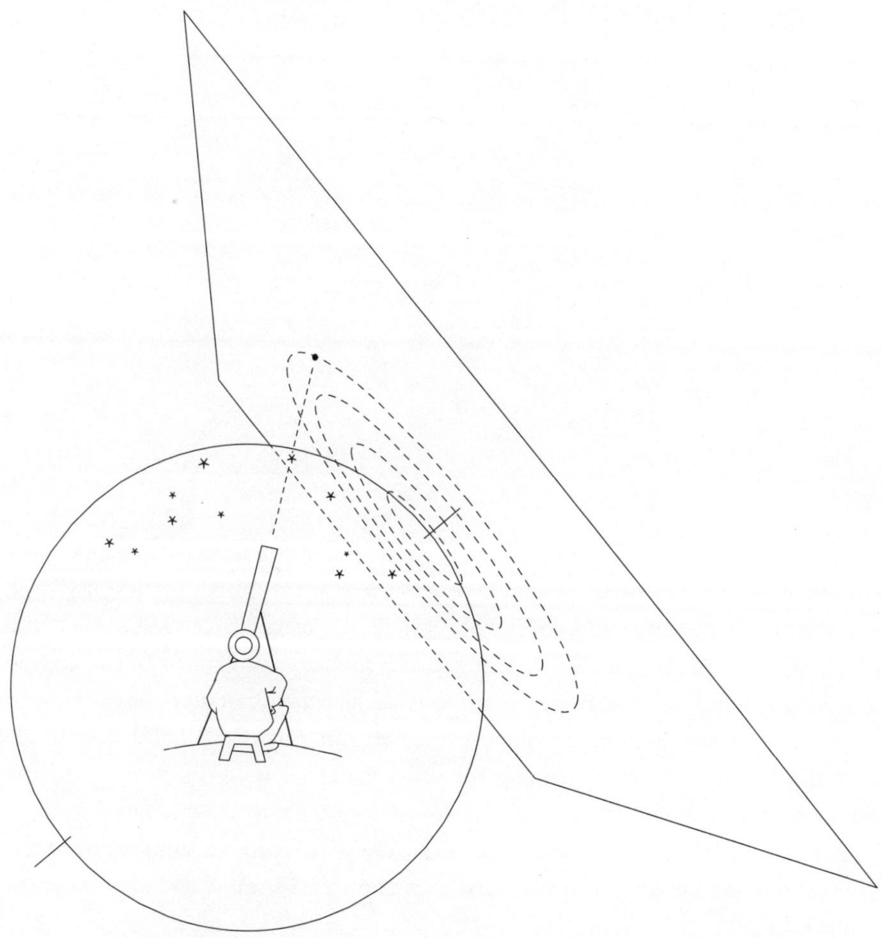

Meanwhile, our observer has another frame of reference which, from his point of view, is fixed and permanent. It is defined by the landscape on Earth that surrounds him, that is, by the circle of the horizon projected onto the celestial sphere, by the circles parallel to it rising up to the zenith and by the arcs of circles perpendicular to them, descending from the zenith down to the horizon. These are the lines that he primarily uses to record his observations. This imaginary grid of lines against which the celestial sphere will rotate, can be projected onto our plane as well, so that now we have two superimposed images; on one hand these reference lines that are fixed relative to the observer, all of which are segments of circles as we have just seen and, on the other, the image of the celestial sphere, of the Sun's path, the Ecliptic and all the stars that complete a rotation every day around the northern celestial pole.

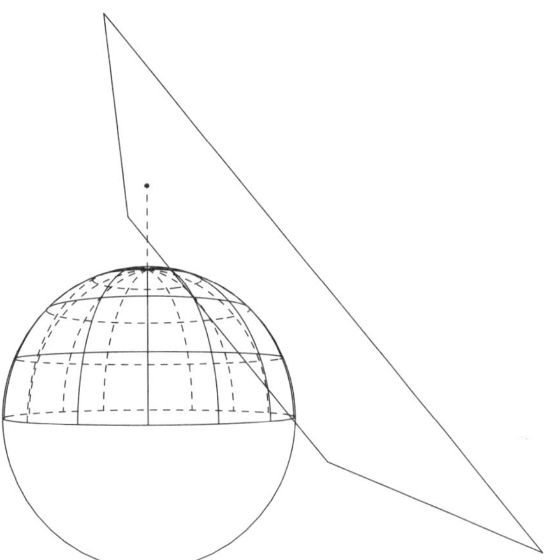

The astrolabe is constructed by cutting out a circle of this plane, sufficiently large so as to contain a good portion of the observable sky, usually including the ecliptic. In fact, we want to cut two copies of this circle, each containing one of the images we have projected, so that we can rotate one against the other and adjust them to the moment of observation. In practice the two superimposed circles were made of metal; the one on top, containing the ecliptic and a few of the main stars, was pierced on much of its surface so as to allow seeing the coordinate lines drawn on the bottom circle. Clearly, these coordinates were peculiar to the latitude where the observations were made and many astrolabes contained a set of alternate circles to be used at different locations.

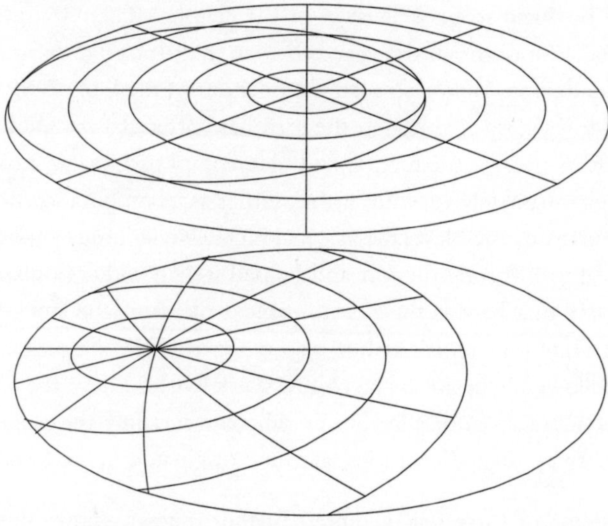

On the obverse face, most astrolabes displayed a graduated scale in degrees and had an alidade, or sighting rule, pinned at the center of the circle, so that the observer, holding the astrolabe with one hand, could measure the altitude of a star or the Sun at any given time.

Then, returning to the front side of his instrument and rotating the rete or top circle in order to place the star at the appropriate location relative to the coordinates on the bottom circle, he could tell the time of day or night or predict the rising of stars above the horizon, the number of hours of nighttime or daylight and such matters. On the obverse face the division of the circumference into 360 degrees could be taken to be a division of the ecliptic. Then an inner circle was divided into the 12 months of the year and each month into days so that with the help of the alidade one could also read the position of the Sun on the ecliptic that corresponded to each day. Although the astrolabe offered a complete representation of the sky, its most fundamental and elementary use was as a clock. Due to its size it was unfortunately very unreliable, either as a computational tool or as an observational instrument. Nevertheless, its practicality and the simplicity of its use made it an ideal tool for instruction and helped it gain wide popularity. Synesius of Ptolemais says that he was the first to discuss the astrolabe since Ptolemy, but we know that Theon, Hypatia's father, also composed a book on the subject and he may have followed Synesius by just a few years. We also have the commentaries of a few other scholars from a later time whereas certainly the earliest surviving astrolabes date from several centuries later and originated in the Muslim world.

Plebeius: Synesius of Ptolemais acquired further renown within the Church because, being married, he was all the same appointed bishop of Ptolemais in Cyrenaica. It is said that he had resisted the appointment, arguing that he could not withdraw the devotion he had to his wife in order to serve the Church, but all reports indicate that he dispatched the duties of his office with dedication and was a well-liked citizen in his native town. To my knowledge, he was one those scholars who advocated the incorporation of classical learning into the Christian conception of life. His reputation must have been considerable, for one of his writings contains a homily he delivered before the emperor condemning the corruption that was rampant in his court. I believe that he died at a rather young age, even before Hypatia's murder.

Albertus: When we consider the development of science during this very extended period, it is not surprising that we tend to return to Ptolemy as a milestone, not only in the development of astronomy, but also as representative of a type of thinking that we identify as our own. In him we recognize first the desire to provide an objective and natural account of the construction of the Heavens, not a supernatural interpretation, but a description of it as an artifact or a machinery devoid as much as possible of fanciful ideas; secondly, the confidence that all the motions of the Heavens could best be described in a mathematical language and, lastly, the need to carry out accurate observations in order to test the validity of the abstract theory. After Hypatia's time we see a decline of interest in mathematics and also a decline in the confidence of using it as a tool for discovery. Even when we find an interest in understanding the contributions of the great scientists from Euclid to Ptolemy, it is not so much in the spirit of contributing to an unfinished enterprise but more in deference to ancient wisdom and expertise. The philosopher Proclus studied Ptolemy's works and criticized them for being too involved, perhaps because he was more interested in philosophical questions than in actual observations; nevertheless, he must have been one of the few amongst his contemporaries who had some familiarity with the methods of mathematical astronomy.

Plebeius: Mathematics must have had some appeal for him, since even on theological subjects he presented his arguments in mathematical form, with propositions followed by demonstrations. Proclus was a a prodigious writer and his interests ranged far and wide, but he also fell under the spell of Plato's doctrines and he contributed greatly to promote them. He was also an excellent expositor and by being diligent and anxious to understand the ideas of others and make them clear to himself, he opens the door for us to understand his age and its intellectual concerns on his own terms and not as we might reconstruct them starting from our own preconceptions. Furthermore, he awakens our sympathy when we learn that his interest in philosophy was not just an intellectual enterprise, but he allowed it to impose certain norms in the conduct of his life, norms of temperance and sobriety and also of generosity, for he was said to be very generous with his own considerable wealth. His father had been an advisor at the Court in Constantinople and Proclus was sent as a youth to Alexandria for his education, expecting that he would follow in his father's footsteps, but he soon became more interested in philosophy and, by the age of twenty, already disenchanted with the climate he had found in Alexandria, moved to Athens, where he remained for more than fifty years rising to be the leader of the Academy. Despite his paganism, Proclus' theology influenced considerably that of the Church and his general philosophic ideas remained influential for centuries. Today we would not pay much attention to the fanciful philosophical constructions he built, subdivided into innumerable categories by artificial subtleties, but the search for some sort of illumination, in which the soul would lose its identity and become one with a being transcending this world, is probably as alive today as it was in his day. I suspect that Proclus' emphasis on the harmony existing amongst all things earthly and divine led him down the same path you criticized in Iamblichus, for his belief in a sacred art of philosophy made him also credulous of occult and magical practices. Most teachers in Alexandria at the time were still pagan, although Christian students participated in the courses too. Gradually there was a revival of theological studies, influenced in good measure by the writings of Proclus, until the early years of Justinian's reign, when John Philoponus, originally from Caesarea, was appointed the first Christian head of the Alexandrian school. Philoponus began to oppose the paganism inherent in Proclus' philosophy, still dominant then. This move had the benefit of averting the closure of the school, saving it from the fate that fell on the Academy in Athens. In any event, John Philoponus was still influenced by Plato's way of philosophizing, but he addressed the great theological debates of his day in a peculiar, materialistic manner that was his own, even when he was considering abstract questions such as the qualities of the soul or the difference between the nature of Christ and the person of Jesus. Could any divine attributes be transferred to a human? Were there two natures of Christ after the Incarnation? Was it possible for divine nature to undergo suffering? Did the resurrection entailed a destruction of the material body? and so on. All these questions kept John Philoponus busy, and many other theologians after him, who may not yet have agreed on their answers to this day, although from the point of view of philosophy these concerns have long since ceased to be of any interest. Philosophy is forever chasing the right questions but not necessarily waiting for the answers.

Albertus: Some philosopher might claim that the questions have never changed; it is only the language about which they are arguing. John Philoponus was also interested in scientific matters and wrote on the astrolabe as well. In the natural sciences his name is well remembered for a few very original ideas of his, which no other contemporary seems to have even entertained. Contrary to Aristotle's opinion that by setting an object in motion we are transferring something into this body that will keep it moving, Philoponus was of the opinion that a persistent state of motion was also a form of rest and needed nothing to keep it going. Thus, he seems to have anticipated the notion of inertia. One might say that a few scattered insights of this nature do not amount to more than mere coincidences, but it is rather surprising that he came upon several of them. He also opposed Aristotle's view on the divine nature of the Heavens and maintained that the Sun and the stars were material bodies, burning in the same manner that we see a fire burn here on Earth. He had a curious argument supporting the creation of the world that was much repeated at later times. He argued that the planet Mars, for example, makes several circuits around the sky for every one that Jupiter makes; therefore, if the Universe had existed for an infinite time, Jupiter would have had to make an infinite number of revolutions, but Mars would have had to exceed that number and it was clear to him that a number already infinite could not be exceeded by another. Therefore, the number of revolutions had to be finite and the Universe had been in existence only for a finite time. If all interest in mathematical astronomy had essentially disappeared amongst the Greek scholars of the VIth century, it is curious that this period coincided precisely with the awakening of that same interest amongst astronomers in India. The development of astronomy in India is a fascinating subject, partly because it has remained shrouded in mystery. In its most remote origins, it could be said that most of the attention was directed toward providing an explanation of the creation of the world, of which the Indian imagination produced many fanciful stories, but also toward devising reliable calendars for the purpose of fixing the occurrence of religious festivals. Each culture has produced its own set of interpretations to explain why we see what we see. The early Indian astronomers thought that the Sun turned around after each day to illuminate the stars from behind the Earth. Given that the production of calendars was one of their main objectives, their observations concentrated on following the motions of the Sun and the Moon. At first, they followed a lunar year and divided the starry sky into twenty-eight regions or stations, lunar mansions, as they called them, so that they could follow the Moon through its cycle. This division originated in India and was never adopted elsewhere. At a later time, the Indian astronomers incorporated notions clearly attributable to Babylonian sources, such as the ratio of two thirds between the lengths of the longest and shortest days of the year, even though it is more inaccurate in most of India than it is in Babylon itself. They also made use of Greek ideas, but how this communication was first established is not clear at all. Remnants of Alexander's Greek settlements in northern India seem to have survived for a long time. Early in Roman times the first migrants from northern China, whose descendants would arrive in Europe centuries later, began to move into central Asia and pushed several groups of Indo-european nomads out of their homeland, through the Afghan passes, into India. The Scythians, amongst them, raided Punjab and descended along the Indus valley, later pushed further south by the subsequent invasion of the Kushans, until they occupied parts of Gujarat and Maharashtra. These groups assimilated rather quickly with the native population and for a long time they kept control of the silk trade routes through Central Asia, linking the West with China by land. They also maintained active trade links with Rome by sea, dispatching ships from Broach, near today's Bombay, along the Arabian and Red Seas, with ivory, jewels, spices and returning with an assortment of metals, or simply wine. The earliest astronomical treatises in India date from this era and they betray the influence of Greek astronomy, probably from the time of Hipparchus. There are also horoscopes that appear to be mere translations of Greek lore. One of the difficulties in surveying the evolution of Indian astronomy is the fact that most documents were never dated and their authors remained anonymous. Oftentimes, older documents were rewritten or corrected several times by distant disciples without leaving traces of the original documents themselves. The knowledge that can be attributed to Greek sources is quite selective. We do not find any discussions of Eudoxus method of concentric spheres to describe the motions of the Heavens, but the method of epicycles was readily incorporated. However, they never used some of Ptolemy's innovations, so one would be led to conclude that they never gained access to the Almagest. For this reason, some historians have thought that old Indian documents would provide a good source to learn about Greek astronomy from the time that we know the least, the two centuries that elapsed since the work of Hipparchus until Ptolemy. The true golden age of Indian astronomy did not arrive until the waning days of the great imperial Guptas, toward the end of the Vth century. Aryabhata is perhaps the first astronomer we know by name and whose writings survived in their original form. He was a native of northern India, from Bihar, but he also influenced the development of astronomy in Kerala in the south, where he may also have spent part of his life. One of his innovations was to propose the daily rotation of the Earth, although he was later criticized for this idea, which was not adopted by his followers. Amongst these, two of the more distinguished, a century later, were Bhaskara, of Kerala, and Brahmagupta, of Gujarat, who shone with their mathematical skills in the elaboration of planetary orbits. Indian astronomy developed characteristics that were different from those of its Greek model. The calendar remained one of its major concerns and, perhaps motivated by religious ideas, it placed great emphasis on the conception of great cycles of cosmic history, so that instead of paying much attention to geometric models, it concentrated on the study of the periodic conjunctions of the planets and on devising arithmetical calculations to predict their arrivals. The details of how and when some of the works of the Indian astronomers were transmitted to the West have also remained in the dark. Initially it appears to have involved some translations into Pahlavi, the sacred language of the Persian people. During the several centuries when the Sassanid dynasty ruled over Persia, they do not appear to have had much interest in astronomy beyond the most elementary question of devising calendars, but there certainly was an interest in astrology and, already during the reign of Shapur I, some translations on astrology from the Greek were available. The Persians then concocted some astrological theories of their own involving the conjunctions of the planets Jupiter and Saturn, so that to each recurrence of this cycle they managed to associate some of the great events in history, such as the demise of a dynasty or the coming of a new prophet. We know that in later times, when the invading Arabs began searching for the expert knowledge they lacked, they first came into contact with the works from India through Pahlavi translations and a considerable time passed before any translations were made directly from Sanskrit into Arabic. It is possible that information regarding these works only reached Byzantium by the Xth or XIth Century and only later may they have arrived in the West, mostly through Arab versions.

Plebeius: A good deal of communication may well have taken place directly between Persia and Byzantium much earlier, at the time of the emperor Justinian. Under the rule of Khusro I a great cultural revival took place in Persia and it continued undiminished for almost a century thereafter, into the reign of his grandson Khusro II. Despite the many years of warfare between the two great powers, their links were never entirely severed. Let us recall the welcome reception that the Persian court awarded to the dissident Nestorians, who were fleeing to the East, as well as the invitation on the part of Khusro I to those scholars who had been sacked from the Academy in Athens by Justinian to come to work in Persia. A large segment of the population in Mesopotamia had migrated from the West, they spoke Syriac or perhaps Armenian in the north, but certainly not Persian. There were also large contingents of Christians as well as a sizeable Jewish community. It is within these populations that the declining hellenistic culture was preserved and it is probably true that, for a relatively short span of time, the main language in which scholars communicated was not Greek but Syriac and a large number of translations were made into this language. When we speak of the Sassanid kingdom, we must keep in mind that it had united, as Cyrus the Great had done many centuries before them, Mesopotamia with Persia proper, even though these two lands, the fertile lowlands of the Euphrates and Tigris, all the way to Syria, and the high arid plateau of Iran to the east, remained always quite different countries. Moreover, Mesopotamia had more than felt the proximity of the Roman Empire, being occupied several times. The military campaigns had been countless, two by Trajan, then those of Septimus Severus, Diocletian and Julian followed, and of course, more than a century later, the wars against Byzantium that started under Justinian and did not end until the fatal blow inflicted by Heraclius on the Sassanid forces. Meanwhile, as Byzantium and Persia bled each other to exhaustion in their territorial disputes, little did they suspect that out of the backward deserts of southern Arabia a new people would rise and, in short order, come to intimidate both into submission. One cannot fail to be impressed by the sudden emergence of Islam, the forceful outburst of the Arabs who, up until that time, could hardly have been called a primitive people but was composed, nonetheless, by small town merchants and traders, along with a vast majority of peasants and numerous tribes of nomad bedouins. Their culture, still dominated by a tradition of lyric poetry and folk tales, can only be said to be much less advanced than those of Persia or Byzantium. However, after the death of Mohammed, driven either by the power of his message or, as some have suggested, by successive years of drought in Arabia that forced the overcrowded towns to disperse, or simply driven perhaps by the prospect of profiting from the debilitated empires of the north, in any event, an uneasy alliance was forged between the more affluent merchants and city dwellers and the tribes of bedouins in the desert, accustomed to making a living as marauders, attacking the caravans that linked the Mediterranean to the Red Sea or the Persian Gulf. As a result of their dashing raids and thanks to their temerity more than to the sophistication of their weapons, they began to conquer lands at a dizzying speed. Neither the horses nor the elephants of the mighty Sassanid kingdom were a match to the dexterity or the endurance of the Arabs' camels. It is interesting that most of their early conquests were made in the territories that were favourable to the dromedary, the soft, desert-like and warm lands from North Africa to the plateaus of Persia. Syria fell in the year 634, Egypt in 639 and Persia in 642. By the end of the century the Arabs had amassed an empire that extended from Spain to the Indus valley and, just as importantly, they came to dominate the seas, most of the Mediterranean as well as the Arabian Sea and beyond. During the initial decades of conquest the booty was so fantastic that there was hardly any need to think of having a salaried army. It is also quite remarkable that these early conquests were not as destructive as one might have expected; the new rulers often had the clairvoyance of leaving the local authorities in place and much of life went on as normal, provided everybody paid the levies imposed on them. In fact, the Arabs began copying the modes of operation of the Persian army and soon imitated quite a few other things. By the year 750, after much dissension and strife concerning the succession of Caliphs, the Abbasid dynasty established itself as dominant, moving the center of power from Damascus to Baghdad and marking the beginning of a great period of prosperity, with the foundation of some new cities and the enormous growth of others, amongst them Baghdad, Samarra and Basra in Mesopotamia, Damascus, Cairo, Tunis and Cordoba further west. The old role that the Arabs had played controlling the caravans that carried the trade between East and West expanded greatly, as has been attested by the Arab coins found later from Scandinavia to China. This extensive network required considerable investments and succeeded thanks to a vast capitalist organization that was able to generate enormous wealth. The transition from the Umayadd to the Abbasid dynasty also signaled the demise of a rather dissipated and indolent Arab aristocracy and the increasing acceptance, within the spheres of power in each region, of those better educated and more resourceful amongst the native populations, to the extent that the Caliph's court in Baghdad was increasingly run by a bureaucracy of non-Arab Persians. Given that the ascendancy of Islam coincided with the epoch when the conquering Arabs finally assimilated much of the more advanced cultures of the lands they now ruled, it has been argued that one cannot speak of a truly Islamic civilization. It is true that many of its accomplishments in the arts and sciences belonged to non–Arabs and even to non–Muslims in many cases. Nevertheless, while it would be an exaggeration to say that it was the old marauding bedouins from Arabia who created a culture of their own and put their stamp on this new civilization, it must be acknowledged that the flourishing of Islam took place under Arab rule, when power was in the hands of a rather enlightened class, quite lenient and tolerant in many respects and it is to their credit that a high-minded, deferential spirit prevailed, even welcoming of foreign contributions that they thought would further their prosperity. Furthermore, we must remind ourselves that the ingredient that probably most characterized this new civilization was the language, which was successfully imposed throughout the vast empire, to the extent that we might be more accurate speaking of an Arabic, instead of an Islamic culture. We see this preeminence of the language even in their use of inscriptions to decorate mosques and other buildings, where the beautiful calligraphy forms part of the intricate ornamentation. I have encountered the overall importance of the language within Islam myself in my modest attempts trying to understand the early evolution of their faith while comparing it to the early stages of Christianity. Unable to understand Arabic, the task is made all the more difficult. One is constantly faced with the fact that a language is not merely a vehicle to convey our thoughts but it shapes them to some extent. It is possible that the Romans never escaped a certain sense of inferiority regarding their language in relation to Greek and Greek remained throughout the lengthy period of Roman rule as the more learned and distinguished tongue. The Arabs acted very differently. The Quran was the manifest word of God and it was written in Arabic; it was by necessity a sacred language. If Islam came of age at the same time it assimilated the cultures that it had conquered, it might be argued as well that the very faith that united all Muslims was very much influenced in its origins by Judaism and Christianity. The prophet Mohammed himself had traveled with several trading caravans to Syria and was exposed to their influence. But Islam never expressed the faith of an oppressed minority, it saw itself as a conquering one. Moreover, it always had a social scope that Christianity lacked, it never separated the spiritual from the temporal concerns of life and its secular and religious preoccupations were one and the same. Thus, government has always straddled the roles of civil administration and moral authority and membership in the community is manifested through loyalty as much as through faith. Indeed, in the beginning the simplicity of the faith was specially emphasized, with little interest in elaborating a complicated edifice of dogma; the faith has no sacraments or initiatory rituals and beyond the recognition of the supremacy of Allah, the one and only God, it merely demanded the practice of prayer and fasting. Of course, one might expect that by bringing under the fold of religion all aspects of human life, Islam lent itself to the possibility of being tyrannical toward the individual, but countervailing this tendency there was another more hospitable one, opposed to the establishment of any institutional authority, any church or priestly hierarchy. By the same token, in its concern about the social order, Islam has a central preoccupation with the law, the law on Earth as much as the law of the Heavens, but even in this regard its concern rests with such things as the fulfillment of contracts, the equanimity in judging and its interest is more in sanctioning than in penalizing. I have found it quite absorbing to compare the various strains competing for supremacy within early Islam, in relation to a similar development in early Christianity, the rivalry between those who favoured a literal interpretation of the Quran and an unadulterated version of the doctrine and those who saw in reason and in argumentation the path to elucidate the faith and to know the will of God. The very notion of Islam, of submission to God, does not entail any sense of guilt, as in Christianity, or of yielding to power, as in Judaism, but is born out of humility and of resignation to one's ignorance. Despite its other-worldliness, Islam showed an acceptance of this world too, a favourable view of life, more hopeful than despairing. This gave it a less idealistic and more practical outlook than was possible in Christianity; this is apparent in such attitudes as its rather favourable disposition toward wealth or, at the very least, its reluctance to require poverty in virtue.

We see it even in their treatment of slaves. Slavery was a considerable institution in the Muslim world and the conquering Arabs went to great lengths to provide themselves with a good supply of slaves, Slavs from beyond the Caucasus, but also blacks from the Sudan, Turks from Central Asia or Christians from Italy and Spain. They bought them from the Jews and much later through the intercession of Venetian traders. The slaves were often assimilated and, in some instances, as with the Mamluk soldiers in Egyt, they acquired power and influence. Of course, throughout Islam and over many centuries, there has been plenty of opportunity for it to turn against its own rules, to betray its own promise and to undo much of what good it had accomplished, showing, if proof need be, that mankind is much better at formulating desirable norms of life than in fulfilling them. The promise of an open, respectful and tolerant society often turned intolerant and rigid in its orthodoxy. The members of the Ulema, the wise and learned, the defenders of the faith, would also grow into an isolated class, sometimes acquiescent to an unjust government and complicit in its excesses, sometimes antagonistic and withdrawn, inclined toward a more individualistic, ascetic and indifferent attitude. Thus, we encounter the same perversions that were to be found in the Christian Church to the West.

Albertus: Let us now consider the impact that the rise of Islam had on our subject, perhaps not in as much detail as we should, but at least to evaluate its influence on the study of astronomy, for we must bear in mind that for more than five hundred years the Islamic world was home to the most advanced civilization, a period that is comparable in length to the subsequent dominance of Europe at the frontier of knowledge, after the Muslim world lost much of its original vigour.

Plebeius: Unlike Christianity, Islam was sympathetic to the pursuit of knowledge. In its acceptance of this world, it was more inclined to celebrate the Creation than to bemoan its sins. As a consequence, instead of being an obstacle, it proved to be a stimulus to human curiosity. In its long history, of course, it has had plenty of opportunity to breed its own share of bigotry and narrow-mindedness but, in principle, the Quran is quite explicit, as when it exhorts rather quaintly: Seek knowledge, in China if necessary. It even goes as far as to say that the ink of the scholars is worth more than the blood of the martyrs, an unambiguous and quite daring statement to make, when you pause to think about it.

Albertus: The early Islamic scholars lived up to that exhortation indeed. It never ceases to astonish the enthusiasm with which they pursued all forms of knowledge and the effort they invested in acquiring the expertise and the wisdom that the ancients had accumulated before them. This enterprise was in full swing at the time of the foundation of Baghdad and the establishment of the Abbasid dynasty, certainly during the rule of the Caliph al Mansur, in the first half of the IXth century. Already under the Umayadds, some scholars had began to do translations of Persian books in Pahlavi, sometimes themselves translations of Indian works in Sanskrit, some of which, in turn, may have contained material from older Greek sources. This work became a project of enormous scope under al Mansur, after the foundation of the Library in Baghdad, the so called House of Wisdom. Here teams of translators collaborated in rendering into Arabic many Syriac and Greek works in medicine, astronomy and philosophy, while emissaries were occasionally sent to Byzantium to obtain copies of works that were not otherwise available. Amongst the translators, Hunayn Ibn Ishaq, originally a Nestorian Christian, an indefatigable worker, deserves special mention. He was a polygloth who, in addition to being the Caliph's physician, translated Hippocrates, Galen and many other works. His son also became a translator from the Greek and was responsible for one the first Arabic versions of Euclid's Elements and of Ptolemy's Almagest as well. When we speak of translation we must think that in many cases the subject matter was more foreign than the language in which it was expressed, so that often words had to be invented to translate concepts for which the Arabic language had no equivalent. Many of the translations had primarily an utilitarian goal, as was the case with the books on medicine, but the many translations of astronomical and philosophical works reveal their wide-ranging interests and almost indiscriminate curiosity. Sometimes, it is true, the interest in astronomy was shared by an interest in astrology, as in the conspicuous case of Abu Mashar, probably a very bright and diligent individual, who became acquainted with much of the astronomical lore in the Persian and Indian sources, as well as with some of the Greek works, but having developed an interest in astrology, put together a rather garbled composite of all this knowledge in a number of astrological works that remained very influential for ages to come. Unfortunately, many centuries later, when modern historians have searched for the lost and precious works of some of the more prominent scientists, they have often stumbled only upon manuscript copies of Abu Mashar's books. For the most part, however, the Islamic scholars rejected astrology and it is quite refreshing to see the earnestness and the open-mindedness with which they faced the knowledge and expertise they were importing into the Arab world. You spoke of how the early Christian theologians had also set out to study the works of the Greek philosophers of centuries past. Yet, would it not be fair to say that they did it with the intent of investigating how much these writers had been able to preannounce a truth that only later had become manifest? I have the impression that the studies initiated by the scholars of Islam had by comparison a totally unprejudiced character and were done truly in the spirit of exploration and discovery. The library in Baghdad soon became the best in the world and the community of scientists it supported was comparable only to the one that had existed in Alexandria in the distant past. By the end of the IXth century the mathematicians of Baghdad were fully conversant with the works of Apollonius and

Diophantus and could extend some of their results. Only more than a century later would another scientific school develop in Cairo and another one somewhat later in Cordoba. These centers with their extensive libraries were administered by a professional staff and were comparable in organization to those we would find in modern times. The same can be said about two other types of institutions that the Islamic world pioneered prefiguring their modern counterparts, namely, the hospitals and the astronomical observatories. The copying of books must have generated a very active industry for, in addition to the large official libraries, there were also many private ones, well stocked, often sponsored by wealthy landowners and merchants. Although the best scientists tended to gravitate toward a capital like Baghdad, many of the best physicians and mathematicians had in fact come from the most disparate and remote lands were they had received their early education and training. One of the more distinguished and influential Muslim astronomers was al Battani, who was in his prime in the beginning of the Xth century, when Islamic astronomy had already reached more than a satisfactory level of maturity. al Battani was born in Harran, in northern Syria, and spent most of his fairly long life in the not too distant town of Raqqa, on the Euphrates, were he set up his own observatory and where, for more than forty years, he carried out numerous observations. Although he made some mathematical contributions relevant to astronomy and even put together a small catalogue of stars, his main contribution was his book, the Science of the Stars, consisting of a set of astronomical tables. As an astronomer, al Battani was primarily an observer. In this respect, it may have had some influence the fact that his father is believed to have been a prominent instrument maker. He measured the obliquity of the ecliptic, determined the mean motion of the Moon and the variations in its apparent diameter, as well as the variations in the diameter of the Sun and studied how this affected the occurrence of eclipses. He was one of the first to consider in detail the variations in the length of the seasons and the motion in the sky of the Sun's apogee, along with an improved measurement of the eccentricity of the Sun's orbit. In all these endeavours he took to heart Ptolemy's advice not to take his own measurements as the final word but to repeat them, and al Battani had no compunction in correcting him in several instances. Nevertheless, in most matters regarding theoretical astronomy, he followed Ptolemy's model. It is quite telling that despite his expertise in the subject, his own exposition of the theory is rather careless and marred by many errors, whereas all his effort goes into describing how he performed his observations. From this point of view, al Battani was a very modern astronomer, the first not only to describe the instruments he was using, but also to indicate the possible margins of error in his observations. Perhaps for this reason he gained such a reputation that his work remained a valuable reference for a very long time and his observations were cited by Copernicus, Kepler and many others. Of a younger generation than al Battani, we should mention perhaps al Sufi, who worked out of an observatory in Shiraz, in southern Persia, and whose main accomplishment was a reworking of Ptolemy's catalogue of stars with some improvements in the determination of the stars' coordinates. Even if al Sufi may not have been a scientist of the first rank, we owe to him the fact that to this day many stars are known by their Arabic names. These astronomers lived during a period when the Abbasid caliphate was in the process of desintegrating. From that time on, the vast Islamic empire would remain splintered in separate regions, administered by local rulers, dissident caliphs, new dynasties or military adventurers. However, the study of the sciences had acquired a secure foothold by this time and was pursued across borders and regardless of regional disputes. It may truly be said that science had become an international enterprise and Arabic the agreed language of communication, playing the role that Latin would play in Europe centuries later. I do not think that much is know about how education was organized in the towns throughout the Islamic world, beyond the insistence on a good grounding in grammar and on the teachings of the Quran in the schools that operated within most mosques, but we ought to assume that the conscientious teaching of a few scientific principles must have been fairly widespread and considered to be part of a good education; otherwise, it seems quite astonishing that in the most remote corners of the Muslim world one generation after another could produce scientists of outstanding ability.

Plebeius: It is interesting to compare the general features of life throughout Islam with the conditions prevailing at the time in Europe. In Europe too the order and the institutions that had been forged during Charlemagne's long reign proved too difficult to uphold and the Empire fell into progressive stages of dissolution, broken up into kingdoms, themselves unable to preserve much cohesion. Very soon they were ruled by figure-heads who presided over an array of autonomous fiefs. These were often doing battle against each other for the possession of lands, guided no more than by alliances of convenience, personal rancours and outright greed. The elaborate and efficient collection of taxes maintained during the time of the Empire was replaced by local levies or extortion money for protection. Even some of the vast holdings of land owned by the Church and administered by the Empire were now partitioned, so that the clergy as much as the nobility participated in these land disputes. Manorial lords, bishops and abbots were all pitted against each other in petty fights. A new military generation came to the fore, upsetting all social hierarchies; its chieftains, who had often began as leaders to gangs of brigands, would provide the new class of opportunistic nobles. The breakdown of the social order, the exposure to predators and the general insecurity of life gave greater importance to the fortification of towns, to walled monasteries and the construction of castles. Whereas earlier a local lord might have resided in a country house near an open peasant village, now life gravitated toward the interior of a castle. These were not the solid constructions of a later age, with stone walls, turrets and solid gates, but simpler wooden structures, most commonly with log palisades surrounded by a ditch. The adversity and the hardships of this era were such that its interests were circumscribed to the local environment. The contrast with the cosmopolitanism that prevailed then throughout Islam is all too manifest. When we search for the amenities of life, the range of aspirations or the vigour with which these were pursued, we find very little that catches our attention. It is possible that amidst so much strife and uncertainty, the emergence of the feudal order was seen as providing a minimum of order and stability in everyday life. One must take into consideration the impact of the persistent foreign invasions, the arrival of the Vikings from the North, the Muslim incursions from Spain and along the Mediterranean coast and the attacks launched by the Hungarians from the East. The Vikings had long conducted raids along the coastal towns of the North Sea until they occupied the Channel region of France and proceeded further south making fast incursions inland, taking advantage of its excellent network of rivers, sometimes moving from one river to another by dragging their boats overland. Much of Brittany and the southern Atlantic coast remained deserted and in a chaotic state during the Xth century. The Mediterranean coast was for the most part closed to traffic, controlled as it was by Muslim pirates. The free towns of Provence had hastily resorted to scattered remains from Roman times in order to build defenses of their own. The Muslim forces managed to go as far north as Grenoble and for a considerable period of time they controlled all the Alpine passes between France and Italy. Given the deterioration of roads, the presence of marauders everywhere, the variety of dialects spoken, one could hardly speak of France as a unit yet. The French monarchy was essentially cut off from the region south of the Loire and attempted to assert itself only in the fertile basin of the Seine and Marne rivers. Paris, which was little more than a village on the Ile de la Cité would slowly emerge as the center of political power, while Rheims, an episcopal see, became the ecclesiastical capital. In the north of Italy, the towns of Lombardy suffered as much as the south of Germany from the Hungarian attacks. Most of them had walls built around them. In Germany, the remains of walls dating from Roman times were often refurbished, other times entirely new fortified towns were erected. The Germans, fighting on foot, were typically overridden by the mounted Hungarians and they managed to hold their own thanks to the fact that up until this time Germany remained covered with vast expanses of forests and marshlands or swamps that made the movement on horseback not entirely easy. It is all the more surprising the extent to which the Germans developed a very active trade, even though most of the traffic was confined to the main rivers and the towns established on their shores, such as Cologne or Mainz. If Germany enjoyed a higher level of prosperity during the Xth century than the rest of Europe, it is due perhaps to the organizing power of the strong Saxon kings, but the economy of the country remained entirely agricultural with only an incipient mining industry that did not become greatly profitable until much later. In the south of Italy, the towns along the Tyrrhenian coast, such as Salerno and Amalfi, had to eke out a precarious existence in the proximity of the Muslim fleets that controlled the western Mediterranean. Rome almost fell to their offensive, had it not been for the intervention of Byzantium. Neither the Adriatic nor the Aegean Seas were ever dominated entirely by the Muslim navies, so that the towns of the eastern coast, from Ravenna to Taranto, were able to sustain a more profitable amount of trade. Amongst them Venice, often in close cooperation with Constantinople, emerged as the preeminent power and the gateway for eastern goods to much of Europe. The Venetians, as I mentioned, also participated in an active slave trade, consisting mostly of Slav captives from Central Europe that were then shipped to Spain or Egypt. In this business they were acting as middlemen for Jewish traders that had settled in Germany. The Jews had maintained an amicable relationship with the Arab populations, both in the Middle East and in Spain, and moved with them along the Mediterranean coast, settling down in southern France and Italy, from where they were soon persecuted and pushed further north into Germany. If throughout the territory of Europe most forms of organized life were in a rather precarious state, the Catholic Church found itself in complete disarray. The Papacy went through one of its least decorous moments when the infamous Teodora with her two daughters, Marozia and the younger Teodora, dominated the scene in Rome, apparently making and unmaking Popes at their will, in a long series of intrigues, conspiracies and suspected poisonings. Marozia managed to have one of her sons appointed Pope as John XI, at age twenty-four, only to see him deposed, imprisoned and presumably murdered by an older son of hers. Nevertheless, she later succeeded in having her grandson named Pope, at age eighteen, who took the name John XII in honour of his disgraced uncle. Nepotism and simony in the Church continued undiminished for a long time. Sylvester II, who became Pope for a few years at an old age, at the end of the Xth century, admitted openly that he had paid a good sum of money for his bishopric although, once in that position, he had been able to profit a great deal more through the appointments of priests and deacons. A native of Auvergne, Gerbert of Aurillac, as he was known before ascending to the Papacy, had been headmaster of the famed cathedral school at Rheims, where the future Saxon Emperor Oton III was his pupil and under whose patronage he would later be elected Pope. Before that time, through a good deal of intrigue and opportunism, he managed to be appointed abbot of the monastery at Bobbio in Italy, later archbishop of Rheims itself and then, again in Italy, archbishop of Ravenna. There is no doubt that, for his times, he was a very cultivated man. He lectured on Aristotle and Cicero and took particular interest in mathematics, to the extent that he might be considered to be the only mathematician ever to become Pope. In his youth he had spent some time in

Spain, where the presence of Muslim teachers made it possible to receive a much better education. A good portion of his correspondence has been preserved and he apparently often refers in it to astronomical subjects, or gives instruction on how to solve arithmetical problems. He also mentions books he is trying to obtain or Arabic works he suggests should be translated. He was certainly ambitious, but also intellectually curious and restless. Considering the general atmosphere of the era it would be safe to call him a pioneer, perhaps born before his time.

Albertus: Indeed, the movement to translate Arabic works in large numbers did not start in earnest until a century later at least. Allow me now to return briefly to the accomplishments in astronomy of his more advanced contemporaries throughout the Islamic world. After the decline of the Abbasids rule, a new center of power developed in Egypt under the Fatimid dynasty. The Fatimids were Isma'ili, originally Shi'a Muslim from Yemen who had opposed the Sunni rule from Baghdad. Approximately a century later their rule in Egypt collapsed amidst disastrous famines and great disorder, but for a time they were quite prosperous and diligent, especially during the rule of the caliph al Hakim, who took the initiative to sponsor the construction of a great library and an astronomical observatory. Two great scientists who took advantage of these favourable circumstances were Ibn Yunus and Ibn al Haytham. Ibn Yunus descended from an aristocratic family and he acquired the reputation of an eccentric and absent-minded man who also enjoyed writing poetry. He is said to have catered to al Hakim interests in astrology as well. We know from his writings that he had already been present at the foundation of Cairo in the year 969 and it is probably in Cairo that he carried out the observations he recorded in his astronomical tables. These included several eclipses, conjunction of planets and other typical measurements, such as the precession of the equinoxes or the obliquity of the ecliptic. However, unlike his predecessor al Battani, Ibn Yunus never describes the procedures he followed in his observations nor the instruments he used. Instead, he wrote extensively on how to solve some practical problems, such as the translation of dates between different calendars, how to determine the meridian at any given location from three positions of the Sun during the day and other such problems in spherical geometry. He also became well known for writing the procedures to determine accurately the prescribed times for the daily prayers. Obviously, the need to keep the time throughout the year gave to astronomy a particular significance within Islam. Every mosque had a time-keeper and this necessity must have been the source of employment to many an astronomer. Ibn al Haytham belonged to a generation younger than Ibn Yunus. He concerned himself with astronomical problems of a theoretical nature and wrote a lengthy criticism of Ptolemy's theory of the planets, but his main accomplishment was perhaps in the field of optics and in the study of the visual system. He was far ahead of his time in doing refraction experiments and he even attempted to measure the height of the atmosphere from the effects it produced on the setting Sun. This inaugurated a series of researches in optics in the Islamic world that led, amongst other things, to the elucidation of the phenomenon of the rainbow. A contemporary of al Haytham was another of the greatest scholars of Islam, Abu Rayhan al Biruni, a native of the region of Khwarazm, south of the Aral Sea. As we learn from al Biruni's own writings, all the territories from the Caspian Sea to Transoxania, from Gurgan to Bukhara and Samarkand, were disputed at the time by several dynasties, so that in his younger years, having distinguished himself already as a learned man, he often found himself acting as an emissary and diplomat between them, making astronomical observations on the way and carrying on with his extensive writings too. Eventually he came under the tutelage of the Ghaznavid dynasty, led by the sultan Mahmud. These were Turkish people who took residence around the city of Ghazna, south of Kabul on the road to Kandahar; they became Sunni Moslems, although they were greatly influenced by the revival in the Persian language and culture. Thus, in addition to his native Khwarazmi language, al Biruni also spoke Persian, which he favoured for literary expression, and Arabic, which he found more suitable for scientific matters. He also knew Greek and Syriac. When the sultan Mahmud launched his devastating campaigns in the north of India, al Biruni took advantage of the opportunity to travel extensively through the country and managed to learn enough Sanskrit in order to translate some astronomical texts. One of his lengthier surviving works is a compilation of his observations gathered in India, which has been a wealthy source of information for historians ever since. Unfortunately, the good that al Biruni did could not compensate for the damage wreaked by Mahmud's troops, who each winter descended onto the Punjabi flatlands pillaging towns and destroying Hindu temples. al Biruni settled in the end in Ghazna, where he did a prodigious amount of writing. It is estimated that the list of his works exceeds one hundred and fifty titles. He epitomizes better than anybody else the encyclopaedic scholar of Islam, even above his contemporary Ibn Sinna, who had a greater influence later in medieval Europe, but whose inclination was more philosophical and toward the examination of abstract ideas, even in his medical writings. al Biruni had a much wider range of interests and wrote in a more engaging manner, either as a historical chronicler, as a student of foreign customs and traditions, as a linguist or as a geographer. He wrote also on pharmacology, on minerals, on climate and a lengthy work on astronomy, which he dedicated to the son of Mahmud, the new sultan Mas'ud. According to legend, the sultan offered him an elephant load of silver pieces as reward but al Biruni declined. His astronomical writings are not particularly original in their results, although they were in the range of his speculations, for he discussed the possible rotation of the Earth and the heliocentric hypothesis of Aristarchus too and, quite extraordinarily, he seems to have entertained the possibility that the planets' orbits might trace ellipses instead of circles.

After the death of al Biruni the most active centers pursuing astronomical studies were in the West, in the Iberian peninsula. In fact, almost a century earlier, during the rule of the Umayyads in al Andalus, the Caliphate of Cordoba had seen a period of splendour in the sciences and the arts, principally under the leadership of the caliph al Hakam II, who had worked diligently to build a great library in Cordoba, sending missions to Cairo and Baghdad in order to fetch books or to make copies of those that were unavailable. However, after his death, the most conservative amongst the clerics led a reaction against all forms of learning, they promoted book-burnings and created a climate that was inhospitable to scholars. This was similar to what had happened in Baghdad after the liberal reign of the caliph al Ma'mun, during the decline of the Abbasids' power, when a rather short-lived reaction had also risen, as it is often the case, in defense of the simplicity of the dogma against the presumed corruption bred by excessive learning. In the second half of the XIth century, the Caliphate of Cordoba was dismembered into a few independent and unstable kingdoms that engaged in almost constant quarrels; nevertheless, they inaugurated a period of renewed interest in the sciences as well as in literature and music, a period that lasted for well over a century, until the times of Ibn Rushd, or Averroes, as he came to be known in Spain. The main centers for this activity were Cordoba, Seville and Toledo; astronomy was first promoted in Toledo, although the city would soon fall to the Christian forces of the king of Castille. A good fraction of the work in astronomy was of a theoretical nature and it had a philosophical motivation, for it formed part of an attempt to revive some of the ideas Aristotle had advocated, including the planetary theory of concentric spheres, long discarded. However, there were also a number of important observations carried out during this period and it appears that the astronomers in Toledo were the first to disentangle correctly the two phenomena of the precession of the equinoxes and the motion of the solar apogee, that is, the change in orientation of the Earth's axis and the change in orientation of the Earth's orbit, properly speaking, of its eccentricity. Moreover, the work they carried out in compiling astronomical tables acquired a greater historical significance when the scholars in Europe, many years later, first learned of the achievements of Islamic astronomy through them.

Plebeius: Whether it was in the dismembered caliphate of Cordoba, in Egypt, in the fertile Crescent or in Persia, the incessant disputes and the rivalries between dynasties made power so fragile and the future so unpredictable that one wonders how it was possible to create a climate favourable to sustain scientific studies. Only perhaps during the rule of the Abbasid Caliphate was there enough stability for a sufficiently long time. Aside perhaps from the Iberian peninsula, the external threats to the Islamic world were never as disruptive as the internal dissensions, although the influx of new populations, mostly from Central Asia, also changed its character from time to time. Toward the middle of the XIth century, the Turkish followers of the chieftain Seljuk arrived in large numbers, occupying Persia first and taking control of Mesopotamia and Syria later. These people had already converted to Islam, adopted Arabic and contributed to make the Sunni orthodoxy dominant throughout eastern Islam. They can only be called a foreign usurping power even though, in retrospect, their long rule of more than a century seems to have been rather benign, for they concentrated on administrative and military matters and left the underlying social order more or less intact. They may even have been taken in also by Persian culture, contributing to its wider appreciation. The later Mongol invasions, when the Seljuk power was already in decline, proved to be a great deal more disruptive. After the troops of Gengis Khan crossed the Oxus river, they began a series of devastating incursions to the south that culminated in a massive invasion years later under the command of his grandson Hulagu Khan, leading to the fall of Persia first, then Anatolia followed and concluded with the taking of Baghdad in 1258. They were stopped only in Syria by the fierce Mamluk army of former slaves from Egypt. Baghdad was reduced to a provincial town, never to regain the supremacy that had made it one of the great capitals of the world for almost five centuries. Likewise, the voluminous trade between East and West that had contributed so much to the prosperity of Islam never recovered its former pace. Furthermore, although the Crusades had been by and large only a minor disruption for the populations of the Middle East, they did contribute significantly to loosen the control that the Moslem navies held over much of the Mediterranean and, consequently, they contributed to a sizeable decrease in their revenues. The expansion of the Mongol people was nothing short of colossal. After Hulagu Khan conquered Mesopotamia, another army under the command of his brother Kubilay began to gain control of the north and south of China. Finally, still more devastating for the Islamic world were the Mongol invasions of a century later led by the infamous and lame Timur, or Tamerlane, as he became known in the West, originally only the steward of the local prince in Samarkand, later a rebel and an insubordinate who became the actual ruler at age thirty-three. Thereafter, for more than three decades, he carried out long, murderous campaigns throughout Persia and the Near East, plundering and looting wherever he went. He managed to control a very vast territory, extending at one time from eastern Anatolia and Syria to the north of India. It appears that the only praiseworthy activity he undertook was to embellish the city of Samarkand, some of whose buildings have survived and, however reluctantly, can be appreciated to this day. Unlike his predecessor Hulagu, whose principal wife was a Nestorian Christian, Tamerlane instead persecuted the Christians vigorously and the large community of Nestorians that had remained in Mesopotamia through the centuries ceased to have any noticeable presence there after his time.

Albertus: Curiously enough, the period of the Mongol invasions may also have been the more prosperous for astronomy. However, this may only be a coincidence and due to the presence of a few brilliant individuals who, from a historical point of view, because of their accomplishments give brilliance to the era in which they lived. We are now dashing through a very long era of several centuries during which the peaks of success and excellence are few and far between. All this turmoil and the shifting winds of fortune sweeping through the Islamic world that you spoke about, hurt the scientific enterprise and astronomy in particular, which, in the process of discovery, has always required long and stable periods of systematic observations. Nonetheless, one such peak of excellence was attained under the guidance of the astronomer Nasir al Din al Tusi who, as an Islamic scholar, deserves, by most accounts above anybody else, the Arabic title of hakim, or wise man. He was born in Tus, Khurasan, in northern Persia, and never seems to have strayed too far from his native land. He completed his education in Nishapur, at the time still a thriving center. In the true manner of an Islamic scholar, he interested himself in all fields of knowledge. His talents were recognized early and this gave him some prominence in society, but the circumstances of his life during his most creative years are somewhat obscure. At a time when power shifted hands all too often in Khurasan, it appears that he sought refuge amongst the Isma'ilis, the vehement shiite sect to which he belonged and that had long been entrenched in their forts in the highlands of Khurasan. To say that he was under their protection may be a misrepresentation, for it is not clear that he was not at their service or that he could move at his discretion. During this time he wrote a long list of successful books in a wide variety of subjects. Like al Biruni, he was an encyclopaedic scholar, but the ways in which they went about in the pursuit of knowledge were quite different; al Biruni relied more on the details of observation, whereas Nasir al Din, being more introspective, searched for first principles in the inner workings of the mind. Thus, the subjects on which he wrote range from logic and mathematics to ethics and theology and he even wrote some poetry. Some of his numerous mathematical writings were didactic summaries of the works of Euclid, Archimedes or Ptolemy, but a good part of them were his original contributions; in particular, he wrote the first satisfactory book treating trigonometry as a complete mathematical discipline. His two books on ethics gave him lasting fame and they circulated widely throughout the Moslem world for centuries; also his writings on theology became the canon amongst the Twelver or Imami Shi'a, as they came to be known. At any rate, when the Mongol invasions overran Khurasan, Nasir al Din was already in his fifties, had considerable fame and entered into the service of Hulagu Khan, whom he accompanied to Baghdad. The Mongol leader appears to have had an interest in astronomy and Nasir al Din persuaded him to sponsor the construction of a great observatory in Maragha, not far from Tabriz, in Azerbaijan, with a great adjoining library containing more than a hundred thousand volumes, most of which, in fact, the Mongols would raid from other libraries in Syria and Mesopotamia. Nasir al Din dedicated the last fifteen years of his life mostly to astronomy, first to the establishment of the observatory and then to the compilation of a set of tables derived from the observations made at Maragha. At the observatory he surrounded himself with a team of capable scientists, including his sons as well as al Urdi and, for a time, Qutb al Din al Shiraz, all of whom made substantial contributions of their own, either in the design and construction of instruments or in devising variations of Ptolemy's theories of planetary motions. The reputation of the observatory traveled sufficiently far that it attracted at least one Chinese scholar. Nasir al Din introduced a very ingenious mathematical device that he incorporated in his description of the orbits of the planets. He observed that when one of two circles, having half the diameter of the other one, is allowed to rotate smoothly inside and along the circumference of the larger circle, then any fixed point attached to the edge of the small circle will move along a straight path, up and down along a fixed diameter of the large circle.

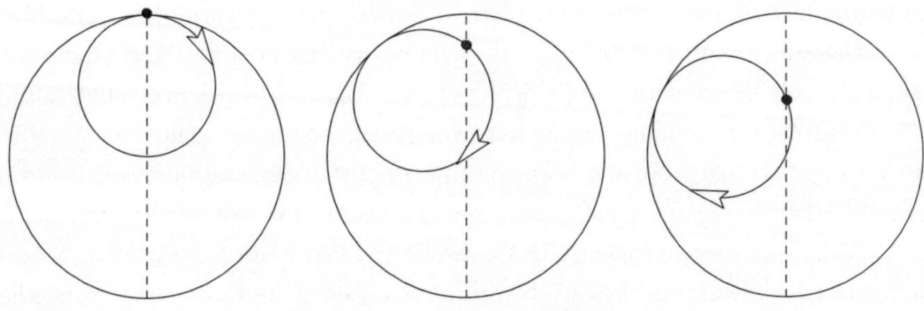

That this is so can be demonstrated using very simple geometry and Nasir al Din would be rather surprised that from all the things he wrote we would choose to single out such a trifle of a contribution, a mere mathematical oddity. Furthermore, he elaborated on this trick, in order to apply it to more general cases, but the essential ingredients and its importance are already manifest in its original form. It is the ability to transform the rotational motion of circles into straight motion along lines. We have seen that, from antiquity onward, there was a desire to reduce all the motions of the sky to rotations of circles or spheres and there was a certain uneasiness whenever circumstances required to appeal to eccentric circles, non-uniform motion along circles and other seemingly unnatural tools. al Tusi incorporated his device into the traditional models of planetary orbits derived from Ptolemy in order to account for such phenomena as the departure of the planets from the plane of the ecliptic, a puzzle that Ptolemy had been unable to solve satisfactorily. It was later used for other purposes by al Tusi's followers and many other astronomers. One of the more distant ones was Ibn al Shatir, perhaps the more accomplished of his era. Working in Damascus almost a hundred years afterwards, he employed al Tusi's ideas to perfect Ptolemy's account of the motion of the planets. His theory of the motion of the Moon, it was found later, is identical to the one Copernicus produced almost two centuries after him. After the time of al Shatir, Islamic astronomy began to decline. It had, however, one last period of glory. You spoke of the devastation caused by Tamerlane and his troops, but some of his descendants were temperate and benevolent rulers. His son, the Shah Rukh, who ruled over Persia, was a patron of the arts and his grandson Ulugh Beg became somewhat of a scientist himself. Unfortunately, when he took the reins of government in Transoxania, he proved to be ineffective and was murdered soon thereafter. Nevertheless, twenty years earlier, around 1420, he had led the way to build a large observatory near Samarkand. One of its instruments was an enormous sextant, namely, an arc of sixty degrees designed to measure altitudes above the horizon, that was built of stone on the slope of a hill, the largest until that time, with forty meters of radius. The observatory was excavated in modern times and the remains of this unique instrument have been found and preserved. Ulugh Beg counted with the assistance of al Kashani, a most skillful mathematician and author of the best work on arithmetic of his time. Here in Samarkand they carried out a number of observations, on the inclination of the ecliptic, on star positions and on the various parameters defining the orbits of the planets. Their astronomical tables, along with excellent trigonometric tables, were the most accurate to date and, together with those of al Tusi from the Maragha observatory, became the more enduring legacy of Islamic observational astronomy. The Moslems' interest in this science did not die altogether. As late as 1547, during the Ottoman rule, an observatory was founded in Istambul, but its contributions during its rather brief existence were minor and by that time astronomy in Europe had already advanced a good deal further.

Plebeius: In Europe the influx of new migrants from Central Asia and the threat of their predatory raids subsided toward the end of the Xth century, much earlier than in Islam. The Magyars or Hungarians were the last nomadic group of Turkish and Mongolic origin to carry out destructive incursions into the territories of Germany, France and Italy. After their defeat they settled in the steppes along the middle course of the Danube and converted to Christianity. The most significant act of Sylvester II during his brief Papacy was perhaps his participation in the coronation of Stephen, their first Christian king, known later as saint Stephen. With the pacification of the eastern frontier the Danube was open once again for German merchants to resume trading with the populations of the Black Sea and with the Levant, even though this route never came to replace the route south through Venice. The disappearance of the Hungarian threat also meant greater prosperity for the independent towns of Lombardy. Here and also in Tuscany, the feudal regime was not as well established as in the rest of Europe; many towns, growing from small rural parishes and staying away from the conflicts amongst the nobles and the Church, progressively developed an industrious class of artisans and merchants that made them quite affluent. Generally speaking, after the year 1000 the well-being of Europeans began to rise. Despite the numerous regional wars, the disputes between feudal lords, between barons and Church, between rivalling cities, many regions enjoyed almost uninterrupted prosperity. Whether one thinks of the fertile plain of Lombardy or the towns along the Rhine valley, of Flanders or Languedoc, commerce and new industries developed at a steady pace. Without the influx of new migrants, the population became more settled. Perhaps this profound transformation was also spurred by the demographic explosion that was then underway and that continued undiminished for three centuries. This is particularly noticeable in Germany, where it subsequently led to the expansion of the Germanic people beyond its eastern border. Throughout this land, an impressive campaign to level as many forests and to drain as many marshes as was possible, opening new lands for agriculture, radically changed its landscape until our own days. Another ingredient of these new economic changes was the emergence of some of the earliest tools and techniques that would slowly transform the rather brutal conditions of labour prevailing at the time, even with such simple things as the introduction of the horse collar in agriculture or the adoption of the water-mill. From the point of view of the legacy that we can appreciate in our own days, aside from the changed natural landscape, the introduction of new stone-cutting techniques is most significant, since a few magnificent buildings and churches is often the only direct testimony we have of that age. Given the slow communications and the difficulties of travel, the enormous growth of commerce appears most striking. After the Normans converted to Christianity and settled in north-western France, they became the most active promoters of trade. Rouen was their most important center and the coastal towns of Normandy were busiest with their fisheries and tanneries. The conquest of England opened up a new market, as well as a rich source of raw materials for the textile industry of Flanders. Throughout France and to some extent in Germany, the more stable and peaceful conditions in the countryside favoured the cultivation of grapes and a profitable wine industry soon developed, with a good portion of the output being sent overseas, sometimes to the Orient but principally to England and occasionally to Scandinavia. The Western Mediterranean was now freed from Muslim domination, mostly thanks to the daring incursions of the Norman mariners who first conquered Sicily and then came to dominate the entire South of Italy. Marseilles became an important trading post, Genoa and Pisa, despite their small size, became considerable maritime forces, while Palermo, the new center of Norman power, occupied a commanding position in the trade between Spain and the Levant. The Norman kingdom of Southern Italy would become in fact the most literate and accomplished of Europe for close to two hundred years. The Normans maintained good relations with Byzantium and with the Muslim world in general, some of whose practices they adopted, including their system of taxation. Beyond the material conditions of life, if we wish to understand the sentiments and the climate of opinion prevailing in Europe at the time, we must cast a glance at the events that have added the word crusades to our vocabulary, in particular the first one, whose preparation and execution caused an enormous commotion, far greater than the effect anyone of them had in the Islamic territories of the Middle East. The widespread cult of relics and the practice of doing penance or giving proof of devotion through pilgrimages are characteristic traits of this era. People not only visited local shrines but undertook long voyages to Saint James of Compostela, Rome or, the more daring ones, to Palestine and the Holy Sepulchre. It is generally accepted that these voyagers were treated quite decently by the local populations throughout the Byzantine Empire and the same can be said when they visited Muslim lands; at most, it seems, they were charged minor taxes for safe passage. However, when the Seljuk Turks took possession of Jerusalem away from the Greeks, Pope Urban II, at a famous convocation in Clermont, was able to stir such passion and resentment describing the atrocities to which the pilgrims were being subjected, that the task of retaking Jerusalem became imperative. It is often said that the true causes of the Crusades, even in the case of the first one, were economic in nature and rooted in a struggle for power and influence, but it is also true that the enterprise was received with widespread enthusiasm and people across social classes joined spontaneously. Above all, the Crusade caused a major economic disruption, for it generated a sudden need for cash with the immediate, although temporary, drop in values. For those with savings or with a desire to stay behind and speculate, it was an opportunity to make a fortune. The Church, in particular, benefited enormously and the Vatican became then one of the strong financial powers in Europe. For many others the Crusade meant ruin. The saddest scene of this great adventure was enacted by the many rural people who were moved by the most vociferous of preachers to abandon their farms and join in the expedition. This Crusade of peasants moved through Germany committing numerous massacres against Jewish communities, only to be decimated themselves when confronted in Hungary. The few that reached Constantinople perished as a consequence of famine and the cold. The Knights' Crusade was better equipped and organized, of course; after a handful of years and with some help from Byzantium they managed to retake Jersualem. Yet, despite the establishment of the Kingdom of Jerusalem and its survival for almost a century, the whole undertaking proved to be very costly and of few benefits to those who had actually achieved it. The true beneficiaries were, without doubt, the maritime republics of Italy that began then to regain control of the eastern Mediterranean, in particular Venice and Genoa, but also Pisa, as well as Marseilles and Montpellier, whose navies could now establish trading posts in the coastal towns of Syria and Palestine. The Crusaders certainly did not fail to notice that these regions, and Byzantium as well, enjoyed greater amenities in life than were possible then in Europe. This also became apparent very soon in the uneven balance of trade with the Near East; while Europe was importing precious stones, spices and silk garments, the ships were often sent forth merely with such things as furs or slaves for sale. But a century later, when Jerusalem had fallen again in Muslim hands and plans were being drawn for a fourth Crusade, the Empire at Constantinople was only a shadow of its former self and the Venetians were at the peak of their power, capable of imposing their own conditions. These turned out to be too onerous, in fact, even though the Venetians were committing most of their resources and an imposing fleet to transport the crusaders. In the end, after considerable wrangling and maneuvering, the Crusade did not steer toward the infidels in Jerusalem but was infamously carried out against the Christian city of Constantinople, which the Venetians had been eyeing from the beginning. Part of the great capital was burnt and several days of murder and plunder ensued. The Venetians, whose culture had been nurtured by their links to Byzantium, were quick enough to collect a large and precious booty; the rest of the troops, instead, laboured arduously to destroy the greater part of the remaining riches of the city.

Albertus: It would not be surprising if the Venetians took advantage to enrich their libraries with many a collection brought from Constantinople. Before this time, as you stated, the Norman kingdom of southern Italy was perhaps the more progressive state in Europe and Palermo one of the more prosperous and learned cities even if it never acquired the rank that Alexandria or Baghdad once had. The other conspicuous center of scholarship during this period was the city of Toledo, where many collections of Arabic texts fell into Christian hands once the city was incorporated into the kingdom of Castille. In Toledo, a community of Arabs, Jews and Christians continued to coexist peacefully for many years, contributing greatly to the transmission of knowledge from the Arab sources to Christian Europe. How this curiosity and urge to investigate the expertise acquired by the Islamic scholars was awaken, or the desire to repossess the long-lost texts of ancient Greece, is not easy to explain in a simple manner. More or less simultaneously, a few scholars scattered throughout Europe began to collect manuscripts and to study Arabic and Greek in order to translate them. This new interest in learning, the search for manuscripts and the effort to translate them into Latin is a mysterious phenomenon, for it was not set into motion deliberately, as was the case with the translations carried out in Baghdad in the IXth Century that we discussed earlier. Instead, the interest seems to have sprung independently in various places and amongst very few individuals. One of the earliest examples is that of Constantine the African, a Benedictine monk, originally from Tunis, who had come to the convent of Monte Cassino, near Naples, and did a number of translations of medical treatises from the Arabic as early as the year 1070. Some sixty years later, in Toledo, John of Seville translated into Latin an Introduction to the Structure of the Spheres and the Movement of the Stars, an early and rather elementary treatise composed by al Farghani, an engineer and astronomer who had been amongst those recruited by al Ma'mum to work in Baghdad. His book, a summary of Ptolemy without any mathematics, remained one of the more frequently used and quoted, even a century later. Perhaps the most prominent amongst the translators was Gerard of Cremona, who arrived in Toledo specifically looking for a copy of Ptolemy's Almagest and stayed on to learn Arabic and to translate Ptolemy's treatise along with many other books. He translated al-Farghani anew and many books on medicine, so many in fact, that it is assumed he worked with a team of translators whom he supervised. It is interesting that some years earlier, around 1160, another translation of the Almagest into Latin was made in Sicily, directly from the Greek, presumably more faithful to the original than the translation Gerard made, although it was the latter that later circulated in Europe. Toledo endured as a center of learning and a magnet for scholars from all over the continent until late in the XIIIth century, into the reign of Alphonse X, the Wise, himself very much respected as a scholar, if not so much as a king. Only to mention the names of some of the translators who initially gathered in Toledo, Adelard of Bath, John of Seville, Robert of Chester, Herman the Slav, Rudolf of Bruges gives us an indication of their most diverse origins.

Plebeius: This movement toward greater learning, the effort to retrieve the ancient literature and the desire to incorporate the science of the Muslim scholars is probably related to another simultaneous phenomenon that also seems difficult to explain at first, namely, the relatively sudden emergence of universities in a great number of cities. Their establishment may have been a consequence of the desire to free education from the constraints imposed by the cathedral and monastic schools, which had had almost a monopoly until then. Teachers as well as students began to form guilds in the manner of artisans and merchants. We can only guess that the growth of the urban population, the greater variety of work opportunities, the increased complexity of commercial transactions and in administration created the need for a better educated citizen. It is not unreasonable to think that also a more affluent population, with more leisure time, would have felt inclined to educate itself better. When we contemplate today the basilica of San Marco in Venice or the buildings of the Camposanto in Pisa, both of which were built during this period, we realize the scope of the enterprises these two wealthy cities of maritime traders were willing to undertake. But these two cities were not the first to establish universities. One of the earliest ones was founded in Salerno, where it had existed for a long time as an informal medical school before becoming a university. The University of Bologna is probably the oldest one to be charted in 1170 and was peculiar in that the students played a significant role in its government. Unlike other later universities, it had its own buildings where the lectures were held. Bologna was principally a school of law, its school of theology did not come into existence until a century later. Its students were somewhat older and had already received an education elsewhere. In later years, when the city of Bologna came under the influence of the German emperor Frederick II, some teachers and students abandoned the University and established new informal schools in other towns such as Reggio, Modena or Arezzo. The University of Padua was also established by students and teachers migrating from Bologna. Those of Siena and Piacenza followed some years later. Most of the Italian universities were free from the ecclesiastical authorities. The universities of Paris and Oxford, founded within the next fifty years, are also amongst the oldest and would become two of the biggest. The University of Paris grew out of the Cathedral School of Notre Dame and acquired prominence as a school of theology. Soon there were universities also at Orleans and Anger. In the south of France the University of Toulouse was established by Papal charter and there was also the school of medicine at Montpellier, a good deal older than all of these. Spain could also boast some of the earliest universities, established by royal decree, such as those of Palencia and Salamanca and later in Seville and Valladolid. In Germany there were no universities until well into the XIVth century; the earliest ones were founded by dissident teachers from Paris who had left the city after the great schism that divided the Church and the allegiances of the majority of scholars. The growth of the University of Paris was very much a reflection of the growth of the city and of the consolidation of the French kingdom. Indeed, when the University came into being, toward the middle of the XIIth century, the first system of walls surrounding the city had just been erected, protecting the new quarters that had grown on each bank of the Seine, outside the original small settlement on the island. The Capetian kings of France controlled then a very limited territory; the county of Toulouse and the fiefs of Languedoc were at the time completely independent and Gascony and Normandy were still in English hands. By the year 1204, King Philippe Auguste II retook Normandy and the very prosperous city of Rouen with it. Although its brisk commerce with England declined considerably, the valley of the Seine itself gave birth to a busy economic life and to some of the first trade guilds, groups of merchants who were doing business throughout northern France, with Flanders and with Burgundy to the west. Meanwhile, the city of Paris had spread further and a new more extensive circle of walls was erected over a period of some twenty years. During this time the major construction project was undoubtedly the imposing Cathedral of Notre Dame that took several more decades to complete. The University settled on the left bank of the Seine. After almost a century of existence its reputation was such that students from all regions in Europe flocked to it, making Latin the shared language spoken in this quarter. Having originated as a Cathedral school, it is natural that theology remained its major field of study for a long time and it was here in Paris that the particular brand of theology or philosophy known as scholasticism took root, which eventually came to dominate the thinking of the Church. This designation refers as much to the teaching method, including lectures and interrogation periods, with time for reflexion and for debates, as to a way of thinking, whose origins are not so easy to discern. Well before the University of Paris had come into existence and preceding also the school of Chartres, which was first to acquire some prominence, the scholastic way of thinking had matured in the writings of Saint Anselm. Anselm of Canterbury was in truth an Italian from Aosta, who joined the Benedictine order and was partly educated in Normandy. William the Conqueror made him archbishop of Canterbury, although their mutual relation later turned sour and Anselm was forced to return for some years of exile to Normandy. Until his time, the doctrine of the Catholic Church had been dominated by the Patristic literature of the early Fathers, most distinctly by the writings of Saint Augustine. Now a new way of thinking seemed to gain in ascendancy. Saint Augustine had said: I believe in order to understand. Clearly, he had meant to say that the world made little sense to him without the sustenance of his faith. In his view, faith was a prerequisite for us to give meaning to our existence. Understanding for him was a fruit of belief. But now this dictum was seen in a different light. I believe in order to understand meant that faith provided the justification or the instigation to seek understanding, perhaps that faith itself was in search of understanding. Thus, man, recognizing within him the gift of reason, responds to this call from a revealed faith and seeks to unfold its meaning through the power of his intellect. Anselm acquired some fame because of his attempts to prove the existence of God through reason, without having to rely on revelation. His argument, in fact, did not amount to more than these bare essentials: Anyone, even the most simple minded, he said, could conceive that being compared to whom no more perfect being can be imagined. Therefore, that being exists at least in thought. But a being that shared the same qualities and existed in reality would be even more perfect, which cannot be for we conceived the other one as the most perfect. Therefore, the most perfect being must exist, not only in thought but also in reality. It is difficult for us to fire up our enthusiasm for such a display of logic; nevertheless, the scholastic theologians must have found it appealing for they soon found themselves immersed in that type of reasoning, getting lost in the most convoluted logical arguments, paying extraordinary attention to the analysis of texts, to the definitions of words, to the subtleties in their use and the various layers in their meanings, never quite elucidating whether their dialectical meanderings disclosed or undermined the mysteries of God. Their ideas never seemed very daring and they seldom let their reasoning run in whatever direction it might, for they knew their final destination in advance, which was to ratify the text of the Scriptures. For this reason, they were accused of being merely weavers of approved opinions and their arguments nothing more than organized common sense. The Parisian school rose to prominence at the same time that a large number of the works of Aristotle became available in Latin translations and his philosophy and outlook had a deep and lasting influence. Unfortunately, it was his metaphysics that had the greater impact. Most of his characteristic language, his distinction between substantial and accidental qualities, between matter and form, between essence and existence, between permanence and change, between the things that manifest themselves in potency or in act, all these subtle and artificial nuances were now appropriated by the scholastic theologians. The most representative and accomplished of the new Aristotelians was certainly Albert the Great, the Dominican friar from Swabia, descendant of a wealthy family, whose father, active in Italy in the army of the Emperor Frederick II, had sent him to study in Padova in his youth. His activities during the first half of his life are somewhat obscure. It is known that he spent a good deal of time traveling and teaching at a number of priories in northern Europe. Then, at age forty, he arrived in Paris and immersed himself in the study of Aristotle. Encouraged by his colleagues and disciples after he had written a commentary on the Physics, he continued on, writing and commenting on all of Aristotle's works for many years, producing a gigantic output of his own hand. In Paris he stayed for less than a decade, for he was then summoned to organize the school at Cologne until, some years later, he was made archbishop of Regensburg. In his very long life and despite his many objections and pleas, the Pope charged him with a myriad tasks, one of which was to tour Germany, exhorting the population to participate in the upcoming Crusade. Albert's writings range far and wide, on Scripture and theology, on logic and the natural sciences and even on astronomy. He was not so interested in ranking the diverse fields of study in a hierarchy, all subordinated to theology, as was the inclination of his scholastic colleagues, but he was ready to consider each science unto itself, with its own methods and pursuits. He represented the true Aristotelian spirit, not so much by reproducing his thoughts, but by being animated by the same voracious appetite for learning. He was unprejudiced and willing to give credit to authority when it was due, or to follow observation or reason when it was not. He never saw any conflict between the sciences that he studied with an open mind and the faith he held by revelation. Mathematics seems to have been a little remote for his taste but, from early in his life, he had been a persistent observer of Nature and he filled his writings with observations he had made or the speculations to which he had been led, whether they pertained to the origin of the rainbow, to the nature of heat or to the constitution of the comets. His comparative study of plants was quite unique for his time and so were his observations on insects, their morphology and behaviour. He also wrote on human anatomy, on medicine and nutrition and on many other subjects. His best known disciple was Saint Thomas, of Aquino, near Naples, who had joined the Dominican order as well despite strong opposition from his family, and had then traveled to Paris to study. After a few years he followed Albert to Cologne, only to return to Paris later to teach, still as a young man, and to do a great deal of writing. Unlike his mentor, by his mid-forties Thomas had completed the majority of his voluminous works; in his last few years he lost much of the vigour that had been characteristic of him and he died when he was not yet fifty, in one of his numerous trips to Italy. Albert had been a philosopher in the widest sense; Thomas, by contrast, was preeminently a theologian. He thought it was man's vocation to reach the ultimate goal of the vision of God and he saw in the human mind, in its highest power of reason, the tool to achieve this goal. Yet, the human mind, he maintained, has been weakened by sin and needs the assistance of faith to achieve its goal. Thomas was keen on distinguishing between the mysteries of the faith, such as the Trinity or the Incarnation, and what he called the preambles of the faith, which were demonstrable through reason, but his desire to show that reason could never arrive at a conclusion in contradiction with Christian doctrine was so vehement and pervades his writings so consistently, that one is tempted to say that this conviction was for him one of the articles of faith, evident in itself and not subject to questioning. In his dialectical method, Thomas would challenge himself with the following question: Given that God is eternal and unchangeable, if the creative act exists in him, it must be identical with his Nature, which is one; therefore, how could the world not have come of necessity from him of all eternity? His answer, of course, was ready in advance. Although the creative act was eternal, its manifestation was not. The Creation, he would say, was not necessitated and God did it only to communicate his perfection and goodness. Only God can be of eternity, things are necessary to the extent that it is necessary for God to will them but, absolutely speaking, it was not necessary for God to will anything but himself and we cannot say that it was necessary for God to will the world, since his will was entirely free.

Albertus: I have never been able to fathom how could Saint Thomas command such great authority, not only in his time but for centuries to come and even outside the Catholic Church. Is it only a matter of taste? That manner of thinking seems so unproductive and barren, even conceited.

Plebeius: Sometimes, when we allow ourselves to embark on his train of thought and follow the many contortions of his arguments, we may feel that he is like a captain steering his ship hopelessly in heavy seas but, as soon as we step back and examine his efforts in a detached manner, the storm and the seas and the ship appear to us as a dream, a concoction of his own imagination in which he was pleased to get lost himself. Aristotle's defense of the eternity of the world always remained one of the more contentious issues for debate. To be more precise, it was not a subject of debate the fact that the Universe had indeed been created, the task was to prove it convincingly, to demonstrate that the opposite view was contradictory.

In fact, when we look now at the arguments used in defense of the eternity of the world, it is difficult to see how to demolish them or, at least, how to demolish them with their own tools. According to Aristotle, the Universe could not have been created because matter is indestructible and uncreated, it can only be transformed through the generation and corruption of forms. Using his language, if the Universe had not existed of all eternity, it was said, it ought to have existed previously as a possibility. Now, matter could have existed in potency of coming into being through the attribute of form or of not coming into being through privation of form. But the Universe has come into being; therefore, it had the attribute of form. However, matter with the attribute of form is the world itself and one had to conclude that the world had existed already before it came into being which is absurd. Thus, the world had existed of all eternity. Saint Thomas' counterarguments were just as convincing. Only from faith do we know that the Universe has not always been, he would say. The Universe could only remain uncreated without the attribute of form; before its coming into being, it had existed not through the passive potency of matter but through the active potency of God. Saint Albert had resorted to a simpler but more translucent argument, to which occasionally Saint Thomas made reference. For him everything had been created, the Heavens, the stars and time as well. Matter and all the rest had been extracted out of nothingness. Only God was eternal, the world was temporal. Eternity preceded time just as the cause precedes the effect. God had preceded the world from all eternity. A lot of ink flowed as a consequence of the confrontation with Aristotle's works. A condemnation issued in 1210 forbade the use of any of his books at the University of Paris, but the prohibition was not followed in all seriousness. Later Pope Gregory IX ordered a purge of the more heretical portions of his writings, so that they could still be used in matters that did not affect the faith. These were high-minded debates amongst scholars; as for the population at large, the Church had not been remiss to prosecute all forms of real or perceived heresies with vigour and determination. For more than a century the battle against the Cathars first, and then against the Albigensian heretics in the south of France, led to some rather horrible scenes. Gregory IX himself is credited with having laid the foundations of that infamous institution that would come to be known as the Inquisition. In any event, some writers continued to interpret Aristotle too liberally, contending that what can be true philosophically may not be true according to theology and conversely. In 1270 the bishop of Paris, Etienne Tempier, issued a list of a dozen items, all heretical doctrines that could not be taught at the university, and returned a few years later issuing a far more detailed list of some two hundred condemnations, amongst which there were even some ideas espoused by Saint Thomas. Similar condemnations were proclaimed at other universities and it took several decades for the Church to come to terms with Aristotle's teachings and to reconcile them with its own doctrine. Another question related to the eternity of the world and much debated by the scholastic philosophers concerned the existence of the infinite. Aristotle had maintained that the infinite could not exist, either in act or in potency; indeed, for him, anything that could exist in potency, necessarily can exist also in act. However, his contention that the Universe had existed from eternity and with it also the human race, led to some unpleasant quandaries. Given that the human soul is immortal, an uncreated world would imply that there exists an infinity of souls in act. It was arguments such as this one that were also used to show, albeit indirectly, that the world must have been created. Otherwise, without a beginning, God could have created a pebble each day, so that gathering all the pebbles they would now form an infinite volume in act, which is impossible; hence, the world was created. The Arab philosophers had followed the same reasoning, accepting Aristotle's contention that an infinity in act was impossible and rejecting the eternity of the world. It is not clear, however, why they had so much difficulty in coming to terms with the infinite. Saint Thomas had been rather uncertain on this subject as well, maintaining that he could conceive God producing an infinitude in act, but if existence was repugnant to the infinite by virtue of its own reason, then God could not, and would not, produce its existence. Such an argument does not appear to have been very helpful. The scholastic philosophers attempted to subdue the infinite into submission through the power of language, but without much success. They spoke of the infinite in act or in potency, the infinite in extension or in number, the infinite by division or by composition and many other such notions that tended to obscure the argument more than illuminate it. The many other questions they debated also tended to end up embroiled in the same type of subtle but sterile logic. Oddly enough, it was the omnipotence of God that made things sometimes even more difficult to understand. Discussing the eternity of the world and the meaning of duration, Etienne Tempier, the bishop of Paris, had asked: Where is the cosmic clock by which we should measure time? Could it be in the daily rotation of the Heavens? Of course not, he answered, for God could, if he so wanted, accelerate or slow the Heavens at his pleasure. Could the world be said to be in a place and if so, was it subject to movement? Was it conceivable to have a plurality of worlds? What was the difference between the firmament and the heaven of the angels? What was the nature of the celestial movers? It is astonishing the number of volumes that were filled discussing these matters. There are two schools of thought that we might say opposed the scholastic movement, although both of them could also be said to have sprung from within its midst. The first is best represented by William of Ockham, the English Franciscan friar who studied at Oxford and Paris and flourished when the scholastic philosophy was already in decline. From another perspective one might argue that he represented the best of the scholastic philosophy; it is only that in his thought we find a first glimpse of a more modern way of philosophizing when, in fact, he was, like his predecessors, a theologian attempting to come to terms with Aristotle's doctrines.

He maintained that anything that is not self-contradictory is always possible and could thus be brought into the world, if God so willed it. This made of logic the most important tool in the hands of the philosopher, but it is through his interest in the analysis of language that William of Ockham really showed his originality, beginning to study the way in which our discourse reflects our thoughts as much as the real world and how it establishes a relationship between the two. For him we could only have an intuitive grasp of individual things; of general and abstract concepts our knowledge was most tentative and conjectural. The categories that the Aristotelians treated so deferentially were only names or short-hand speech, never things in themselves that existed outside our imagination. In his day, these ideas were new and presented a refreshing point of view. William of Ockham is best remembered today for having introduced the principle of economy in language, according to which one should never say in more words what can be said in less. After his days in Paris, William spent some time in Avignon embroiled in a controversy with the Pope on the regulations governing his order. The result of these efforts was an excommunication, so that he decided then to flee to Munich, where he sought the protection of Louis of Bavaria, the future German Emperor, and it was at his court that he spent the last twenty years of his life. It is believed that he fell victim to the great plague of 1348, when he was already well into his sixties. The other school of thought that reacted against the scholastic movement had in Saint Bonaventura one of its most illustrious representatives. A contemporary of Thomas Aquinas, he had also come from his native Viterbo, in Lazio, Italy, to study at the University of Paris, where he remained as a lecturer. Unlike Thomas he had joined the Franciscan order, within which he rose through the ranks to eventually become its Superior General. This led to innumerable trips and to the abandonment of his teaching career, but he never neglected his writings and he had overall a very productive life. His attitude toward philosophy may be said to be one of retrenchment, for he concentrated on the inner life of the soul and its greater capacity to experience than to comprehend. For him it was through the Scriptures that philosophy had to approach the divine. One can almost say that his attitude was that of renouncing knowledge but, in fact, he maintained that the individual could seek knowledge by following on Jesus' path and living his life. He would have said that knowledge was experience and not just a scaffolding of ideas and words in one's intellect. For him the task of the theologian was merely to make the credible intelligible and he warned those who got too involved in the study of philosophy of becoming vain and arrogant by eagerly seeking to know and not to love. In his view, love was able to penetrate beyond what the intellect can reach. Because man had been made in God's image, God was in his memory, in his mind and in his will; therefore, it was the whole of man, not just the intellect alone, that was striving to return to God. Clearly, Bonaventura's ideas cannot really be called a school of thought as much as a road-map; his inclination was that of a mystic, he saw in all things the footprints of God and he could not relinquish his conviction that the knowledge of God through love was more compelling than it was through the restraining force of reasoning.

Albertus: While your theologians were discussing the plurality of worlds or the nature of the celestial movers, there was a small community within the Church that was interested in more practical astronomical matters. During the time that Saint Thomas was teaching in Paris, a book was published by a certain John Sacrobosco, with the title On the sphere; despite his Latinized surname, he was presumably a Scot who had been a student at Oxford and had come now to teach in Paris. This book was quite elementary and merely taught the main ingredients of the celestial sphere, the equatorial and ecliptic planes and the general motions of the planets in a language that did not use any mathematics. Nevertheless, the book became very popular and was used in many schools. It came to replace the book by al Farghani that contained very much the same material. Sacrobosco could hardly claim any originality, except that his presentation was more orderly and didactic. Until his time the knowledge of astronomy in the West had been extremely limited; the more knowledgeable scholars, such as Saint Albert and Roger Bacon, one would suspect, were well aware of the superior science the Arabs possessed, but none could claim to be entirely familiar with it. Those in the Church who were in charge of keeping the calendar were the only ones who had some interest in practical astronomy, for they had to match the civil, Julian calendar, that was a solar year, with the ecclesiastical calendar, which followed the lunar cycle, but in this task the Church had followed for centuries the nineteen year cycle for the Sun and Moon which, although imperfect, had served its purpose well. Anyone who was interested in predicting eclipses or the conjunction of the planets had to seek one of the more elaborate tables that had then become available. This is an interesting aspect of the origins of astronomy in Europe; at least three sets of tables had been translated from the Arabic, one of them were the original tables of al Khwarizmi of the IXth century; then there were the tables of al Battani and the third were the Toledan tables of more recent times, which incorporated several new ingredients, not all of them coherent, some going back to Ptolemy. The original tables of al Khwarizmi, for example, still betraying their Indian origin, were reduced for the location of Ujjain, in central India, and those attempting to use it in the West were forced to do the necessary modifications. As they attempted to come to terms with these discrepant tables and adapted them to their use, some Church scholars saw the need to better understand the theoretical foundations on which they rested and began to take an interest in mathematical astronomy. A few years after Sacrobosco's book appeared, a more advanced, anonymous treatise came into circulation with the title Theory of the Planets. Here Ptolemy's theory of epicycles was explained in mathematical language and with geometrical diagrams. This book also became widely successful at the universities, as attested by the fact that dozens of the original copies have survived to this day. By contrast, Ptolemy's Almagest, which had been translated a hundred years earlier, still remained daunting for the novice astronomers of the XIIIth century and not many copies of it were in circulation. We must remind ourselves that, at this time, in faraway Persia, the great Nasir al Din al Tusi was leading the observatory of Maragha, even though the fruits of his works did not reach Europe until some 200 years later. In the West, the court of the wise king of Leon and Castile, Alphonse X, was the most advanced center for astronomy. Under his own direction, a number of scholars of Arabic, Jewish and Spanish extraction, including also some Italians, worked on a set of tables that were to correct the inaccuracies of the older Toledan tables. This collection also included detailed instructions on the use of the astrolabe, the quadrant and various other instruments as well as discussions on more theoretical subjects, such as the much debated one concerning the precession of the equinoxes. Perhaps it was the decline of the scholastic philosophers that opened the way for a renewed interest in the sciences, but during the first half of the XIVth century Paris led the way in forming a school of astronomers. It was from Paris that after some fifty years the Alphonsine tables began to be disseminated throughout Europe. John of Lignières, a native of Picardy and associated with the University of Paris for many years, was one of the more prominent and diligent astronomers of this new school. Now we can speak properly of astronomers, for these men carried out observations of their own. John of Lignières determined to obliquity of the ecliptic with some precision and composed a catalog of 48 stars from his own personal observations, a rather unusual endeavour in his time. Much of his work was devoted to composing sets of tables for the conjunction and opposition of the Moon as well as the planets, tables which, in their later versions, incorporated some ingredients from the Alphonsine Tables, but with further simplifications. In addition, he wrote a number of treatises on the use of instruments, an astrolabe amongst them that he had designed with a slightly different stereographic projection. His work was continued within a few years by a number of disciples and associates, in particular by John of Meurs, a Norman who acquired greater fame as a theorist of music but who also composed a set of tables of his own and was prudent enough to verify their accuracy by carrying out a number of observations. He was also called upon by Pope Clement VI to Avignon to advise him on the long-delayed reform of the calendar although nothing came of it. Albert of Saxony was one more amongst these disciples who worked on a set of astronomical tables in Paris. In fact, much of the material in these various tables overlapped; in the end it was only Albert's tables that were published and gained wider diffusion throughout Europe. Around 1340 some version of the Parisian tables made its way to Oxford, where they were adapted and further improved. At Oxford's Merton College a number of scientists built up a school only second to Paris. The Oxford tables were widely used in England, for the most part with an astrological purpose; yet, these tables also found their way back to the Continent and were used by numerous scholars for over a century. One of the more remarkable personalities to emerge from Oxford was Richard of Wallingford. Being a Benedictine monk it is unlikely that he was associated with Merton College, although he may have studied there. Still in his early thirties, he left Oxford to become abbot at the monastery at Saint Albans. It was during a trip he made to Avignon to seek papal confirmation that he contracted leprosy, the disease from which he would die a decade later. Nevertheless, he remained as abbot during all this time and managed to have a very active life. Already before leaving Oxford he had written a book on spherical trigonometry, the first of its kind in the Latin West. Although he was interested in astronomy, he was neither an observer nor a theoretician. His main interest was in the design and construction of instruments, both for observation and for computation. Of the latter, the albion, for the purpose of computing planetary orbits, was one of the most ingenious and achieved some popularity. Its assembly of graduated discs was not designed to reproduce graphically the orbits of the planets as in similar earlier instruments; instead, it aimed at reproducing the computations done with tables. Its main objective was practicality, perhaps ingenuity too, but certainly not precision. Finally, while at Saint Albans, Richard worked on the design of one of the earliest and also one of the most ambitious mechanical clocks. It was supposed to display, on a single dial, the motions of the Sun and Moon and all the planets, besides signaling the hours, the tides and an assortment of other things. The escapement mechanism, used to control the uniform rotation of a wheel, had been discovered some decades earlier and the construction of church clocks became somewhat of a fashion. Richard spent so much money on the clock for Saint Albans that he was asked to defend the project before the King. In the end his death prevented him from seeing it to its completion. Amongst the contemporaries of these scholars there are two I wish to mention, namely, Thomas Bradwardine and Jean Buridan, because the problems they addressed, although not directly related to traditional astronomy, reflect the broadened intellectual landscape of the philosophers of this era. Bradwardine was a product of the Oxford school; having joined the Church, he rose to some prominence in the ecclesiastical hierarchy to become Chaplain to king Edward III. It is said that he was present at the capture of the city of Calais, during the early stages of the hundred year war against France. Eventually, he was made Archbishop of Canterbury, but it was in the year of the Black Death and he fell victim to it within a month. In fact, he was succeeding two other archbishops who had also died in less than a year. In any event, although Bradwardine's main interest was probably theology, he also studied problems in geometry and mechanics; in particular, he wrote a book on Proportions, as he called it, in which he sought to elucidate the relationship between the ratios of speeds of motions and the ratios of their causes, or the ratios of their resistance. Clearly, he was after an algebraic relation between the magnitudes expressing such things as force, velocity and resistance. The results he obtained proved to be quite erroneous, but the boldness of his attempt and, reading his arguments, the difficulties he faced because of the lack of appropriate conceptual tools, make his efforts all the more praiseworthy. He also wrote a treatise on the continuum, defending Aristotle's position against the atomists, who thought the continuum to be composed of discrete material indivisibles. Here Bradwardine was treading on safer ground; he would explain, for instance, that when you have two concentric circles, by drawing all the radii emanating from the center, you establish a correspondence linking each and every point in the interior circle with each and every point on the circle outside; therefore, if the atomists were right, both circles would have to be composed of the same number of atoms and have the same length, which is absurd. The other scholar I mentioned, Jean Buridan, a native of Calais, was a secular cleric who rose at a rather young age to be rector of the University of Paris, a post he relinquished some years afterwards, but to which he returned later in his life. He was also interested in problems of mechanics and introduced the notion of impetus, the power a body acquires and preserves when set into motion and that is dependent on the quantity of matter and its velocity, a notion very similar to what we would call momentum. Here he clearly departed from Aristotle's ideas, for he was postulating that the impetus would endure forever, were it not corrupted by an opposing resistance. He also speculated that no mover might be required for the celestial spheres if there is no resistance against their motion and, perhaps driven by a desire of economy of effort, he even suggested that it might be the Earth that rotates daily and not all the spheres of the Heavens. Buridan approached his scientific problems, not as a mathematician or an experimentalist, but as a philosopher; nevertheless, he did not argue as the scholastics had done, but in a lively, idiosyncratic way that we would recognize as belonging to a later age. He made a clear distinction between the laws in science that are valid because they are deduced through logic and those that are proved empirically, on the supposition of the natural course of Nature, as he said. He insisted that the principles of physics can only be shown empirically and not metaphysically, as the scholastics had held, that is, by merely showing their opposites to be contradictory. His interest in logic led him to investigate the laws of deduction and the way in which they, in turn, could be deduced; likewise, he showed an interest for paradoxes in logic, such as the one raised by the man who says: What I am saying is false. All these problems, in which language explores the world as much as itself, he addressed with insight and common sense. The most distinguished disciple of Buridan in Paris was Nicholas Oresme, a native and resident of Normandy, first in Rouen and then in Lisieux, where he was appointed to the bishop see. Oresme had been the tutor of the dauphin to the French crown and remained later as an advisor to the new king Charles V. The scope of his interests is quite remarkable and he often addressed them in novel and original ways. Part of his fame rests on the treatise he wrote on money, the first compendium of its kind, where he discusses the need for a healthy currency, the theory of value, the nature of speculation and other topics. I wish to mention a small but momentous scientific contribution of his that appears in a small book entitled Treatise of Configurations of Qualities and Motions. He thought of representing graphically the intensity of a certain quality along the extension of a rod by drawing a perpendicular segment at each point along its extension, in such a way that the length of this segment would correspond to the magnitude of the intensity at that point. Despite the simplicity of the idea, no one until his time had used such two dimensional charts. Moreover, Oresme proceeded to make this idea more general, for he noticed that the instantaneous velocity of a moving body along a line was such a quality susceptible of representation; for the horizontal extension he could merely take the position occupied by the moving body or, even more abstractly, he could use time as the extension. Thus, in his graphs, moving horizontally indicated the uniform passage of time, whereas moving vertically indicated the variation in the magnitude of a certain quality such as velocity at a given time. He then observed that a horizontal line would represent a motion with constant velocity, whereas a rising line would correspond to a uniformly accelerated motion, all properties that we now learn as school children and we find it difficult to perceive the arduous path of abstraction that led to their formulation. It must have been most delightful then to discover for the first time that having displayed the velocity as an intensity in such a graph by means of a rising or declining line, the quantity of motion, in other words, the distance traveled, was represented by the area enclosed under the line. Oresme also discussed the concept of impulse introduced by Buridan, but in this matter he was more conservative, more inclined toward the idea that the very motion of a body exhausts its impulse, thus moving away from the notion of inertia and preserving the requirement of movers that impel and sustain motion. Nonetheless, he was one of the first to draw a clear distinction between absolute and relative motion, pointing out that most of what we call motion is of the latter kind. Like Buridan, he discussed the possibility of the rotation of the Earth but preferred to reject the idea. Allow me to mention one last idea of Oresme, which is interesting in itself. He was a fierce enemy of astrology, perhaps because king Charles V was captivated by it and importuned him often with much astrological nonsense. Being as diligent as he was, he wrote extensively on the subject, always pointing out how detrimental astrology was for philosophy. In the process, he came across the following argument. Since astrology presumes to base its predictions on the configuration and relative positions of the planets, it was necessary to take into consideration that the periods of their orbits are very likely incommensurate and that, consequently, it was impossible to speak of anything resembling a cosmic cycle, for the planets would always find themselves in a novel configuration that had never existed before and that would never repeat itself again. Hence, any attempt at predicting the future in this way was futile, he maintained. Of course, it is very unlikely that such a reasonable and subtle argument would have deterred any convinced astrologer from his efforts. Oresme is the last prominent scholar belonging to the French school. At the end of the XIVth century Paris ceased to be the dominant center for scientific progress, especially in astronomy.

Plebeius: That is hardly surprising; France as a whole had suffered considerably from the ravages of the long war against England. Secondly, regarding the prevailing climate for scholars, the schism within the Catholic Church and the recognition in France of a dissident Pope, shunned almost unanimously by the rest of Europe, divided the allegiances at the University of Paris and led to a large exodus. This coincided with the period when some of the first universities were being founded east of the Rhine, so that those scholars in Paris who where of German extraction returned to teach in their native land. After the fall of the Hohenstaufen, for more than a century, Germany had remained in political disarray with incessant disputes amongst local knights and princes vying for power, but these troubles did not deter the growth of a certain number of cities which, proceeding almost as independent states, became rather prosperous and wealthy. The universities of Heidelberg and Cologne, founded shortly after 1380, came about as a consequence of this prosperity. The few universities already established a few decades earlier had benefited from the patronage of powerful and resourceful donors, as was the case with the University of Prague founded by the king of Bohemia in 1348 or the University of Kracow established by the Polish king Casimir I in 1364 and also the University of Vienna founded in 1365. A contributing factor was the fact that central and eastern Europe had been spared to some extent the devastating effects of the great plague that struck at mid-century.

Albertus: The teaching of astronomy at the universities still remained quite elementary for some time. Only a few of the textbooks available in Latin explained the basic theoretical methods followed by Ptolemy to construct the orbits of the planets or how the elements of these orbits were retrieved from direct observations, so that the students could have some understanding of how the astronomical tables then in use had been composed, beyond being able to use them as an arcane computational device. For the first time we witness a more widespread desire to construct mechanical instruments, some of them for observation while others were gadgets that reproduced the computations prescribed by the tables. All of them were hand-held instruments, not designed for precision measurements, and one may suspect that their popularity had more to do with their use for astrological predictions. The simplest instrument was perhaps the cross-staff, in which the observer placed his eye at one end of a long rod, while another rod perpendicular to it, was allowed to slide along its length until its endpoints coincided with the two objects in the sky whose separation one intended to measure. Their angular separation was then read off from a graduated scale marked on the long staff. Other more sophisticated instruments, such as the armillary sphere or the torquetum, also introduced during this time, enabled the observer to relate the various angular coordinates of an object in the sky, depending on whether they were measured in relation to the horizon, to the ecliptic or to the celestial axis. Quadrants and astrolabes, even sundials, were also built in great numbers. It is interesting that with the development of the mechanical clock, the sundials, instead of becoming obsolete, were more necessary than before, for they were used to monitor the behaviour of the quite unreliable clocks. It must be said, however, that throughout this time, nowhere do we find a sustained effort to conduct long term observations, nor a persistent desire to attain greater accuracy in the measurements. The interest in astronomy was of a more qualitative nature. The variations in the Moon's apparent velocity across the sky or a determination of the eccentricity of the Sun's orbit did not attract the attention that far more philosophical questions seemed to command, concerning the plurality of worlds, the substance that made the outer spheres, the nature of the prime movers, the possibility of an infinite void and many more. The influence of Aristotle remained decisive; some followed the lead of the Arab astronomers of Toledo in an attempt to revive the system of concentric spheres he had favoured to explain the motions of the Heavens, in opposition to the Ptolemaic method of deferent circles and epicycles. An old idea of Ibn al Haytham was resuscitated, according to which, two concentric spheres were assigned to each planet so that a third sphere, with its diameter equal to their separation, was allowed to slide and rotate in the intervening space; it was this sphere that carried the planet on its surface. Depending on its position, the distance from the planet to the Earth, at the center of this configuration, could vary. As usual, the system of nested spheres was arranged in such a way that the outer one was in contact with the inner sphere of the next planet, until one reached the sphere of the stars, not leaving any space unaccounted for between any two of them. Since the time when al Farghani's book was translated into Latin in 1137, the actual sizes assigned to the planetary spheres, which he had reproduced more or less faithfully from Ptolemy's computations, became widely known and were further reproduced, sometimes with minor variations. Some scholars took care to add to the thickness of these spheres the diameters of the corresponding planet, something Ptolemy had not done, but this was truly a dainty detail and did not change substantially the dimensions of the Cosmos. The distance to the sphere of the stars was still held to be of the order of 20,000 Earth radii, which is not quite the distance to the Sun, of course. Nonetheless, the Universe was held to be enormously vast. In the popular literature it was pointed out that if Adam had set out on a journey toward the stars, walking 20 miles a day, even if he had started on the day of the Creation, he would still have had several thousand years to walk in order to reach his destination. At the time, the day of the Creation was generally placed some six thousand years in the past. Dante Alighieri seems to have been fascinated by the astronomical ideas that were being entertained during his lifetime. In his fantastic vision of the Commedia, as he explores the underground circles of Hell, the rising cliffs of Purgatory in the sublunar world or the various realms of Paradise, he repeatedly makes reference to the configuration of the planets and the stars at each move. He also attempts to give some meaning to the tides, to meteors, to the lunar spots and to the reddening of Mars. If I am not mistaken, he mentions having read the book by Al Farghani himself. Thus, in his description of the hierarchy of Paradise, when he assigns a place in the third Heaven to those spirits who had been most disturbed by their worldly passions, he decides on the sphere of Venus, because the cone of the Earth's shadow from the Sun was believed to reach up to that distance. There is no doubt that Dante attempts to present a thorough and full explanation of the cosmic drama and the truth that he lays before us as final and conclusive. In his almost admonishing style, he does not invite us to inquire about the meaning of things but to submit to an inexorable order. Perhaps it is appropriate to see in Dante's great poem the last and most lucid statement of an era that was then disappearing. Throughout his life, his advocacy for the return of a German Emperor to Italy and for the restoration of the Papal authority seems to express a longing for an order and institutions that were then fading, if not vanishing altogether. In this regard, it is all too tempting to compare him to his younger contemporary Francesco Petrarca, who seems, by contrast, the early herald of a new epoch, less dogmatic, less interested in giving a complete account of the architecture of the Universe and more inclined toward expressing all the uncertain and ambivalent facets of human life. Intensity of experience and eloquence of expression become more important for him than contemplation and devotion. I suppose that religion was no less important to him, but for Dante Christianity offered a closed theory of the Universe and the life beyond this life occupies the center of his attention, whereas Petrarca sees the Christian faith essentially as a moral doctrine, as an ideal to guide each individual in this life.

Plebeius: Yes, it is characteristic of Petrarca that he always returned to the writings of Saint Augustine amongst the Church Fathers for inspiration. Nevertheless, although there may be some truth in the manner you have contrasted his attitude against Dante's, I would be inclined to adopt a contrarian point of view. In some respects, it is in Dante that we may find the qualities that we would identify as modern and ahead of his time. Dante remained always a lay scholar, something that was very unusual in his day. Petrarca, more conventionally, took vows and became a member of the Church. No doubt, the two shared a great many things. Both of them had to lead a life in exile; in fact, Dante and Petrarca's father had had to leave Florence almost simultaneously. The two shared also a love of literature, a desire to lead a reclusive, studious life, at the same time that they were unable to stay away from the major contemporary debates affecting public life and felt compelled to participate forcefully in them. Both deplored the disarray and corruption in the society to which they belonged, but Petrarca thinks of himself as a stranger in his own times and looks with nostalgia toward the distant past of antiquity that he finds a great deal more congenial. Dante, on the other hand, is at home in his own world, which makes him all the more bitter and angry; we might say that his eyes are in the future, for he has the zeal of the prophet and the reformer. Petrarca had developed a great animosity toward the scholastic philosophers, he thought of them as engaging in sophistry and arcane debates that did not touch on what for him had become the most fundamental concern, namely, how is man to find his way in the world and lead a virtuous life. His animosity was directed toward some of the commentators of Aristotle, in particular Ibn Rushd, whom he disliked so much that it turned him away from all things Arabic. His reaction was more temperamental than reasoned; yet, it made him also disinterested in scientific matters. He preferred self-examination to seeking knowledge for its own sake. In this respect, I think that Dante's attitude is more curious and open-minded, he seeks and welcomes new knowledge and wishes to disseminate it. All the vehemence we find in him is due more to his didactic instinct than to any compulsion to adhere to a strict doctrine. When we consider their attitudes toward astrology, Dante may give the impression of being somewhat credulous. He insists on recording the position of the planets at each stage in his poem and he tells us that by following one's star one cannot fail to reach a glorious port, but his references are always cautious and measured. Stars influence but do not compel, he tells us. He probably believed, like many of his contemporaries, that the order of the Heavens could sway events on Earth, but his allusions remain elusive and vague, as in the much quoted verse which identifies the love that moves the Sun and the other stars with the love that moves the human soul. Petrarca, to the contrary, very much like Nicholas Oresme, could not stop himself from decrying astrology as a hoax and its practitioners as charlatans, warning his readers not to be lured or deceived by the pretensions of those impostors. However, this was also the orthodox attitude that the Church had always taken since the early Fathers. One could not advocate astrology and remain a good Christian at the same time. The notion that one could predict the future based on the configuration of the planets carried with it implicitly, except for Oresme's subtle argument, the possibility of a recurrent history, repeating itself after each celestial cycle. This was a horrible idea, as Saint Albert had said. Clearly, it is central to Christian doctrine that the drama of universal history plays itself only once, with a single, final denouement. On the other hand, astrology's presumed ability to foretell the future contradicted the freedom of the will, as had already been emphasized by Saint Augustine. For him it was most important that the human soul had been made in the image of God and its capacity to choose between alternatives had to be free and unrestrained.

Albertus: Should we then hold the Church as an ally of the sciences and think that it held an enlightened position against the unworthy and deceitful fabrications of the astrologers, or was the Church itself guilty of obscurantism, demanding strict adherence to its dogma and refusing to contemplate that some of the theses of the astronomers and astrologers might be subject to verification? Ptolemy had already sketched a distinction between two types of astrology. Firstly, there was the study of the possible influence of the planets and stars on the sublunar world. This is what we might call the science of astrology and, not knowing better, we might think of it as an over-ambitious form of meteorology, to use the vocabulary of our own day. However, there is also the aspect of astrology that attempts to fix the propitious moments for the performance of certain rites or determine the personal traits of an individual from the moment of his birth. There is also the casting of horoscopes, the association with divination and demonology, with magic and the adoration of stars. All these fringe and more fanciful concerns are what has always given astrology its dubious reputation. However, to complicate things further, there are those who contend that during this period astrology gave a helping hand to astronomy, if only unwittingly. They argue that most scholars at the universities were largely concerned with philosophical questions, whereas those who were interested in prophecy and divination wished to have reliable ways of charting the course of the planets, so that in this way they stimulated and promoted the development of mathematical astronomy. At all events, one of the factors that may have contributed to making astrology popular, well before Dante's time, is the fact that amongst the many books in philosophy, astronomy or medicine that were being translated from the Arabic there was a great number of astrological works, many of them of Greek or Indian origin, avidly read by the population at large. The interest in astrology did not subside for a long time. I understand that after the introduction of the printing press, many publishers could not operate their businesses at a profit without producing a good number of booklets on astrology, horoscopes and the like. It is quite possible that many a good piece of literature saw the light of day only thanks to the books on astrology that were being produced by the side.

Plebeius: Perhaps the single event that contributed most decisively to the popularity of astrology was the great plague that devastated Europe in the middle of the XIVth century. We have made reference a few times to some of its victims. The Black Death, or the Great Dying as it was called at the time, was indeed a cataclysmic event of the greatest proportions, not only because of the many lives it took, but also due to the manner in which it altered the consciousness of those who survived it. Its ferocity and the speed with which it propagated could only be explained as an extreme punishment brought about by the wrath of God. Europe had been free from plagues for many centuries and the population had increased enormously. The last plague pandemic had occurred during the reign of Justinian, when the pestilence had descended along the Nile river and quickly spread through the Empire and to the West. As it was perceived at the time, the whole human race had come close to being annihilated. In the XIVth century the plague is believed to have originated in China, probably in Yunan, one of its native areas, like Siberia or certain regions of East Africa and the Arabian Peninsula, places where it has lain dormant for long periods of time in a fragile equilibrium with the general environment and its climate. From Yunan the plague is thought to have been brought to the Gobi desert and then carried west by Mongol horsemen upon their migrations. The conditions that have to occur for an outbreak of bubonic or pneumonic plague are quite specific and interesting to consider. The bacillus lives in the digestive track of a flea which is hosted by rats. Sometimes the bacillus reproduces itself uncontrolably and is regurgitated by the flea which deposits it on the rat's skin. If the rat has some kind of skin injury, the bacillus is certain to penetrate the blood stream and eventually kill the rat. Only then will the flea very likely migrate to another host, perhaps a human, something more likely to occur when humans and rats live in proximity and frequent contact. Now, once again, the bacillus has to find a way to penetrate the blood stream of its new victim. However, as far as I know, what actually causes an outbreak of plague has never been entirely explained, whether it could be due, for example, to climatic changes that disrupt the quiet but fragile coexistence of bacillus, flea and rat. What is unquestionably true is the fact that climate plays a crucial role in the propagation of the plague for the bacillus finds it difficult to survive on its own unless the weather is relatively warm and humid. Then, carried either by man or through the displacement of the rodent population itself affected by drought and climate changes, the plague began to move westward through the steppes of Central Asia. It is known that a community of Nestorians in Tien Shan was decimated by it in 1339; in a few years it had arrived at Samarkand and not before long it was spreading through the Caucasus and into Crimea. In 1347 it arrived in Constantinople and the next year at Antioch and Alexandria. It is sometimes said that the plague first arrived in the West in october of 1347, carried by a Genoese ship that was returning from one of their trading posts in the Crimea. With an already afflicted and despairing crew, it was allowed to dock briefly in Messina, Sicily, but that was enough to infect the local population and the plague began its trip northward with astonishing speed. Within a year it had swept through France and the Low Countries and it had arrived in England. Sailors brought it further to Scandinavia and victims were recorded even in Greenland. Not all regions were equally affected. In Spain, the Mediterranean coast of Catalonia was more severely affected than Portugal.

In France and Italy, the towns in particular were ravaged, losing in some cases between half and two thirds of their populations. In Flanders, the countryside suffered a deadlier blow than the towns. In England, the advance of the plague was unyielding, perhaps because having trade relations with France, Italy and Holland, its ports were exposed to more than one strain. Within two months, the country was in great disarray and not even the Parliament managed to convene for the entire year. In London, the need for sanitation was perceived and even a quarantine was instituted, but the city was not spared in the least. The quarantine had first been attempted by the Venetians, but their city too fell all the same. Germany was less affected and even less the territories to the east from Hungary to Poland. In Germany, however, there was another calamity that was entirely man made. The Jewish population was made responsible for the spread of the plague, even when it was being decimated in the same manner as the general population, but this reasoning had no effect. The Jews were persecuted and thousands were killed. In the city of Basel, they were rounded up and immolated together on an island of the river Rhine. There were many and varied reactions to the plague. In Germany, the groups of flagellanti became quite common. The strange practice had originated in Perugia, Italy, long before. These were people who believed in doing severe penance to expiate the scourge that had fallen upon them. They traveled from town to town recruiting new members and congregating in public places in order to endure collective floggings that took place amidst screams of pain and chants of exultation. It must have been a sobering sight. The Church initially tolerated with some complicity the practice of this overly pious groups but, in the end, turned against them and declared they were heretics. Most people believed the plague was some form of divine punishment. Many became extremely devotional and pilgrimages were quite popular. People spoke of the daily baptism of tears, as if there was a certain redeeming quality in the suffering itself. The view that life was all misery and distress did not appear far-fetched. There were others, of course, who adopted an opposite point of view; seeing the futility of fighting against the plague, through a virtuous life or by any other means, they adopted a careless and Epicurean attitude toward life, taking advantage of it before its impending end. Above all, the plague broke the general fabric of society. Families broke up under strain, despairing priests abandoned their parishes, doctors charged exorbitant fees to unwitting victims. It is not easy to imagine the effect of seeing scores of relatives and acquaintances, one after the other, develop within hours the dark skins caused by internal hemorrhages and then perish within two or three days. The notion of the dance macabre is not entirely a product of the imagination; the victims of the plague often suffered seizures to their nervous system that led to trembling and convulsions. Petrarca, who described the scenes he had witnessed in Florence, insinuates that an incredulous posterity might take his testimony as fable. At the time, the true cause of the plague was not discovered. Those who favoured astrology speculated that a conjunction of Saturn, Jupiter and Mars a few years earlier had unsettled the good order of the world. Others believed that some earthquakes that had taken place in Asia had liberated poisonous fumes from the core of the Earth and that these had spread the pestilence far and wide. The fact that the transmission of the plague took place through rats was not discovered either. In fact, it was believed that rats were as much victims as humans.

Albertus: Certainly, the plague of 1348 must have been one of the greatest up-heavals civilization has ever seen and we should do well to study how the institutions, the beliefs and the customs of civilized life were shaken by it, not out of any morbid curiosity, but to remind ourselves of the fragility of our own condition. We have become so arrogant of our capacity to know that we cannot imagine ourselves exorcising demons and ghosts in vain, while harm in one form or another might be descending surrepticiously aiming at our destruction. We could well be as ignorant of its origin as oblivious to our own ignorance.

Plebeius: The material conditions of life were disrupted for many years after the plague. There was as much a rural crisis as there was urban disarray. The old manorial system with a stable peasant population began to disappear. There was such a shortage of manpower that labourers felt free to move and seek better wages. Initially, the prices kept pace with the rising wages, but after a few years the reduced demand brought the prices down, which in turn produced a deceptive sense of prosperity. Nevertheless, the considerable rural depopulation also led to the concentration of property into fewer hands and numerous peasant uprisings followed. The cities had also lost a considerable fraction of their populations and the universities, in particular, were decimated. The decline in production and the labour shortage accelerated the introduction of new manufacturing methods along with some new mechanical tools. These innovations brought some unexpected benefits at the same time that the general hardships of life seemed to have become so much greater. What made the effects of the plague so long lasting was the fact that the plague had not departed entirely. It returned in very frequent cycles; barely a decade after the Black Death a second plague struck, the so-called children's plague for the disproportionate number of children that it killed. Throughout Europe, for roughly a century, the plague struck two or three times per generation, so that the population had very little chance to recover. The prosperous city of Florence, for instance, that had had a population of almost one hundred thousand at the beginning of the XIVth century could count only forty thousand a hundred years later. Thereafter, the plagues became more sporadic, although they did not disappear altogether for yet another hundred years. In Italy, the social and economic decline was accompanied by the devastating campaigns carried out by free standing armies, often led by ruthless condotiere, mercenaries serving one lord or another, often pursuing just their own interests, ravaging and pillaging any town on their path. It was only the larger cities that could remain autonomous and assert their power. Thus, in addition to the progressive concentration of wealth within the rich bourgeoisie, political and military power also started to concentrate around a few powerful states, dominated either by a city, such as Milan or Venice, or by a dynasty, as was the case with the house of Savoy or, most conspicuously, by the Church in Rome. The flourishing life of the self-governing Italian towns of a century earlier did not survive unscathed and what had been a wonderful civic experiment in urban development was to a large extent discontinued. It seems so puzzling that, amidst these circumstances, a period of intense interest in learning and the arts got under way in northern Italy, producing an unprecedented number of craftsmen, scholars and artists, whose reputations have in many cases endured to this day. Their accomplishments were generally sponsored and supported by the new oligarchies, sometimes by what we can only charitably call enlightened despots. Nevertheless, although the work of these writers, sculptors and painters was often directed toward celebrating the life of those who hired them or toward decorating their palaces and residences, those who participated in this new cultural movement were soon pursuing their own objectives and they gave expression to sentiments and aspirations that were their own. It is ironic that, along with the rediscovery of antiquity and its values, a renewed interest in the role of the citizen became manifest, at the very same time that democracy was being undermined with the rise of these new oligarchies and the myriad principalities that kept themselves busy in endless rivalries, misalliances and wars. These contradictions we perceive only in retrospect and after a long time. It is impossible not to notice, for example, that while the artistic legacy of this era is overwhelmingly religious in its subject matter, the culture it emerged from was predominantly secular. Likewise, we find today that despite so much turmoil, in a society that was rapidly transforming itself, the written testimony shows that men saw themselves quite at ease in their condition, eager to explore and with the desire to cultivate their own capacities. The gloomy view that saw life merely as a transitory stage in which the endurance of one's soul was tested, gave way rather suddenly to a view that saw it as a noble adventure, besieged by tribulations perhaps, but also worthy of being exalted and celebrated.

Albertus: I recall that we have discussed these matters before, Plebeius, and I believe that my objections were and still are that this fantastic movement of renewal, despite its spontaneity and its sincere curiosity and optimism confronting the world, was not in fact very much interested in the sciences. Perhaps it was too literary and artistic in its orientation and had more interest in cultivating style and expression than in seeking new ideas. All the inventiveness it stimulated seems to have resulted at best in the development of new techniques and not so much in fundamental or lasting theories. It is possible that its rebellion against dogmas or its own distrust of all knowledge received from authority may have contributed to its eclecticism and its distaste for systematic thought.

Plebeius: I have considered your objections but I still suspect that you give to this era less credit than it deserves. Its curiosity about the natural world may have been confined to a desire to tinker and experiment more than to produce elaborate theories, but in its very eclecticism lay part of its strength. It is undeniable that it produced some splendid individuals and this was not coincidental at all, for it formed part of their outlook in life. In this regard they succeeded perhaps better than we have in our own age; instead of molding a person forcefully so as to pursue a specialized and well defined objective, they tended to see each individual as a composite of diverse aptitudes and gifts, to be cultivated and nurtured into a harmonious whole. I am thinking of such men as Paolo Toscanelli or the great Filippo Brunelleschi. Brunelleschi was not even a learned man, in the sense it had at the time, for he had not studied either Latin or Greek. Having started as a sculptor, he gradually turned to architecture, where he made his most celebrated contributions. But as a draftsman, he was perhaps the first to explore the art of representing depth on a two dimensional surface and he came to discover the rules of linear perspective, despite the fact that he had not had any substantial mathematical training. The same may be said of the vastest project he undertook, the construction of the dome of the cathedral in Florence. All his innovations in design, the technique of building the vault without armatures, the many tools he created, hoisting devices, rotating cranes, load positioners, were not the product of any systematic study but the outcome of a long apprenticeship, a good deal of experimentation together with his daring genius. Would you not consider Paolo Toscanelli, this other remarkable Florentine, equally representative of this era? Master Paul the physician, as he was known, had an enormous range of interests outside medicine, including mathematics and astronomy. You may say that he confounded his observations of comets with a good dose of astrological speculation, but he was just as inquisitive and restless as the scientists of a later age. He was in contact with the most knowledgeable scholars of his day and received their praise for his expertise and good judgment. Although he was twenty years younger than Brunelleschi, they established a very close friendship of mutual respect and admiration. When the lantern at the top of the dome of the cathedral was installed, after Brunelleschi's death, I think it was Toscanelli who had a hole drilled through it, so as to be able to carry out measurements of the Sun's motion, following its projection on the cathedral's floor. Toscanelli, like Brunelleschi, left almost no writings of his own and his legacy can only be measured by the way he influenced his contemporaries. Geography was another of his interests and he had a hand in bringing to life again an interest in cartography. He was known to interrogate at length the many travelers that passed through Florence. A famous letter of his written toward the end of his life and addressed to the canon of Lisbon suggests the feasibility of a trip to China by sailing west across the Atlantic. By all accounts, Paolo Toscanelli was a singular character, quiet and unpretentious, despite the preeminence he occupied amongst his colleagues. He devoted a good deal of time to administer his family business, which were considerable, but it is said that he was almost ascetic in his private life and that he took the pursuit of knowledge as a serious calling, not so much in a contemplative way but with an eye toward the benefits that mankind might draw from it. He had studied medicine at the University of Padua, which was at the time perhaps the preeminent University in Europe and where a number of other distinguished contemporaries also studied, such as Nicholas of Cusa, whom Toscanelli also befriended. The Faculties of Medicine and Rhetoric had been added to the Faculties of Canon and Civil Law by the Venetians in 1405, once they had wrestled control of the city from the Carrara family from Verona after a few years of struggle. The Venetians sought to enhance the rank of the university and profit from it as well; they did not allow the citizens of Venice to study anywhere else and provided a favourable climate for students from the rest of Europe to enroll in it. I cannot resist mentioning also Vittorino da Feltre, originally from one of the small towns at the foothills of the Alps the Venetians had acquired in their expansion. He was also a product of the University of Padua, to which he remained associated for quite some time. He had enrolled in the Faculty of Rhetoric and later took time to study mathematics, although at the time there were no teachers of mathematics proper at the University and one imagines the instruction was carried out by the teachers of astrology or moral philosophy. Because his means were quite meager, Vittorino found himself obliged to take up students of his own to pay for his studies and this forced occupation turned then into a lifelong concern with education. For a while he taught in Venice, then returned to Padua, where he began to take students in his house, rich and poor alike, each of which would pay according to his means. It was during this time that he began to give shape to his educational methods, which later brought him renown and a large following. Eventually he was appointed to the chair of rhetoric at the University of Padua, but soon thereafter the ruler of Mantua Gianfrancesco Gonzaga, with his wife Paola Malatesta, invited Vittorino to become the tutor of their children, a position he accepted gladly given that the Gonzaga granted him total freedom in designing the course of studies and choosing the type of activities the children would engage in. It was at Mantua that he spent the last twenty years of his life teaching long hours every day, not only with the Gonzaga children but also with other youngsters from various sectors of the community that he adopted in his small school and who were able to receive the same superior education. Like his contemporaries Brunelleschi and Toscanelli, Vittorino left no writings of his own hand. He used to say that enough had been written already in the past and that it was more worthwhile to put one's thoughts into action. Indeed, he devoted most of his time to testing his own ideas on education, seeking to find the proper role to be played by physical exercise, the role of punishments, the depth to which certain subjects should be studied and their diversity, the extent to which memory should be trained or the place that games should occupy in a child's life, in short, all the ingredients that should contribute to best develop his mind, his body and his character. Vittorino is a most inspiring figure. By my account, the excellence of an era or a civilization is measured by the extent to which it concerns itself with education and by how it argues what should be the aims of education, when we do not see ourselves just as the product or the victims of the arbitrary circumstances of history, but society as a whole pauses to consider what it wants to become, how it aspires to be perceived and what tools it has at its disposal to achieve it. One of the peculiarities of Vittorino and the educators of his era in relation to their predecessors is the fact that they remained pious men, willing to preserve religion as an ingredient of a good education, but religion was seen now as concerning the private, spiritual life of the individual, so that they saw it appropriate to turn the main focus of education on its public spirit. How to achieve perfection as a Christian concerned each man or woman, how to achieve perfection as a citizen and how to educate for good citizenry, was a collective enterprise. All type of relationships and commerce amongst the members of a society were not supposed to resolve themselves in friction and discord but had to be cultivated as a craft, an art called civility. Logic had to be taught for clarity of thought and expression, not for sophistry and vain ostentation; the aim of the art of persuasion was not trickery or deceit but to make debates fruitful and expeditious in arriving at the truth. Education had in their view a new set of concerns, such as the balance between self-interest and public good or between authority and freedom. It was no longer aimed exclusively at the individual, providing him with a safeguard against the adversities in life, but it wanted also to help us define our mutual relations, between the sexes, toward children or the elderly, also to discern what makes our conduct worthy of emulation and a myriad such questions.

Albertus: If I were to search within the field of astronomy for individuals that were as enterprising and high-minded, I would have to turn without question to Georg Peurbach and Johannes Regiomontanus, certainly the most distinguished of their day. You mentioned earlier that, after the great schism within the Church and with the support the French theologians gave to the Pope in Avignon, many scholars from the University of Paris moved to Germany or further east. One of their destinations was the University of Vienna that had been founded only recently. It was here that Henry of Langenstein came to teach, an acerbic critic of Ptolemy, who established in this city a modest but uninterrupted tradition in astronomy for almost a century. Georg Peurbach, born near Linz in 1423, initiated his studies in Vienna although, by the age of thirty, when he obtained his master's degree, he had already traveled through Germany, France and Italy. He seems to have been offered the possibility of teaching at the universities of Padua and Bologna but he preferred to return to Vienna. In the following years he wrote a book entitled New Theory of the Planets, which was a reformulation of texts of a similar nature that had been in circulation for almost two hundred years and that were meant to be an introduction to Ptolemy's theory. It is interesting to note the other activities that Peurbach was engaged in during this time, for they may reveal, perhaps not his own personal inclinations, but the requirements placed on a brilliant scholar at the university. During some years we find him lecturing on Virgil or Horace and, at other times, he was serving as court astrologer to the king of Bohemia and Hungary, who presumably needed very much his help. He was only fifteen years old and throughout the territories that had been under the influence of Germany, from Saxony to Bohemia, from Bavaria to Hungary, the various branches of all the aspiring royal families were intermittently ruling one region or another and constantly quarreling over the spoils like children. The king also saw the need to spend some time doing battle against the Turks who had by then advanced to Serbia. In any event, his life met an abrupt end at age seventeen and Peurbach came then under the service of Frederick III, the nominal emperor. In the year 1456 he observed the passage of the comet now known after Halley; by this time he was being assisted by Johannes Regiomontanus, a young student who, like himself, had adopted as his own the name of his home town, Könisberg, in Franconia, albeit, in its Latinized form. Regiomontanus was then twenty-years-old and the next year, at the allowed age, joined the faculty at the University of Vienna. In 1460, Cardinal Bessarion, the Papal envoy who was of Greek origin, visited Vienna and encouraged Peurbach to translate Ptolemy anew. According to Regiomontanus, Peurbach knew the somewhat deficient translation from the Arabic of Gerard of Cremona almost by heart. With the assistance of Bessarion, Peurbach was able to complete only a fraction of the work; in less than a year he was dead at the age of thirty-seven. Regiomontanus then determined to learn Greek and accompanied Bessarion back to Rome, where he completed the translation that came to be known as the Epitome to the Almagest. It was a shorter version than the original but in some respects an improved version, of greater clarity and precision, with added commentary, references to newer observations and to discrepancies between theory and observation. For example, it calls attention to the fact that Ptolemy's prediction for the variations in the apparent size of the Moon were larger than observed, something that seems never to have been of paramount concern to their predecessors. With Peurbach and Regiomontanus we can say that for the first time in Europe there were astronomers who were competent enough to understand Ptolemy's work in all its intricacies and make also their own original contributions. In the ensuing years Regiomontanus first lectured at Padua, for

112

he had accompanied Bessarion when the Cardinal was appointed Papal legate in Venice. During this time he composed a book On Triangles, as he called it, which he dedicated to his patron and that was the first complete treatment of planar and spherical trigonometry in the West, comparable in accomplishment to the similar and much earlier book of al-Tusi. Later, he moved to Hungary under the service of the new king Mathias Hunyadi, who had also acceded to the throne at age eighteen and who, unlike his predecessor, went on to have a long and successful reign. He was somewhat unusual in that he excelled as a general but proved to be a reliable patron of the arts and sciences at the same time. One of his greatest accomplishments was the foundation of the great Library in Buda. He also succeeded in attracting a good number of scholars and Regiomontanus, who had already a solid reputation, helped lead the effort to establish a new Academy at Pozsony, on the Danube not far from Vienna, later to become the city of Bratislava. All the while he managed to compose several astronomical and mathematical tables and a set of ephemerides amongst them, which was said to have been taken years later by Colombus in one of his trips. After three years, however, Regiomontanus decided to move to the city of Nuremberg in Germany, which he found more congenial and closer to the other centers of learning in Europe. It was here that he started a new pioneering career as a publisher of scientific works, starting with the New Theory of the Planets of Peurbach, which now became well known and widely used. Unfortunately, he was unable to carry out the extensive plan of publications he had anticipated. Within four years, not yet forty years of age, he succumbed to an outburst of the plague during a visit to Rome, where he had traveled summoned by the Pope to make suggestions on the reform of the calendar. Those who came after him lacked his enthusiasm and dedication and the valuable Epitome to the Almagest, which he had written with Peurbach, was only published some twenty years later in 1496.

Plebeius: Bessarion had been the young and bright archbishop of Nicaea, who came to Italy as a member of the delegation of the Emperor of Byzantium in his visit to Florence in one of the many attempts to reconcile the Eastern and Western Churches and also to seek support in the West against the Turkish threat. But Bessarion decided to remain in Italy, mastering Latin and Italian and becoming a distinguished scholar, translating some Aristotle himself and amassing a great collection of books that was the foundation of the Biblioteca Marziana, in Venice, still standing to this day. He rose through the ranks within the Catholic Church and was rewarded with a number of diplomatic missions, such as the ones you mentioned to Vienna and Venice and he was eventually made Patriarch of Constantinople in exile. I do not think he had any passion for astronomy but, for that reason, it is all the more fascinating the interest he took in the work of Peurbach and Regiomontanus. Nicholas of Cusa, to whom I referred earlier, also participated in the synod between the Eastern and Western Churches in Florence and had been sent to Constantinople earlier in preparation. His career, as a man of action and of learning, conducted from within the Church, parallels that of Bessarion and was equally illustrious. He was a native of Germany, form the Moselle region, near Treves, and he went on to study at the University of Padua during Vittorino's time. His degree was in Canon Law but his interests ranged widely. Later he embarked on a path that we might call of ecclesiastical diplomacy. The Church during this epoch was in a state of stupendous confusion. For thirty years there had been a French Pope and an Italian Pope. In 1409, when a Council was convened in Pisa to solve the disputes, it ended up adding a third Pope. Only ten years later was it possible to unite the Church under one single authority. Meanwhile, a number of Councils were convened and dissolved in rapid succession, attempting without much success some degree of reform, always with the aim of curtailing the dissolution and excesses found throughout the Church hierarchy, the intrigue, the nepotism, the business of indulgences and other such niceties. The Council that was convened at Basel dragged on for some twenty years and it was here that Nicholas of Cusa started his career. Like his friend Aeneas Sylvius Piccolomini, who would later become Pope, he took up the councilar cause that sought to assert the supremacy of the Council as the Church authority, in opposition to the Roman Pope. For a time the Council, in exasperation, nominated its own dissident Pope and Piccolomini became his secretary. Nicholas of Cusa soon turned to the cause of the Roman Pope and he became, in Piccolomini's words, his most forceful defender. Having become a priest by this time, Cusanus was rewarded with the Bishopric of Bressanone, in the Tyrol region of Italy, a post he did not assume immediately because he was also sent to Germany as Papal legate, a mission that involved a great deal of traveling and that he undertook with particular zeal, delivering numerous sermons and preaching the need for reform within the Church. All this ecclesiastical activity took a good deal of his time, but he never abandoned his philosophical pursuits and managed to write meanwhile a number of books. One must conclude the he was very much a dilettante, his thinking is very unsystematic and cannot be said to follow any school. He believed that true divine knowledge was inaccessible to man and that all we can attempt are tentative approaches. For him, each individual had to find his own road to knowledge and he emphasized knowledge from experience more than bookish learning. Knowledge was conquered not so much through scholastic training, but by means of brief illuminations. Convinced in this way of the inability to convey true wisdom, he resorted to the use of metaphors and a varied imagery, sometimes mathematical, all of which makes for a colourful writing style. Later in his life, when he attempted to assert the land rights that, according to medieval custom, were assigned to him as prince of his diocese, he ran into difficulty with the Austrian duke who, as count of Tyrol, was unwilling to grant them. Cusanus, eager to do battle, hired some mercenaries but eventually was cornered and had to accept a not very favourable settlement. Meanwhile, his friend Piccolomini, who had achieved some fame through the brilliance of his writing and oratory and who had been decorated by the German Emperor, in whose service he was, had also turned around by now to the cause of the Roman Pope. He had become a priest and received as reward the appointment to Bishop of Trieste. Piccolomini rose quickly through the ranks of the Church and, within a decade, he was nominated to the Papacy, a post he occupied under the name Pius II. He could now make some attempt to help his old friend. He excommunicated the Austrian Duke and appointed Nicholas of Cusa to the college of Cardinals, although this was not of much help and Cusanus remained for a good portion of the last years of his life in Rome, writing a quick series of some of his best books. Pius II was one of the more learned and certainly one of the best writers amongst the Popes. However, once he was appointed to this office in 1458, and for the next six years until his death, he turned against the humanists and scholars, of whom he had been a prominent member or, at the very least, we might say that he neglected their interests. His main obsession became the Crusade against the Turks. The fall of Constantinople in 1453 was truly felt as a cataclysmic event in most of Europe. Pius II sent emissaries to all the courts of Europe to rally their support so that all could join in a common cause, but his great efforts had very little success. The end of his life can only be said to have been rather sad, if not pathetic. Being already ill and weakened when he gained the support of the Venetians to launch the Crusade, he decided to take charge of the great enterprise in person. He traveled to the city of Ancona to meet their approaching fleet, but he died as the fleet was entering the harbor. Very soon all the forces that he had gathered were dispersed and nothing came of the Crusade. Nicholas of Cusa had died only a few days earlier in the nearby city of Todi as he prepared to join his old friend in Ancona. Nicholas of Cusa was a very singular man and because of his vigour and originality I find him to be one of the more endearing characters of this entire era. His writings contain many imaginative and unconventional ideas, although they are often presented in a rather convoluted and obscure way. But it would be unfair to accuse him of being vain about what he was attempting to accomplish with his philosophic studies. In one of his essays he presents an entertaining conversation between a philosopher and a fool, the latter being a maker of spoons, and in the end it is the fool who leads the philosopher to the true knowledge of the divine, for the fool is a crafts-man who creates something that does not exist in Nature and, as such, is closer to the original creative activity of God than the studious but only imitative efforts of the philosopher. I am certain you would contend that his flirting with mathematical ideas were only the musings of a philosopher overly fond of Plato's mode of philosophizing and not very representative of a truly scientific spirit.

Albertus: If I remember correctly, it was Peurbach himself who had examined his mathematics and found it wanting. I do not wish to overstate my case or to insist too forcefully on the dearth of truly original scientific thinking during this period. You reminded me appropriately of a few individuals who expressed their originality in other remarkable ways. I believe that Paolo Toscanelli, without being a mathematician, was amongst the first who attempted to study Archimedes, at a time when very few knew of his work. The renewed interest in cartography was undoubtedly motivated by the advances in navigation and the first explorations of distant seas, but it failed to incorporate any new scientific principles or make much progress for many years. The great Geography of Ptolemy had remained unknown or ignored in the Latin West and was only translated from the Greek in the beginning of the XVth century. Its first printing was done much later in 1472, with many editions following in subsequent years, always with the addition of new maps since the original ones of Ptolemy had long been lost. These new maps were still quite inaccurate, just as those of Ptolemy had been, and remained so for some time. This is not to say that they were not more descriptive of the true geography than the old portolan charts that mariners had employed until then, charts that merely attempted to describe a coastline or provide compass orientation in setting the course from one port to another. On the other hand, in order to represent a sizeable segment of the known world on a map, there was first the problem of how to map the curved surface of a sphere on a plane; secondly, it was necessary to know the latitude and longitude of a sufficiently large number of the localities that were going to be plotted. Progress on either front was very slow to come. Ptolemy had solved the first problem by placing the sphere inside the cavity of a cone and projecting then from the center of the sphere, along straight lines, through each point on its surface toward its corresponding point on the cone, thus making use of the fact that a cone can be cut along an edge and then rolled out onto a plane. It was not until almost a hundred years later that cartographers incorporated new geometric procedures in the design of maps different from those of Ptolemy. Likewise, it took a long time before a concerted effort was made to devise efficient methods of measuring the latitude and longitude of the places that were being represented in the maps. Nevertheless, as it happened with the introduction of perspective in painting, the needs of cartographers may have provided an additional stimulus to incorporate further mathematical subjects in a general education. The same can even be said regarding the needs of bankers and accountants and their increasingly complicated array of commercial transactions. Toward the end of the century we see the proliferation of mathematical manuals describing simple, efficient ways of carrying out arithmetical operations, mostly for the benefit of bookkeepers and traders. All these advances were very modest indeed. Despite the notable progress in general education, it seems to me that there was not a concurrent advancement in the sciences themselves, either in the originality of devising new methods to treat old problems or in the imagination of conceiving new ones. For all the prosperity we witness in this age, first in the wealthy courts of Italy, but then also in Germany and Flanders, with the generous patronage of artists and philosophers that resulted in so many lasting works, we do not find a similar effort in the scientific disciplines, at least nothing comparable to what we are able to witness only fifty years later. Regiomontanus was certainly a pioneer, not only in his dedication to master Greek with the intention of translating a scientific classic, but also in having become an accomplished mathematician, as well as by having understood very early what opportunities had been made available by the arrival of the printing press. It is curious that this technique, which the Chinese had developed centuries earlier, seems never to have made its way to the West. Its rediscovery in the middle of the XVth century must have had a considerable impact throughout society, not just amongst the many scribes who made their living in the laborious production of manuscript copies of books. Initially, much of what was printed must have been either religious works or shorter tracts, calendars and astrological charts. The publication of books that could have an influence on the diffusion of scientific knowledge took more time to develop. For this reason, the long list of works that Regiomontanus had envisioned publishing seems all the more remarkable and it is unlikely that he would have been able to carry out this ambitious project, even if he had lived a long and healthy life.

Plebeius: Much of what was published were indeed religious texts and those scribes who had been employed producing manuscript copies of them, seeing their source of income disappear, attempted to argue that it was unbefitting to let machines do a work that required greater respect and devotion. I will not dispute your assessment that an era that was so prosperous in other respects was rather poor in its scientific accomplishments. But it is quite a different matter to argue that the intellectual climate created by the humanists of the XVth century was in fact detrimental to the development of science or, as some suggest, that it placed too great an emphasis on literature. There are those who go as far as to say that this movement, by being somewhat rambling and erratic, hampered the development of the type of methodical thought that is required in the sciences. They contend that those who opposed the movement and remained faithful to the tradition of scholastic learning, many of whom still held on to their academic chairs, at the University of Padua for example, they were the ones who paved the way for a true scientific revival later on. This can only be true in a very contorted way. Those who were enthusiastically attempting to restore the great texts of antiquity were only in part motivated by literary considerations. You mentioned Cardinal Bessarion who contributed significantly to secure a new translation of Ptolemy, despite the fact that his direct interest in astronomy must have been rather limited at best. It is noteworthy that the interest in ancient Latin and Greek works also stirred an interest in the text of the Bible and its old versions in Greek and Hebrew. These studies, such as those of Lorenzo Valla initially, and Erasmus later, reveal not only a literary or linguistic curiosity but also a changed outlook within Christianity, a more critical and introspective attitude, with a desire to place itself within the course of history more than to interpret history dogmatically according to the Scriptures. The old theological disputations concerning doctrine ceased to attract much attention. It is true that Christianity saw itself now immersed in a more secular culture, but the role of religion in the life of the individual or of society as a whole did not diminish. A withdrawal from the world was no longer seen necessarily as an example of virtue. The faithful could now pursue a dignified and virtuous life within the context of their civic responsibilities. Wealth and prosperity were not seen as hostile to a pious life, as long as they were a stimulus to contribute to the public good. The Church itself, typically the custodian of the more conservative views, tolerated some dissent within its ranks. Nicholas of Cusa, a prominent cardinal, advocated at one time an encounter between Christianity and Islam to seek their common roots and further their mutual understanding, something that would have seemed unthinkable a few generations earlier, when purity of the doctrine was held at a premium. One finds signs of this increased open-mindedness even in minute details. The small booklets printed for those who undertook long trips of pilgrimage were no longer limited to admonishing the pilgrims regarding the observance of their mission, but also tried to satisfy their curiosity, informing them on the customs and traditions of the lands they would visit, very much like a modern travel guide. If these things point to a more tolerant and receptive attitude and, more generally, if we can speak of an acceptance and a reconciliation with the material world, then I think that, even within the confines of the Christian community, we witness the same inquisitiveness, the same disposition to learn that was animating the society as a whole. Thus, extending our view, when we observe the increased dedication to many arts and crafts, the interest in cataloging plants and herbs, as it is found later in countless printed versions, the curiosity about foreign lands or the desire to study the human body, we realize that all these things were the early manifestations of what later was pursued in a systematic way. Perhaps there was not yet the attitude of seeing the world as a riddle and the desire to interpret it according to general laws, but it was moved by something just as significant, the desire to observe and the pleasure of discovery. No one illustrates this inclination better than Leonardo da Vinci, who must have been only a generation younger than Regiomontanus. Leonardo had very little formal schooling, since he was made an apprentice to a sculptor at the age of twelve, but his imagination and originality enabled him to make of every experience an adventurous exploration. Learning about linear perspective led him to study the process of perception, optics and the visual system. Likewise, from an artist's concern with human musculature, he went on to investigate thoroughly the human anatomy and to carry out a long series of remarkable drawings. We cannot have much doubt of his passion and determination when we observe that he did not shun away from spending many a night in the nauseating atmosphere of the mortuary amidst frayed and dismembered corpses, at a time when these activities were still frown upon and censured. Being interested also in all sorts of mechanical devices and machines, Leonardo combined these interests to investigate the mechanics of locomotion in man and in animals. With his insatiable curiosity and the ability of a great draftsman, he left scores of sketches of plants, animals, rocks and fossils. His activities as a designer of machines were also extensive and far-reaching, he studied aerodynamics and the possibility of flight; he concerned himself with watermills, offered his services as a military engineer and devoted his attention to a host of technological problems in a variety of branches, from masonry and tanneries to foundries and the art of casting bronze. On all manner of subjects he left a great abundance of notes, always bursting with new ideas, but these notes also remained scattered and rather formless at the end of his life. Leonardo was not a contemplative man, he was certainly a man of ideas, but he cannot be considered a man of action either. He was too speculative to be called an engineer and too practical to be called a philosopher. Despite the singular quality of his genius, he failed to have a great impact on his contemporaries, partly due to his inability to bring his original conceptions to fruition, but also due to his reticence and his preference for privacy. Looking at his legacy from afar, we can certainly consider him a scientist before his time, always engaged in a most imaginative dialogue with Nature, sometimes speculating daringly, other times reasoning cautiously from the simplest tools and examples that lay before him.

Albertus: Let us not stray too far afield and turn now to a man of another mold and with an outlook very different from Leonardo's. I am speaking of Nicolaus Copernicus. Some twenty years younger than Leonardo, he belonged to the same generation of Michelangelo and Dürer, of Erasmus and Machiavelli and, of course, of Martin Luther too. He was born on the confines of West Prussia, in the town of Torun on the Vistula, in a territory that had recently been ceded to Poland, but Copernicus spent most of his youth away from home and, thanks to his education, he can also be said to be the product of the enterprising and optimistic intellectual climate of his era, even if, by temperament, he seems to have been more circumspect and withdrawn. His father was a prosperous businessman and also his mother belonged to a wealthy family of merchants; a brother of hers, a prominent clergyman and bishop of Varmia, took charge of guiding his nephew's studies after the death of the father. We can safely say that Nicolaus was not bound to a destitute future. Initially he was sent along with his brother to the University of Krakow, where he already showed an interest in astronomy; his teacher had held the first chair of mathematics established at Krakow. Later he may have visited several German universities, until he finally enrolled at the University of Bologna.

In Bologna he became assistant to a distinguished local astronomer, although, ostensibly, he was enrolled in studies of canon and civil law. In any event, he did not remain there to obtain a degree, but traveled then to Rome where he stayed during the year of the jubilee in 1500, and moved later to Padova where he also studied medicine, once again without obtaining a degree. In the meantime, his uncle had secured for him a position of canon at the cathedral chapter of Frauenberg and Copernicus took up this position when he returned to Poland, already in his thirties. At first he stayed at the bishop's see, acting also as his personal physician, but after a few years he moved to Frauenberg itself, where he was to remain for the rest of his life. The position of canon was mostly administrative, the duties were light and it offered quite a decent income. It is possible that Copernicus' family expected of him to pursue a more illustrious ecclesiastical or diplomatic career, but his interest in astronomy sent him off course to a more quiet and private life. By the time he settled in Frauenberg, where he built a modest observatory for personal use, he was already embarked in his project of recasting the mathematical astronomy of Ptolemy into a heliocentric framework. During those years a manuscript of his entitled Brief Commentary was circulated privately; it contained, in a sketchy form and without any proofs, most of the ideas that were to appear thirty years later in his great treatise On the Revolutions of the Celestial Spheres, published shortly after his death. We do not know how Copernicus became persuaded of the motion of the Earth, rotating on its axis as well as orbiting around the Sun. He never discussed the genesis of his ideas in his writings; instead, he presents the finished results without any clue that might reveal his train of thought. It is probable that the hypothesis of conceiving the Earth as one more amongst the planets, making them all orbit around the Sun, appeared plausible to him already during his student days in Italy. He may then have decided that it was worthwhile to investigate whether the theory of Ptolemy could be reformulated adopting that point of view. During his stay in Padua, for instance, there was a young professor named Gerolamo Fracastoro, also interested in astronomy, who years later developed his own system of the world, largely an unsuccessful attempt at resuscitating the old model of concentric spheres of Eudoxus. It seems so very likely that the two men met to discuss the shortcomings of Ptolemy's astronomy and the ways to amend it, as must have done also other mathematicians and astronomers at the time. In the first book of his treatise, Copernicus gives a general outline of his theory. He begins by presenting the arguments in favour of the motion of the Earth. Ptolemy had argued that it was preposterous to imagine that the bulky Earth could be rotating on an axis, since it would certainly fly into pieces if it were to do so. Copernicus turned the argument around. Was it not more demanding to ask of the immense vault of the Heavens to rotate daily? Why did it not fly into pieces? Likewise, to the argument that the rotation of the Earth would force it to leave clouds and birds behind, Copernicus could respond that it was equally reasonable to assume that the entire sublunar world, with the air and everything it contained, participated in the same motion. Lastly, as far as the possibility of the Earth being hurled about in space, Ptolemy had argued that the tendency of all objects to fall toward the center of the Earth was sound evidence of its occupying the center of the world, but Copernicus could now turn the same argument against itself. The tendency of all things to be as close as possible to the center of the Earth had given it its spherical shape, but we also see that the Moon and Sun are spherical as may well be the case with all other heavenly bodies. Was is not plausible that they also enjoyed the same power of gravity? Then, if the Earth turns out to be like the other luminaries in the sky and we see them all in motion, why could we not assume that it too is moving? Nevertheless, we must suspect that these were not the arguments that set him off in his investigation; he is much more likely to have been moved by the ideas of Martianus Capella, for example, a commentator of the Vth century whom he quotes and who, amidst the disputes trying to elucidate whether Mercury and Venus were below or above the Sun, had argued that it was more reasonable to suppose that they rotated around the Sun and not the Earth. As far as the superior planets was concerned, you may recall that in Ptolemy's description of their orbits, as the Sun rotates around the Earth, the radius of the epicycle guiding the planet around must always march parallel to the radius going from the Earth to the Sun. Now, in the simplified case in which we make the Earth the center of all the eccentric circles, you immediately convince yourself that, if those two radii, that of the epicycle and the one leading from the Earth to the Sun, have the same length, then it follows that the distance from the planet to the Sun remains constant. Therefore, just like a sailor trying to unravel the movements of various ships when his own is also moving, why not jump ship, so to speak, place ourselves in the Sun and, given that both distances from the Sun, to the Earth and to the planet, remain constant, describe then their orbits as circles around the Sun. Copernicus himself, alluding to the relativity of all motion, quotes the words that Virgil puts in the mouth of Aeneas when he says that we sail out from the harbour and see the land and cities recede away from us. Unfortunately, we must consider all this speculation to be idle talk, since we will never be able to retrace Copernicus' steps. Moreover, I do not want to simplify beyond recognition what, at the time, must have looked like an entangled plot of competing ideas and hypotheses. But it is clear that, once Copernicus understood the bare outlines of his new thesis that put the Earth in motion around the Sun, the entire scheme of a planetary system centered at the Sun must have appeared compelling to him. He could now explain in a simple manner the variations in brightness of the planets and not only their retrogressions along their orbits but also the reason for their occurrence, which appeared to be shorter the more distant the planet was. He could also explained in a simple manner why the retrogressions took place only when the planets were in opposition to the Sun. Fracastoro, defending his own model of the orbits, had attempted to explain the variations in brightness by postulating that the spheres carrying the planets were filled with a certain aether of uneven distribution, so that their light was dimmed when traveling through the denser regions. This was possible but seemed ad hoc and rather conspiratorial. Copernicus must have felt that his interpretation was much more economical and elegant. The suspicion or the confidence that he was on the right track must have given him the energy and determination to undertake the laborious task of overhauling Ptolemy's Almagest in its entirety and to attempt to show that he could predict the orbits of the planets with greater accuracy. I do not think we should stop to consider in detail the contents of Copernicus' treatise. It would be naive to assume that he managed to arrange the planets in circular orbits around the Sun, dismissing all consideration of epicycles and such complications. In many regards, Copernicus' account is more involved than the one he set out to replace. For the Earth's orbit around the Sun, Copernicus could have adopted Ptolemy's simple model of a eccentric circle, but he gave too much credit to the accuracy of Ptolemy's observations, as well as to those performed centuries later by Arab astronomers, and this induced him to believe that the inclination of the ecliptic had changed, that the rate of precession of the equinoxes was not constant and that even the eccentricity of the Earth's orbit had varied, as well as the orientation of the eccentricity relative to the stars, or that the line of apsides, as it is called, had moved along the ecliptic. Thus, he determined that the center of the Earth's orbit was not in some fixed position away from the Sun, but itself rotated around it on a circle with an epicycle. Copernicus was certainly the first to recognize the motion of the line of apsides, but the change in the eccentricity was fictitious and a product of inexact observations. The irregular precession of the equinoxes that he thought to have confirmed was obviously caused by the oscillations of the axis of rotation of the Earth and he adapted the old trick that had been first discovered by Nasir al din al Tusi and that we have already discussed, of converting a double circular motion into a rectilinear one. There is little doubt that much of the earlier work by Arab astronomers was relatively well known in Italy during that time and Copernicus may well have adopted several of their ideas. In fact, to account for the Moon's motion he discarded Ptolemy's model that required such a large variation in the Moon's distance to the Earth and was much more successful adopting the idea of double epicycles that had been used by al Shatir, the disciple of al Tusi at the Maragha Observatory. As far as the planets were concerned, he dismissed the use of the infamous equant introduced by Ptolemy, the point off the Earth from where the planets' motion was seen as uniform. Nevertheless, he also used orbits composed of eccentrics and epicycles, now revolving around the Sun instead of the Earth, in a manner that was reminiscent of the way they had been used by the astronomers of al Tusi's school. I am not implying that Copernicus adapted these old ideas in a simple and straightforward way. To the contrary, he laboured anew on all of the elements of the planets' orbits, their eccentricities and orientations, using observations that he carried out over many years and by comparing them to ancient observations through lengthy and difficult computations. His great work concludes with the description of the motion of the planets in latitude. This had caused considerable trouble to Ptolemy, which is quite understandable given that his orbits were centered at the Earth and not at the Sun. Unfortunately, Copernicus missed here the opportunity to simplify the theory to any considerable degree, in part because he remained very close to Ptolemy's methods, but also because he referred all motions to the center of the Earth's orbit and not the Sun itself, which are not quite the same thing, so that his theory cannot be said to be fully heliocentric yet. This is one aspect of Copernicus' work that has always remained puzzling. On one hand he knew that his ideas, if they proved to be true, would cause a revolution and would upset a view of the world that had prevailed for more than a thousand years; on the other hand, he follows Ptolemy so closely in purpose, in outlook and in his methods, that his writing may sometimes be confused with the Almagest. As Kepler would say much later, Copernicus often appears more eager to explain Ptolemy than the real world. Furthermore, all throughout his arguments we never find the excitement of a man who has discovered a golden path, a truth that he wishes to reveal to the world. We have no doubts of his determination and conviction, but he never impresses us as someone on a mission. He did not work secretly, but neither did he advertise his views, and the fact that he only acquiesced to publish them after much prodding is further proof of his reticence.

Plebeius: This reticence of his cannot be attributed to fear of censure or reprisals by the Church. Martin Luther called him a fool, intent on turning the science of astronomy upside down, but Copernicus himself seems to have always sided with the orthodoxy and, to my knowledge, the Church was sympathetic to him, at least in the beginning. His work was not placed on the Index of forbidden books until half a century later.

Albertus: Indeed, Copernicus dedicated his book to Pope Paul III. All the same, he must have felt very wary of advocating openly his views regarding the motion of the Earth. He says so much in his dedication to the Pope, speaking of the difficulties that he might face and the scorn, in his words, which I had reason to fear, on account of the novelty and preposterousness of my opinion, nearly persuaded me that I should entirely abandon the work I had already begun. This did not happen thanks to the encouragement of some friends, amongst them the bishop of Kulm and most of all Georg Rheticus, a young mathematician from Wittemberg, who at age twenty-five came to Frauenburg and stayed for two years, studying and collaborating with Copernicus, at the time already well in his sixties. It was Rheticus who wrote a first account of Copernicus' theory and its favourable reception gave Copernicus the reassurance and vigour to write a final and polished version of his work in the few remaining years of his life. One of the more important consequences of Copernicus' work was the fact that man no longer found himself looking at the Heavens as if he was at rest at the center of a sphere, observing the luminaries dashing through on the surface of the sky; instead, the Earth's roaming through space enabled him to observe the relative distances and the general disposition of the planets and the stars. Once he understood that the Earth was in motion, he possessed a new tool that gave access to knowledge that had not been available before. One would be tempted to say that this observation escaped Copernicus, but this is far from the truth. Indeed, he states quite clearly that his new description of the orbits of the planets, being a combination of the Earth's and the planets' motions, linked them all together in such a constrained but consistent way that one could not alter some part of it without disrupting the order of the entire construction. Nevertheless, Copernicus did not dwell very long on these matters. It is true that, in order to learn the true dimensions of the planets' orbits, he needed to know at least one absolute distance, such as the distance to the Sun, and this he could not determine any better than his predecessors. Another observation that seems to have attracted less attention than it deserved concerned the distance to the stars. Traditionally, the distance to the Sun had been estimated to be somewhat in excess of a thousand Earth radii and the distance to Saturn, the most distant planet, was approximately ten times that distance. The sphere corresponding to the stars was expected to have a radius twice the size of Saturn's orbit or, at most, a few times larger. However, it was now realized that the stars did not move appreciably as the Earth rotated around the Sun. Of course, if the stars were minute sources of light affixed on the surface of a large sphere, it was difficult to make any assessment of their distance. Copernicus only ventured to say that the sphere of the stars was immense. He did not pause to observe that, if the sphere on which the stars were distributed had some measurable thickness, then the relative positions of the stars would be seen to vary during the course of a year. This required that the radius of this sphere was at least a thousand times the radius of the Earth's orbit and not just a few times, as it had been assumed until then. It was only after Copernicus' time that questions of these nature began to be asked more often. Copernicus' theory was not seen merely as a description of the paths followed by the planets on the sky, but it was part of a fuller account of the physical world we inhabit. The notion of the Earth being hurled through space provoked questions that had not arisen until then. Our planet had long been held to be at the center of the Universe and all objects had a natural tendency to gravitate toward its center. If this proved not to be the case, how could one account for the attraction exerted by the Earth? Could other planets be endowed with similar powers and contain inhabited worlds similar to our own? And what power guided the Earth and the other planets in their orbits around the Sun? These questions had now acquired a new meaning. The periodic apparition of comets was yet one more mystery of the observable Universe that needed to be addressed. A number of them were closely scrutinized during the XVth century; quite exceptionally, there were two of them in 1457. Arriving on the heels of the fall of Constantinople, they were seen as the portent of some tragic events to come. Then there was the comet of 1577, studied with great interest by the great astronomer Tycho Brahe. From a rather abundant harvest of comets, it was noted first that their tails were always pointing away from the Sun, making it quite plain that these objects were not meteorological phenomena of the sublunar world, as had been maintained since antiquity. Tycho Brahe proceeded very carefully in his study of the comet's orbit and concluded, not only that comets resided outside of the terrestrial world in the aethereal region, but that their orbits around the Sun had to be very elongated, or else they were changing their speeds as they advanced along their paths. In the first instance, it was clear that the orbits had to cut through the spheres assigned to the planets, providing new evidence that these spheres could not have any material existence but were an artifice of the astronomers' theories. In 1572 a new star appeared in the constellation of Cassiopeia, readily noticeable to anyone vaguely familiar with the constellations. The star was as bright as Jupiter and it was said that at its peak it could be seen in daylight. The young Tycho Brahe, like many other observers, studied it assiduously, monitoring its changes in color and brightness until it faded completely a few months later. It became clear from the observations that the visiting star could not be an atmospheric phenomenon but that it had to reside above the lunar sphere. Moreover, it did not exhibit a tail like most comets or an orbit similar to theirs. It was in fact very much like all the other stars and just as distant, for its position did not change appreciably in the background of the rest of the stars as the Earth moved around the Sun during a year. One was compelled to admit now that the celestial sphere was not as immutable as it had once been thought to be and that it might perhaps be subject to the same processes of transformation and decay occurring in the terrestrial world. It is interesting that the main focus of astronomy in the West, ever since Babylonian times, had been to study the permanent features of the sky and the regularity of its motions. The Chinese astronomers, in contrast, had for millenia paid far more attention to meteors, comets and sudden new stars, which they dutifully recorded as meaningful, even portentous events in the life of the Empire. In England, the new star of 1572 caught the attention of Thomas Digges, a well-to-do gentleman with an interest in astronomy who published a brief tract recording the results of his observations. He assumed, erroneously, that the dimming of the star would be a periodic phenomenon associated to its motion. In fact, the star never brightened again. Thomas Digges had been educated by his father and by his father's friend John Dee. One could say all three belonged to a first generation of English men of science, interested in astronomy and astrology, but also in an assortment of problems of applied mathematics, such as surveying and ballistics, subjects on which they wrote extensively. Some years later, when Thomas Digges was editing some of his fathers works for publication, and having become acquainted with Copernicus's work in the meantime, he decided to add an appendix containing a rather free translation of the first book of Copernicus' treatise. It is this appendix which has given Thomas Digges some lasting fame, for in it he went beyond Copernicus and postulated for the first time that the sphere of the fixed stars was infinite in extent and so he thought was the number of stars, that the differences in luminosity amongst them probably reflected the fact that they were at various distances from the Earth, the dimmer ones progressively escaping our sight, which was capable only of reaching a finite number. Thus, an infinite void, populated with an infinite number of luminaries, Digges contended, how could one conceive of a better court for the great God, in his infinite power and majesty? From 1512, when the preliminary Brief Commentary of Copernicus' ideas had first circulated, to well into the XVIIth century, his theory of celestial motions gradually and slowly became accepted within wider circles, until it was eventually the only theory taught at the universities. The principal aim of his work had been to offer a faithful representation of the motions of the Moon and planets, capable of providing accurate predictions. One could argue, it seems to me, that it was only circumstantial that his theory came into being at the same time that a new conception of the celestial sphere was taking shape, more akin to the terrestrial world, a conception that united both worlds in one common Universe of space and matter. We can say at least that Copernicus' theory was coincident or symptomatic of this new conception but hardly its only originator. And just as the Heavens were losing their pristine and immutable qualities, so too the terrestrial world became accessible and comprehensible in an objective and matter-of-fact way as it had not been before. It is always difficult to reconstruct the climate of opinion of a distant era, but we must remind ourselves that, up to this time, the learned opinion as much as the common folk understood our world to be populated by spirits, demonic aquatic spirits that stirred up the tempests in the seas, terrestrial spirits that caused the Earth to shake and shed volcanic emanations from its womb, as well as the astral powers that had the capacity to seal the course of one's life and with which astrology concerned itself. When the nature of fossil rocks had not yet been understood and coral reefs were considered to be minerals, it was reasonable to think that not only animals and plants were endowed with life but rocks too were possessed by spirits which, so it was said, enabled them to rise into mountains. In the course of the XVIth century man began to examine the material world in a more detailed and systematic way, sometimes out of mere curiosity, attempting to understand Nature and its ways, sometimes with an eye toward developing a technique, imitating the ways of Nature itself. A few years after Copernicus' treatise was published, a similarly remarkable book on mineralogy and mining was published by the German

126

scientist Georgius Agricola, a book that, for its time, was a model of clarity and thoroughness, describing the methods of prospecting and of cave excavation, the tools and machines used in assaying and smelting and even taking an interest in the illnesses associated with mining. During those years Paracelsus was publishing his works on Alchemy as well. Despite all the extravagance and iconoclastic gestures of this singular man, one must admit that the way in which he went about searching for medications to treat a variety of diseases was rather sensible and practical. It is true that his theories were coloured by a belief in the supernatural power of the stars which, according to their composition, acted through the medications by an affinity of the elements involved. Nevertheless, his emphasis on direct observation and experimentation, motivated perhaps by his disrespect for any form of authority, led him to try all manner of combinations of herbs and mineral composites and, even when they seldom achieved their goal, they were the basis for other more fruitful and effective attempts in the future.

Plebeius: It is hopeless, however, to make any attempt to read his writings which can be utterly arcane and unintelligible. Nevertheless, I agree that the objectives he pursued had nothing to do with the search for the transmutation of the elements and other such practices, in which so many charlatans of his time indulged so easily. He was amongst the first to advocate the humane treatment of insane people on the grounds that they were ill and not possessed by spirits. Reading quite recently on how the concept of illness has evolved throughout the centuries, I have found that the slow and difficult progress of the medical arts in general is simply fascinating. I have to concede that, in so far as it concerns medicine, your earlier observation about the dearth of scientific achievements until this era applies thoroughly. One early accomplishment was the publication of the illustrated book on the Fabric of the Human Body, by the great Flemish anatomist Vesalius, which represented the first substantial advance since the time of Galen. This book must have been printed very close to the date when Copernicus' treatise was published, although its immediate fate turned out to be quite different. Vesalius' book was very much celebrated and readily adopted in most schools of medicine. Until that time, the knowledge of human anatomy and physiology was very limited and so were the tools at the disposal of a doctor. We have spoken of all these gentlemen who were either practicing physicians like Paolo Toscanelli and Fracastoro, whom you mentioned, later also William Gilbert, others had been students of medicine, like Copernicus himself. But medicine was in a rather sorry state at the time. All theories of diseases were based on an imbalance amongst the basic humours circulating in the body, the blood, the phlegm and both biles, the same that determined whether the character of an individual was phlegmatic, choleric, melancholic or sanguine. In addition, these humours were associated with the qualities of humidity, dryness, coldness and heat, each of which corresponded in turn to one of the basic elements: water, air, earth and fire. And, of course, there were further astrological associations between the positions and conjunctions of the planets and certain human conditions. How could then a physician treat successfully his patients with this jumble of principles? He could sense the patient's pulse, examine the colour or texture of his blood, but diagnosis was most often reduced to an exercise in physiognomy. As far as corrective action was concerned, bloodletting was one of the most common prescriptions, or perhaps one of a handful of purgatives or diuretic substances could be indicated but the stock of medications at his disposal was paltry. This was not for lack of trying. Physicians had resorted to anything and everything in order to palliate the cruel ravages of disease, they had brought plants from the tropics, they would even recommend sheep droppings or wolves feces for some illnesses, but seldom could they claim any success. Physicians did not shy away either from practicing surgery, but here, once again, one shudders to consider the methods and manners they employed; it is enough to think of the barbarous practice of cauterizing the wounds of battle with burning oil as it had been practiced until then. The discovery of effective therapeutic methods proceeded at a very slow pace indeed.

Albertus: Yet, the willingness of the alchemists to persist, by trial and error, contributed more to the development of a truly experimental science than the fantastic and capricious speculations of the astrologers, despite the fact that their theories were also burdened often by mysterious interpretations of supernatural powers. Little by little astrology began to be discredited amongst the more educated members of society, although this applies only to its divinatory pretensions and the casting of horoscopes and not so much to the belief of a true kinship between the order in the Heavens and the unfolding of history on Earth. The criticism of astrology had more to do with a belief in the free will of the individual and with the impossibility of a direct, immediate effect of the planets on his or her decisions and actions. By contrast, the notion of a true affinity between the celestial sphere and the human sphere remained very much alive and was pursued with even greater delectation. The belief that somehow the planets dwell in us never really disappeared, although it was now interpreted perhaps in a more recondite, subtle and indirect way. In this climate, there was also a widespread interest in magic, not particularly in black magic, associated to sorcery and evil spirits, but in what was known as natural magic, addressed to awaken the hidden sympathies of the Universe. Assuming an underlying harmony between the world above and the world below, it was believed that through incantations, by singing certain hymns or chants, by using scents or talismans and following certain rituals, it was indeed possible to invoke the passive forces of Nature and bring them into action. It is impossible not to see in all of this a great deal of naive superstition; yet, it is also true that it was pervaded by a sense of awe and wonder, by the conviction that the divine had descended to participate in the world, with our hands and through our knowledge. Therefore, the sense was awaken of a hidden but impending solution to the riddle of the Universe and this is not all that different from what animates any student of science to this day. In any event, it is quite astonishing to see to what lengths they went seeking ancient sources of wisdom, either in Egyptian mysteries or in Chaldean oracles, in the utterances of Asclepius or in the presumed revelations of Hermes Trimegistus. Everything seemed to be pregnant with meaning, anywhere might the key to this cosmic riddle lie dormant. Islamic occult writings never became very popular but the Jewish Kabala, supposedly transmitted by Moses, was studied assiduously, with its hodgepodge of incantations, magical numbers and propitiatory use of Hebrew letters and words. Many of those who wrote about these things considered themselves devout Christians and attempted to see all these ancient doctrines as progressive stages in the ultimate revelation of the Christian message. The Church as an institution, being a good deal more conservative, must have viewed with suspicion the activities of those whose intellectual curiosity appeared to be too lively and vigorous for their own good.

Plebeius: Of course, throughout all this period the Church was busy trying to avert its own dissolution and was confronting the far greater problem of the protestant insurrection. Nevertheless, they also made a futile attempt at controlling what the faithful could and should read, instituting the Index of forbidden works, an infamous invention that has been discontinued only in very recent times. It is interesting that Luther, who had dismissed and ridiculed Copernicus, was much more receptive to the work of the alchemists, but this only because he interpreted their efforts in an allegorical, religious way. He saw in the purification of metals or the distillation of liquors a mysterious, secret significance, evocative of the work of God on the day of Judgment when, through fire, he would separate the just from the impious. Calvin, on the other hand, with an excessive desire for the good that is typical of so much zealotry, condemned all sciences as diabolical and held any manifestation of intellectual curiosity to be, in his words, audacious, impudent and an enemy of religion. It is sad to say that neither the Protestant Churches nor the Catholic Church felt very much inclined toward tolerating any dissent and, for one reason or another, a number of men and women were put to death, merely on account of their ideas. One thinks of some prominent cases, such as those of Thomas Moore in London and Michel Servet in Geneva, but there were many more inconspicuous cases of persecution and many lives were ruined because of presumed heresies or sorcery. Giordano Bruno epitomizes the tumultuous spirit of this era more than anyone else and his condemnation and public execution at the stakes at Campo dei Fiori in Rome made his figure all the more imposing. As a very young man he had joined the Dominican order, but was forced to leave it after a few years due to his dissenting opinions and conduct. He then started a nomadic life, first visiting several Italian cities and later traveling through much of Europe, giving lectures, making a living as a private teacher and stirring up controversy wherever he went. By the time he was arrested by the Inquisition in Venice in 1592, at age forty-two, he had been to Lyons and Geneva, where he temporarily became a Calvinist, and later to Toulouse and Paris. From Paris he went to London and Oxford for a while and, after a sojourn in Paris again, he departed to Germany, spending some time in Marbourg and Wittenberg, during which time he turned Lutheran, finally moving further to Prague and then Frankfurt. Bruno was interested in the most varied subjects, in magic and logic, in memory and mathematics. He had an efervescent mind and he wrote in a style that is as animated and exuberant as it is obscure and enigmatic. The subjects he discussed fall within that vague territory that lies between the philosophic and the religious and they certainly betray his scholastic training. Our world, he says, is composed of form and matter, the form being understood as something akin to the soul of the world. Both form and matter are accidents of an underlying substance that is divine in origin. Only in God do they manifest themselves as undifferentiated. Man attempts to understand God by studying the natural world and Giordano Bruno speaks of this attempt as a heroic love, love in its desire to rise to the divine and heroic because it remains unfulfilled and is futile during this life. Nevertheless, he contends that the world of the senses is made in the image of the divine world and sometimes he comes very close to insinuate that it is the very thing itself, which no doubt must have brought upon him accusations of holding the most blasphemous of opinions. Of course, Bruno was also very receptive to Copernicus' ideas; not only did he adhere to the new model of planetary orbits but he also became its active propagandist. This dissent from orthodoxy, however, did not play a major part in his subsequent trial by the Inquisition. In any event, Bruno went further, he believed that Copernicus had conceived his model only mathematically, whereas he could now understand it in a vaster philosophic framework. Unlike Copernicus, he proposed that the world was infinite in extent, reflecting the nature of an infinite God. It is possible that he adopted this view having learnt it from Thomas Digges during his stay in England, or he may have revived the views of the Roman poet Lucretius, whom he had also read. The Heavens, Bruno continued, were populated with an infinite number of stars at various distances from the Earth and these stars are surrounded by planets, which are inhabited just as the Earth is. The stars are all in motion; as they are consumed by the fire burning in them, further fuel gravitates toward them and revives their flames. In his view, it was impossible to speak of the center of the Universe, for any inhabitant in one of these populated worlds could consider himself to be at its center. In retrospect, we can understand how Giordano Bruno would feel compelled at times to maintain that this world of the senses we inhabit was not only made in the image of the divine world, but that they both were one and the same. Until his time, it was customary to assign to our world a place in space, just as the Empireum or divine world occupied another place, somewhere beyond the last sphere. Now, instead, in this new conception of an infinite space, pervaded with worlds such as our own, there was no there, so to speak, that was not like here, and the Universe was only an image of itself. It seems that Bruno borrowed some of his ideas from Nicholas of Cusa and in some respects the latter's ideas were even more peculiar. Cusanus had taken as a point of departure to his arguments the necessary imperfection of all human knowledge and the inability for a material object to attain perfection of form. Then, reasoning from this premise, he concluded that it could not make sense to speak of the center of the world, given that the world could not be a perfect sphere. Therefore, since no single point could be the center, the center could be anywhere and was everywhere. At the same time, he did not abandon the notion of a finite Universe, but not being able to speak of a Universe bounded by the precise contour of a sphere, he made the Universe finite but boundless, speaking of a sphere of which the center is everywhere and the circumference nowhere. Using this type of of mathematical imagery, he had been struck by the fact that two opposites, such as the curving of a circle and the rectitude of a line become the same thing in the limit of an infinite circle. Since, in his view, only God was infinite, it was in him that opposites came to coincide, but this lay beyond the grasp of the human intellect, just as the notion of a sphere that has no circumference and a center everywhere cannot be fully comprehended. His ideas, I suppose, lacked the necessary precision but he took great pleasure in getting lost in the labyrinth of his own imagination.

Albertus: One could have made some sense of those ideas, provided one kept them confined within the field of mathematics, but it is futile to drag them elsewhere and interpret them arbitrarily. Before we digress into discussing these matters, I wish to return to the subject of astronomy and concentrate for a moment on the work of Tycho Brahe, whose name I have already mentioned, one of the most influential astronomers in history. He was born into a noble Danish family of some wealth and influence. During his time, Denmark as a whole was enjoying considerable prosperity and, not unlike Germany, a sizeable segment of its population was educated, curious and had the means to keep abreast of the ideas being advanced in the rest of Europe. Tycho was interested in scientific matters from an early age and a conjunction of Saturn and Jupiter, whose occurrence fell several days off the date he had computed using the Alphonsine and the Prutenic Tables, may have been a decisive factor in his determination to devote himself to astronomy. As a youth, and with the intention of smothering this pernicious interest, he was sent to Leipzig to study law, but he used his money and time to acquire and study scientific books, as well as to travel and establish contacts with learned men across Germany. Back in Denmark, he busied himself with astronomy and alchemy, showing no interest whatsoever in a career at the court as was expected of a man of his class. Throughout his life Tycho was to show much resolve and stubbornness in the things he pursued. He alienated himself from his peers still further by marrying a commoner, a marriage that, according to tradition, could not be sanctioned officially. As it was so common in those times, his wife never elicits a comment in his writings and letters, so that we do not know the tenor of their relationship or how she may have influenced him. All we know is that the marriage did last to the end of his life. In any event, an extraordinary occurrence in November of 1572 led Tycho to concentrate all of his efforts into astronomy. This was the apparition of the new star in the constellation of Cassiopeia. As I mentioned earlier, he studied it in detail, measured its position relative to other stars and failed to detect any motion in relation to them, leading to the conclusion that the new star belonged indeed to the sphere of the stars and was not an atmospheric or sublunar phenomenon. A similar conclusion he would reach a few years later in regard to the comet of 1577, whose orbit, he determined, intersected the planetary spheres, thus contradicting Aristotle's view that comets were produced in the atmosphere. Meanwhile, Tycho continued to travel throughout Europe, visiting other astronomers like himself and, upon his return to Copenhagen, delivered a set of lectures on the current state of the science that proved to be quite successful and aroused a lot of enthusiasm in his audience. During this time Tycho had been contemplating the idea of emigrating to a place that he might find more congenial to his scientific pursuits when, at age thirty and thanks to some benevolent recommendations, the King of Denmark granted him in perpetuity the use of the small and isolated island of Hven, just at the southern entrance to the Sound, then in Danish hands and an important source of income from the levies imposed on the intense sea traffic. In addition Tycho was awarded some grants in order to support the establishment of an astronomical observatory. After surveying the feasibility of the enterprise, Tycho threw himself wholeheartedly into it and for the next twenty years resided on the island, first erecting a building, which he called Uraniborg, the first to be fully dedicated to astronomy in Europe and, in succeeding years, conducting observations, training assistants, writing and otherwise managing the operations of the observatory, which in the end included also a paper mill and a printing press. I do not know whether we can say that Tycho Brahe belongs amongst the scientists of the greatest originality. We tend to rank as such those who originate new ideas and make us think in ways we had not thought before. Tycho was not original in this sense; we might even say that he was a man of a rather simple minded idea, namely, he wished to reach the greatest accuracy possible in his astronomical measurements. But the determination with which he pursued this goal and the manner in which he allowed himself to be guided and instructed by experience, which led him eventually to some remarkable scientific discoveries, was indeed original and deserves our utmost admiration. Perhaps some other of his contemporaries shared his belief that the road for advancement in astronomy lay in more precise measurements, but he had the resolve and dedication to prove it in deeds. Throughout the years, much of his time and energy were devoted to his instruments, to their design, construction and testing. Here, of course, he encountered the obstacle that the artisans of his day, carpenters, smiths, engravers and even the clock makers did not have the means or the skills to satisfy the requirements to which he aspired. He utilized a great variety of instruments, sextants and quadrants, armillaries, torquetums and rulers, some of them portable, some mounted and revolving, others fixed, instruments of different sizes, sometimes made of wood, other times of steel. He experimented at length with clocks but found in the end that they were too unreliable for precision astronomy. He introduced a new sighting mechanism with double alignments and adjustable slits to increase the accuracy and he also introduced the markings along diagonal lines to increase the fineness of his graduated instruments.

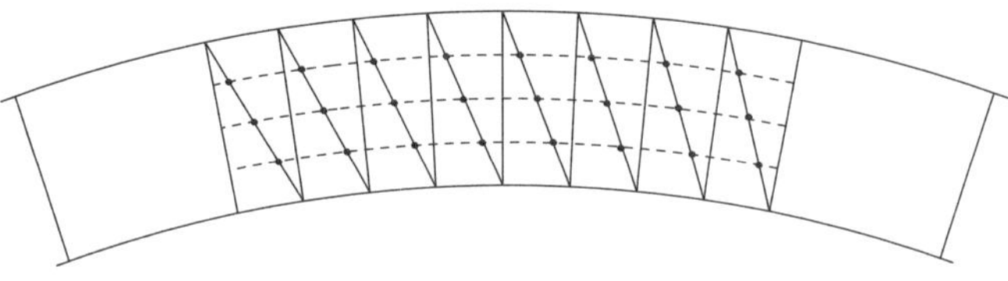

Tycho's goal was to obtain measurements that were accurate to just one arc minute, which, at a distance of 100 meters, amounts to an ambiguity of less than 3 centimeters across, an accuracy that, after many efforts, he occasionally attained. His great mural quadrant, built at Uraniborg along a north-south direction with an opening high on a facing wall for observations along the meridian, was two meters in radius, so that one arc minute precision would have corresponded to markings half a millimeter apart. Tycho was not only an innovator in the design of his instruments but also in the way he carried out his observations, using several instruments simultaneously and occupying several assistants, usually more than one to operate an instrument, while others, by candlelight, not needing to adjust their eyes to darkness, would write down the records of the observations or monitor the clocks. These amounted to simple rules of operation but they were essential to reach the best performance attainable with the naked eye. The task of obtaining accurate observations was always dependent on the reliability of the instruments used, but to test these instruments one was forced to turn them to the matrix of the sky, far more precise than any man made precision instrument, so that the conclusions drawn from astronomical observations were always in the process of disentangling themselves from the fact that the things one wanted to measure set the standards of the instruments used to measure them. Tycho's success relied, above all, on his carrying out observations repeatedly and methodically over many years. In the end he was not able to put to use all the observations he recorded, in particular regarding the orbits of the planets, and his great project of overhauling the entire field of astronomy and establishing it on solid foundations remained unfinished. Nonetheless, his accomplishments were not few. He compiled the first star catalogue to supersede entirely the one that had survived since the times of Hipparchus and Ptolemy. He also carried out all the customary observations of the Sun, measuring such things as the inclination of the Ecliptic and the lengths of the sidereal and equinoctial year, but his theory of the Sun's motion was marred, surprisingly, by his acceptance of the erroneous value of the solar parallax, equal to 3 arc minutes, that had been handed down since Ptolemy's time. The parallax reflects the displacement of the Sun's position in relation to the stars when seen from opposite poles of the Earth, or from the same position at sunrise and sunset. To put it in different words, the solar parallax is nothing but the angle corresponding to the Earth's diameter as seen from the Sun. In reality, this effect is much smaller than the effect caused by atmospheric refraction, all the more so when the Sun is observed close to the horizon, so that Tycho was never able to separate one effect from the other. His theory of the motion of the Moon was, by contrast, one of his great achievements. From his exhaustive observations, he first noticed that the inclination of the plane carrying the Moon's orbit had changed since Ptolemy's time, namely, the inclination of the axis of this orbit with respect to the ecliptic had varied slightly. You will recall that we spoke of this axis sweeping a cone as it changes its orientation while keeping its angle to the ecliptic fixed. On further scrutiny Tycho realized that the measurement of the inclination depended on the face of the Moon, or on its position along the orbit at the moment of the measurement. In this way he was led to the thesis that the axis of the Moon's orbit describes another small circle twice every orbit. That is, in addition to the general precession that reorients the plane of the orbit very slowly with a period of eighteen years, the orbit's plane jitters slightly, it wobbles minutely like a spinning top when it begins to lose its balance, or you might say that the axis of the orbit travel around its cone like a drunken man, unable to keep a steady course. This added motion should in turn affect the times and positions where the Moon arrives at the nodes, the all important crossings of the ecliptic, something that Tycho Brahe immediately set out to verify. Here we have a conspicuous instance in which the observations led to a theoretical hypothesis and this hypothesis in turn guided the observations that were to follow. Another of his remarkable discoveries concerning the motion of the Moon was the observation that its orbit around the Earth proceeded faster in the summer than it did in the winter, when the Earth and the Moon are closer to the Sun; this led him to conclude that the Sun had a retarding effect on the Moon's orbit. Although Tycho was primarily an observer, he developed a theoretical model of planetary orbits that was a hybrid of the models of Ptolemy and Copernicus. According to it, the Sun rotated around the Earth, which was at rest at the center of the Universe, while all the planets circled in orbits around the Sun, in addition to being carried along by the Sun itself in its rotation around the Earth. Tycho became very fond of his theory, but it never attracted a great following. From a geometric point of view, it was indistinguishable from Copernicus' model and it is difficult to explain why, when the latter ran into difficulties because it had set the Earth in motion, the expert opinion did not seek a refuge in Tycho's model that accomplished the same result without contradicting the Scriptures. In 1588 King Frederick II of Denmark died; his son Christian was then a minor and would not accede to the throne for another eight years. There is no doubt that Tycho's great enterprise was made possible by the benevolent disposition of the king, who may not always have seen the benefits of investing in the astronomy of precision, and also, it must be said, it was made possible by the benevolent disposition, if perhaps unwilling, of the peasants and tenants in the fiefs from whose income Tycho derived his grants. It is quite possible that he was somewhat profligate in disposing of the monies awarded to him and he may also have been obstinate and difficult to deal with. It is notable, for instance, that despite the many assistants that collaborated with him, he did not form any real disciples. In any event, after the coronation of Christian, Tycho's fortune turned against him. The new king was much more engaged and active than his father had been, but also prone to a good deal of mischief. Tycho lost a portion of his income and, after some fruitless negotiations, opted to go into exile. His reputation abroad was considerable and he was soon invited to Prague by the Austrian Emperor Rudolph II. Tycho moved most of his instruments to Prague but hardly made any use of them, for the last few years of his life were beset by a number of personal difficulties. When he died, at age fifty-five, his funeral in the cathedral of Prague and the procession afterwards were accompanied by great pomp, something quite unprecedented for a man of his profession. Let us now return to our digression on mathematical matters, but I wish to begin by relating to you some more general mathematical ideas that were first explored during the period that we are discussing. Although at the time they may not have seemed to bring anything to bear on such debates as to whether the Universe is finite or infinite, much less on general astronomical questions, now in retrospect we see that they have helped us greatly to elucidate these matters. Allow me to explain how this came about. Overall, a large number of books on mathematics were published during the XVIth century, which allow us in particular to assess without too much difficulty the progress made in mathematics during that time. In the early years, the vast majority of books were merely computational manuals addressed to those needing to make elementary calculation in their daily lives. Some of these manuals still explained the advantages of using the Indian instead of the Roman numerals and of carrying out one's sums with the pen instead of the abacus. They explained further the various ways of carrying out a subtraction or a long division, the means of verifying the exactness of one's results, how to extract roots, as well as a smattering of geometry. Occasionally, we find the more advanced textbook, following on the footsteps of Regiomontanus and addressed to scholars and specialists. Such was the case, for instance, of Johannes Werner's work, in which he attempted to revive the study of conic sections that had lain dormant for so many centuries. Werner was originally from Nuremberg, a city whose schools had the best instruction in mathematics to be found in Germany. He later continued his education in Italy and returned finally to Nuremberg, where he became a priest as well as an occasional advisor to Albert Dürer in his own mathematical studies. The study of conic sections was also pursued by a few Italian geometers, but their expertise did not match yet that of Apollonius, whose works, like those of Archimedes, would still not be translated completely for a few more decades. The recovery of these texts contributed significantly to stimulate the study of mathematics and to its advancement. These geometers were interested in mathematics largely out of intellectual curiosity. There were also, of course, the many practitioners of other disciplines who often contributed interesting problems of a mathematical nature, above all the astronomers, but also the cartographers and surveyors, the civil engineers and ship-builders, the military engineers too, interested in ballistics and the designers of machines, often interested in problems of statics, studying the equilibrium of forces or the center of gravity of bodies. Specifically in relation to astronomy, there were advances in trigonometry and in the development of methods to reduce the complexity of computations. These led eventually to the invention of logarithms, perhaps not of great theoretical importance but certainly a panacea for the calculators of that era. A measure of the sophistication of the scientists of this age is the fact that, for the first time, they speak of estimating the possible errors made in measurements. Although we cannot yet speak of a calculus of errors, nevertheless, from this time on, scientists would be concerned with sizing up their own ignorance or the scale of their own mistakes and Tycho Brahe was certainly a prominent pioneer in this field. Now let us turn to the specific mathematical problems that I wish to discuss. Although algebraic equations greatly challenged the abilities of the best mathematicians of this era, many applied themselves to their solution with enthusiasm and largely only for sport, given that the applications of this subject to practical problems were extremely limited. You are given, for instance, the perimeter and the area of a rectangle and you are supposed to find the length of the sides. This leads to an equation of the second degree which today any child can solve. In general, one performs a number of predetermined algebraic operations on a certain magnitude and knows the outcome, the task then is to find the original magnitude. The great difficulty in dealing with these problems was that there was no mathematical language to express them properly and only the most skillful could keep track of the maze of calculations that needed to be carried out. A further complication resulted from the fact that each equation had to be treated separately; one could become an expert in a set of tricks applicable to a few cases but no to others and these tricks were often kept secret, so that a mathematician could challenge his colleagues in open competition and test their expertise. They would say, for example: Give me a number such that when multiplied by itself three times to which you subtract five time the number multiplied by itself only twice gives you unity. Anyone who dealt with these sort of problems became known as a cosista, from the Italian language, an expert in the thing, the thing referring to the magnitude that in each equation was the object of their search. It was a group of Italian mathematicians, active in the northern cities of Bologna, Milan and Padua, who first made progress in the solution of algebraic equations, but it was primarily in Germany where the first steps were taken to develop an abbreviated language and eventually a symbolism to deal with these complicated mathematical procedures more effectively. The development of mathematical notation is not a profound subject in itself, it slowly evolved through the years and the symbols that we use today owe their adoption more to chance than to any intrinsic value. Nonetheless, the very use of a specific mathematical language deserves attention and respect. Clearly, when we introduce symbols to denote things, we are merely assigning names and substituting words for ideas that we already have in our minds, just as we do in any other language. In mathematics, however, things are slightly different and anyone who has even the least familiarity with it has had this experience, namely, that the language takes a power of its own and we are often led, not by the ideas we want to represent, but by the logic of the very language that we have ourselves created. The equation I mentioned before was written at one time as

$$1\,^3\,m.\,5\,^2 \text{ equal } 1.$$

The modern $+$ and $-$ signs, adopted for the operations of addition and subtraction, were borrowed from German merchants who used them to denote a surplus or a deficit of goods. Our sign for equality was a deliberate creation of a British calculator, who thought that nothing could be more equal than two parallel lines. Thus, that same equation would be written today in a standard form as

$$x^3 - 5x^2 = 1.$$

The use of letters to denote unknown quantities can be found occasionally in antique sources but only became widespread during the XVIth century. Let us not discuss the methods to solve these equations. Those of second degree had been studied since Babylonian times and the Greeks had developed a geometric method to solve them. It is not different from what today we would call, in algebraic terms, the method of completing the square. That is, if the purpose is to find the number x which satisfies the equation

$$x^2 + 2x - 3 = 0$$

we wish unity to appear alongside the first two terms, so we increase each side by one

$$x^2 + 2x + 1 - 3 = 1.$$

Combining the first three terms we obtain

$$(x + 1)^2 - 3 = 1$$

or, equivalently,

$$(x + 1)^2 = 4$$

and the equation is now readily solved, since we know which numbers squared give us 4 and from them we can deduce the possible values of x. The equations containing third and fourth powers of the unknown caused much greater difficulty. The ancient problem of doubling a cube led to an equation of third degree and had been solved by geometric means. It was only in specific cases that these equations could be solved. There was also no way of knowing when an equation had indeed solutions or how many, since in those days the solutions were only conceived as positive numbers. The solutions were generally obtained in progressive steps, first solving some simpler cases, then reducing those more complicated to the simpler ones and so forth. More interesting perhaps than the procedures and tricks employed are the disputes and acrimony amongst the characters involved in this chase. Gerolamo Cardano was very much at the center of all this controversy and perhaps justifiably so. He counts no doubt amongst the most extraordinary men of science, not perhaps by the perfection and permanence of his accomplishments, but certainly because of the sheer scope of his interests and the indefatigable energy with which he pursued them. His father had been a friend of Leonardo da Vinci and it seems that he encouraged his son to be just as universal and unrestricted in his studies. He was above all a mathematician, but also one of the more successful physicians in Milan. On these subjects he wrote extensively and also on everything else, in an unending series of volumes, from music to astrology, from the art of gambling to moral philosophy. At age seventy he was imprisoned for heresy by the Inquisition but, after a trip to Rome, he won the favour of the Pope along with a pension. His autobiography is quite remarkable in that he gives a careful account of all his great virtues that is as exhaustive as the account he provides of his many vices. He was the first to publicize the general solution to algebraic equations of the third degree, breaking an oath to secrecy he had made to the colleague who informed him of it, although it must be said that he did so only several years afterwards, when he learned that this colleague was not the true originator of that coveted solution. The equations of the fourth degree were first solved by a disciple of Cardano who, as was the usage then, had become a member of his household. In fact, this young man had been only a servant boy in the beginning but, by being quite capable and studious, he eventually graduated to become a mathematician on his own right. The elucidation, at last, of those algebraic equations had a considerable impact and led, indeed, to further attempts to find solutions to equations of higher degree. As it turned out, these attempts lasted for centuries and did contribute to advances in various directions, one of them being the proof of the impossibility of finding such solutions. After the middle of the XVIth century the Italian school ceased to lead the way in the study of mathematics and the subject of algebraic equations was taken up in France, in particular by François Viète. Viète was a native of Vendée and he never ventured too far from his homeland, studied law in Poitiers, held some legislative posts in Rennes, later in Tours in the service of King Henri III and eventually at the court in Paris. His lifespan coincides quite precisely with that of Tycho Brahe but, just as the latter's main contributions are easily recognized and helped define the science of his time, those of Viète are rather more elusive. There is little doubt that he was the most capable and learned mathematician of that period, even more so considering that he was an amateur scientist and seldom had the time to devote himself entirely to his studies. His service at the court was of an administrative nature and had nothing to do with his mathematical interests or even with astrology, as would have been the case at other courts. At one time, however, the ambassador to France from the Netherlands informed the King of a challenge proposed by a Dutch mathematician regarding an algebraic equation of the 45th degree that had not been solved, at which point François Viète was summoned and in relatively short order produced all the solutions, at least all those that were then conceived as possible solutions, being positive numbers. His contributions to this subject are amongst the more lasting. His view of algebraic equations resembles more our own than that of his predecessors. Until his time each equation was treated in its own individuality, but Viète began to substitute the parameters in an equation by letters, treating a class of equations instead of any one in particular, using what he called the logic of the species rather than the logic of the number, so that a quadratic equation would be written in general terms as

$$ax^2 + bx + c = 0,$$

where a, b, c are taken to be known numbers and x is the number to be sought. Viète, in fact, used vowels and consonants and not, as we do today, the first and last letters of the alphabet to distinguish one class of numbers from the other. Moreover, he would not have yet used our language, saying that the numbers a, b, c are fixed parameters, whereas x denotes a variable, a number that we let roam amongst all numbers, seeking to make the expression on the left hand side equal to zero, but his thinking is nonetheless rather similar. It is interesting that most of Viète's mathematical interests, such as his contributions to trigonometry, were related to his interest in astronomy and to a general theory of the Cosmos he intended to develop, critical of Ptolemy and of Copernicus as well, both of whom he had studied, but very little of this saw the light of day. For this reason I said that it is difficult to assess him as a scientist. He shone with the brilliance of his mathematical contributions, but one could argue that these were the happy outcome of an effort that in its more ambitious goals was probably misguided. In later years he was involved in a pair of controversies, one of them against a Dutch mathematician who claimed to have squared the circle and here Viète was certainly right; the second controversy concerned the reform of the Julian Calendar undertaken by the Church and here he was unfair, if not wrong. Being amongst the first to become thoroughly familiar with the works of Diophantus and Apollonius, he also led the way in expressing and solving geometric problems in an algebraic language and, conversely, translating algebraic problems back into geometry, ideas that would be later continued and perfected by his successors, Pierre Fermat and René Descartes. These two men had a few more things in common besides the mathematical problems that attracted their attention and that led them to similar discoveries. They had both grown in rather privileged surroundings in the French provinces and each received an excellent education. Their fathers had achieved some distinction in their professions, but it was the wealth of their mothers that provided in part for their comfortable life in the years to come. They both studied law but rejected it as a profession and, despite their scientific interests, neither of the two was lured to settle in Paris. There are also clear differences between the two men. Descartes opted for a rather solitary and nomadic life. Fermat remained in the south of France instead, became an administrator in the Languedoc and formed a family. One might say that this distinction, between a nomadic and a sedentary life, applied also to their intellectual development, for Descartes pursued a number of subjects in physics and physiology besides mathematics, whereas Fermat remained exclusively a mathematician throughout his life. But I do not think they would see this distinction as very pertinent, because Descartes saw all his endeavours as different aspects of one single investigation that derived from his all-encompassing philosophical outlook, while Fermat could very well argue that the variety of problems he pursued in mathematics often required different principles, tools and strategies to confront them. Here we are only concerned with the birth of a new manner of addressing problems in geometry and algebra, which, as I have tried to indicate, was in gestation for a long period of time and required an appropriate language to be developed successfully. It is only after the work of Fermat and Descartes that this subject acquired a form the we recognize as very similar to the one we have come to adopt. At the time, many mathematicians did not feel they were creating something new; they thought they were reconstructing some of the results that had been known in antiquity to such men as Apollonius and Archimedes, knowledge that was later forgotten when some of their works were lost. But this is true only in a vague and indirect way. In his brief dissertation on geometry, Descartes presents his results with refreshing clarity in a very modern language, a language, of course, that we have learnt from him. However, the problems he addresses and the perspective he adopts are still bound to an older tradition and it is Fermat who introduces new ideas and adopts a changed outlook. Descartes sought to provide geometric constructions to all manner of algebraic operations. Thus, for instance, when it comes to multiplying two numbers

$$a.b = x,$$

Descartes rewrote this in the form of proportions

$$a/1 = x/b$$

so that, using similar triangles, the quantity x could be realized as the leg in one triangle and be measured.

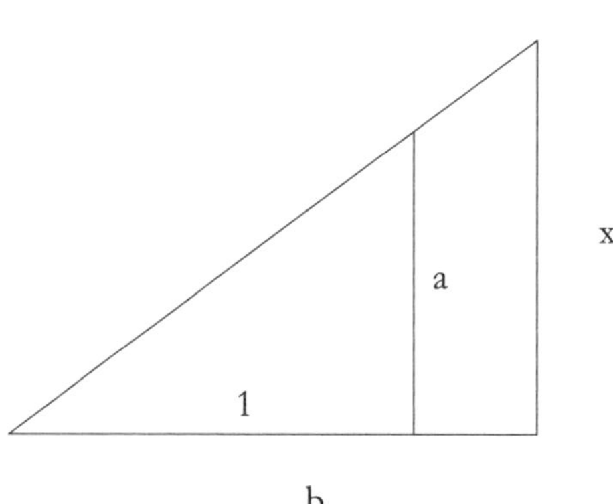

It is noteworthy that Descartes broke with the tradition of interpreting the product of numbers as areas. In his representation, all quantities are regarded as lengths of segments. He proceeded to device other similar constructions, to extract the square root of a number or to solve more general equations. Let us consider an equation of second degree, which I will write, for convenience, in the form

$$x^2 - 2ax - b^2 = 0.$$

Proceeding algebraically, the unknown quantity x is found by completing the square, that is, adding the same term on both sides of the equality

$$x^2 - 2ax + a^2 - b^2 = a^2,$$

so that it can now be written

$$(x - a)^2 = a^2 + b^2,$$

which is easy to solve as we saw earlier. Seen from a geometric point of view, this last equality is an expression of Pythagoras' theorem, where the lengths a and b corresponding to the two legs of a right triangle are given to us, so that we only need to add the length of the side a to the hypothenuse of the triangle to obtain the unknown x, something we can easily accomplish using a compass.

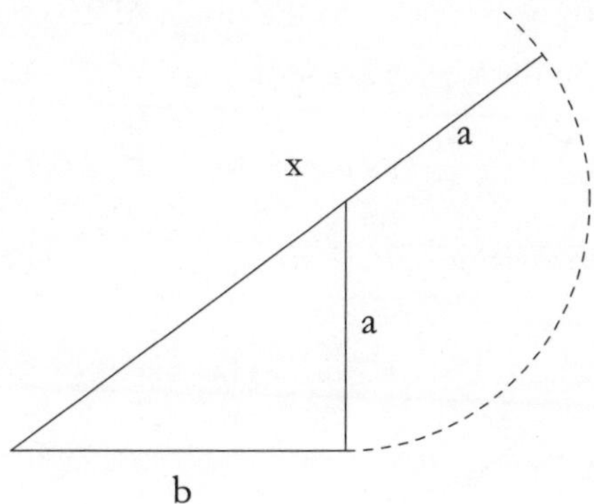

Contemplating the same equation, Fermat would have considered that the expression $x^2 - 2ax - b^2$ could adopt various values depending on the value of x. Hence, seeing x as a length, he would have allowed x to roam along a line, starting from an original position at length 0 and, for each such value, he would have plotted the other quantity $y = x^2 - 2ax - b^2$ along a line perpendicular to the original one, so as to keep track of it until it vanished. Of course, Fermat was not the first one to introduce such plots. Nicholas Oresme had done it much earlier. However, he noticed now that he had these two quantities x and y, representing the distances to two perpendicular directions and the algebraic expression he was considering and for which he had drawn a graph, could be written in the form

$$y - x^2 + 2ax + b^2 = 0.$$

He realized then that any such algebraic constraint containing two unknown quantities represented a curve on a plane. Thus, the original problem of finding the roots to an equation was now embedded into a larger context and one could not only speak of solving algebraic problems geometrically, but also study the properties of curves with the tools of algebra. Fermat was truly an amateur mathematician who pursued his studies for pleasure and not to advance his career; thus, he felt no need to publish his results, although they became known through the correspondence he maintained with a few scientists in Paris. Having identified a point on the plane with two numbers x and y, the coordinates representing the distances to two perpendicular lines, one could identify the curve produced by the constraint

$$x^2 + y^2 = 1$$

with a circle, for this is nothing but a restatement of Pythagoras' theorem, requiring that the distance from our point to the origin of coordinates be identical to one.

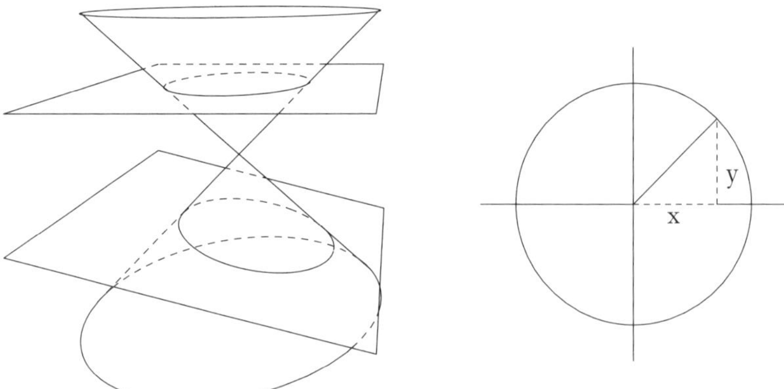

Fermat realized that all such constraints which did not involve higher powers of the quantities x and y than the second would produce curves that coincided with the conic sections long studied by Apollonius. These were the ellipse, the parabola and the hyperbola, and one could take as a representative of each species the curves corresponding to the equations

$$x^2 + y^2 = 1$$

for the circle,

$$x - y^2 = 0$$

for the parabola,

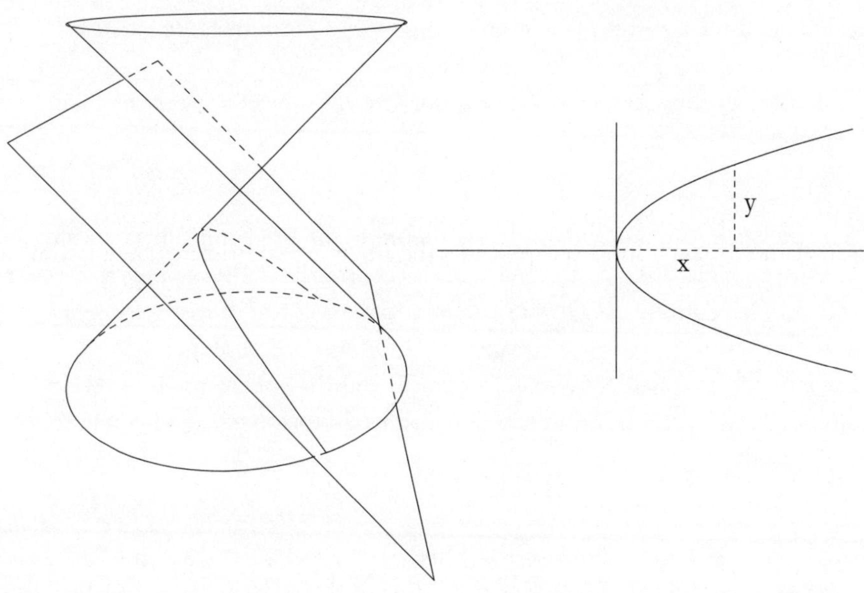

and

$$x^2 - y^2 = 1$$

for the hyperbola

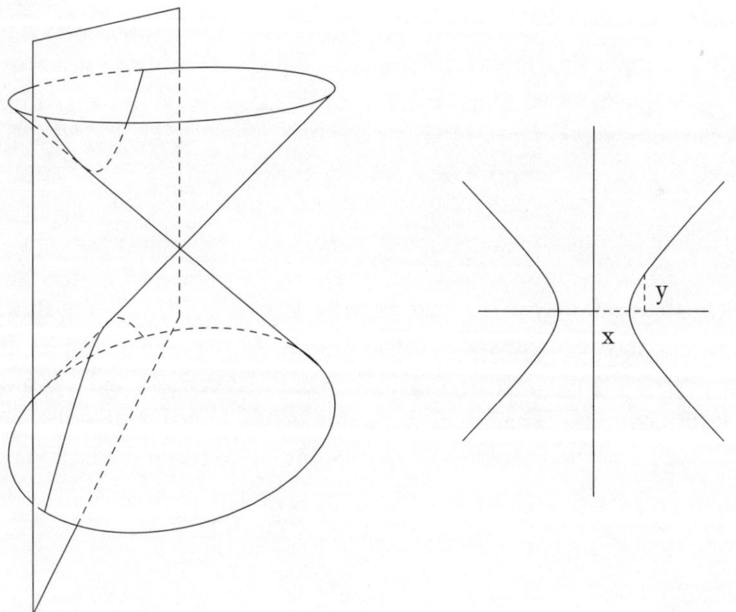

It was soon realized that one was not restricted by these methods to planar geometry. Introducing a third axis and three quantities x, y and z one could describe

surfaces in three dimensions with the same type of equations one had used to describe curves on a plane. Thus, the surface of a sphere, for example, would be represented by those points whose coordinates are subject to the constraint

$$x^2 + y^2 + z^2 = 1$$

for, once again, extending Pythagoras' theorem, the left hand side is nothing but the square of the distance to the origin of coordinates. Proceeding in the same manner, it is relatively easy for a mathematician to conceive that we might live in a spherical world. Just as the equation $x^2 + y^2 = 1$ represents a spherical world in one dimension, namely, a circumference, and the sphere $x^2 + y^2 + z^2 = 1$ represents a spherical world in two dimensions, so too an equation of the form

$$x^2 + y^2 + z^2 + w^2 = 1$$

would represent a three-dimensional spherical world in which one would go around in circles, regardless of the direction one followed, up or down, forward or backward, left or right. As in the previous cases we speak here of three dimensions because we can choose three of the coordinates arbitrarily, the fourth one being determined by the constraint. You may argue that this is merely an equation and, furthermore, that it is difficult to picture in one's mind how it could represent the world we inhabit. Nevertheless, the mathematician has now the advantage that he can carry out his geometric computations in an algebraic language, using the same methods that applied to the circumference or the sphere, without strictly requiring to have an accurate visual representation of the space he is dealing with. All the same, let me now digress for a moment and make some comments that may seem at first rather childish. Suppose we start with the equation

$$x^2 = 1.$$

Of course, there are only two solutions, $x = 1$, $x = -1$. If we plot these points along a segment, you might say, by analogy with our previous examples, that they form a sphere of dimension zero. The segment between -1 and 1 is a disc or a ball of dimension one. Therefore, the condition $x^2 < 1$ defines a disc or ball of dimension one and the condition $x^2 = 1$ defines a sphere of dimension zero.

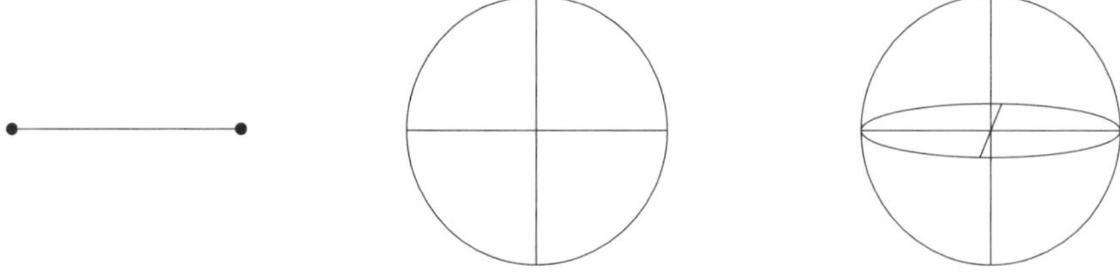

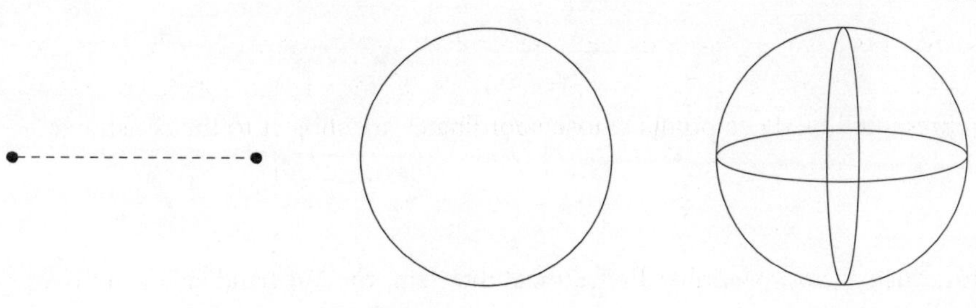

We now proceed upwards, $x^2 + y^2 < 1$ defines a disc or a ball of dimension two and $x^2 + y^2 = 1$ defines a sphere of dimension one. We use the names disc or ball to refer to a solid interior and sphere to refer to the outer layer, regardless of the specific dimension. The same description will apply when we move still one dimension upward. Notice now the following peculiar phenomenon. If we take two copies of discs of dimension one, bend them and join them along their edges, that is, along their corresponding spheres of dimension zero, we end up forming a sphere of dimension one. Likewise, if we take two copies of discs of dimension two, bend them and glue them along their spheres, their circumferences in this case, we end up forming a sphere of dimension two. Thus, gluing two discs or balls of a given dimension results in a sphere of that same dimension.

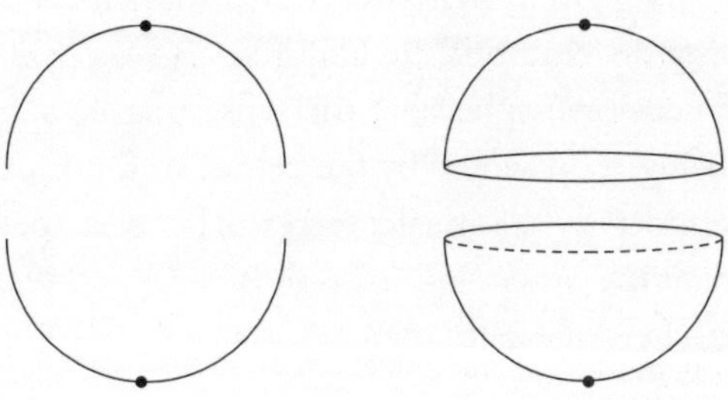

Notice also that in each case the centers of the two discs become two poles in the resulting sphere and that the path along which we have joined the discs could be called the equator of the sphere. We can do now exactly the same thing with two copies of discs or balls of dimension three and glue them along their outer spheres. Of course, we cannot do this in practice, no matter how much bending we exert, but the procedure gives us an abstract representation of how a three dimensional spherical space ought to be understood.

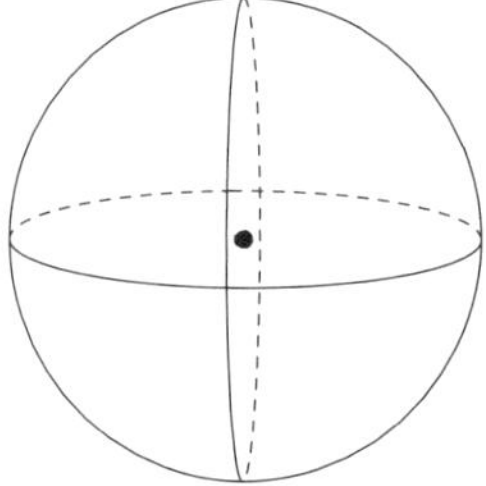 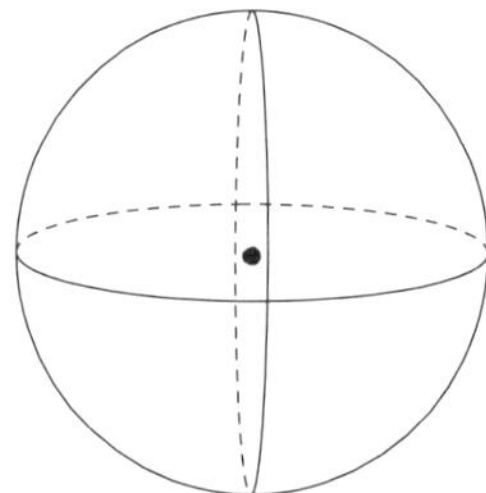

Each ball represents what we should call a hemisphere with the center of the ball standing for the pole of that hemisphere. Starting from this point and heading in any direction, we eventually come to the outer sphere, which I have called the equator and, after crossing it, we find ourselves in the interior of the other hemisphere, assuming the gluing had been accomplished, so that we descend now toward the other pole. Once we go past this other pole, of course, we find ourselves returning to the point from where we started. Thus, if we were to think that the Earth sits at one pole in the Universe, the larger and larger spheres in the sky surrounding us suddenly begin to become smaller, converging eventually to the opposite pole in the Universe.

Plebeius: I am reminded of that canto in the Paradise of Dante's Commedy where Beatrice explains to him that the spheres of the Heavens, which rotate ever so faster the farther away they are from the center of the Earth, instead of becoming larger, they progressively shrink into a series of luminous spheres of increasing velocities but diminishing radius as they converge to their common center, the site of the Prime Mover and Creator of the Universe. It is as if Dante resorted to the image of a three dimensional sphere, in the sense that he speaks of two poles in the Universe, one being represented by the center of the Earth, the abode of the devil, a center surrounded by the smaller terrestrial circles, the circles of the Moon and the planets and, at the other end, the pole where God resides, surrounded by the circles of angels, cherubims and their associates.

Albertus: One could argue that Nicholas of Cusa had also apprehended the true nature of a three dimensional spherical world when he said that it was a sphere with the center everywhere and the circumference nowhere, but it seems certain that he lacked the mathematical tools to make these notions precise. One can always concoct such vague expression that ring well to the ear but can then be distorted to mean anything we want. Mathematics is often misunderstood in this regard. We can now say that the algebraic language of Descartes and Fermat enables us to describe the three dimensional sphere as the quadruplets of numbers that satisfy the constraint

$$x^2 + y^2 + z^2 + w^2 = 1,$$

but those who refuse to be introduced to the language of mathematics find this to be vague and subject to interpretation. Quite to the contrary, unlike philosophy or poetry, where allusions and free associations are part of their charm, mathematics must create its own language with its own rules of operations, so that everything said means exactly what one intends and nothing else, at the cost of appearing sometimes too terse, removed or mysterious. At any rate, the methods introduced by Descartes and Fermat had their most immediate impact in the study of conic sections. The results of Apollonius regarding these curves, obtained by purely geometric arguments, could now be reproduced and extended when expressed in the language of algebraic equations. Quite coincidentally, the conic sections, the ellipse in particular, had made a renewed appearance in a most remarkable context some twenty years earlier when Johannes Kepler had demonstrated that the orbit of Mars described such a curve, although at the time his accomplishment received little attention. In the year 1600, shortly after Tycho Brahe had arrived in Prague, Kepler became his assistant. As it was, their collaboration lasted less than two years. Moreover, despite the fact that both seem to have had great respect for one another from the outset, which was in part the reason for their coming together, during their first year of collaboration the relationship was somewhat disagreeable, if not acrimonious. Nonetheless, Kepler soon became the most important member of Tycho's circle and, in his deathbed, the great astronomer urged Kepler to carry to completion his projected Rudolphine Tables, in accordance with Tycho's own planetary model rather than Copernicus'. At the time, Kepler was already exploiting the wealth of observational records that Tycho had assembled on Mars and was attempting to elucidate the shape of its orbit. When Tycho Brahe died, Kepler, then thirty-years-old, was immediately appointed to occupy the position of Imperial Mathematician, a post he held for ten years until the Emperor's own death. For the first six years of that period Kepler laboured exhaustively and exhaustingly on the orbit of Mars, writing at the same time a detailed account of his investigations which he presented in his Astronomia Nova, a book that, with some justification, aspired to lay a new foundation for the science of astronomy. He had begun by establishing that the orbit of Mars lies on a plane inclined less than two degrees with respect to the orbit of the Earth and, furthermore, that this plane goes through the Sun and is fixed, contrary to Copernicus' assumption that the plane oscillated. Always using Tycho's observations, he then attempted to model Mars' orbit in the traditional way with an eccentric circle, although, unlike Ptolemy, he did not require that the Sun lay opposite the equant and at the same distance from the center, the equant being the point from where the motion of the planet is seen as uniform. Allowing these two distances to vary independently, he was able to obtain a more satisfactory representation of the orbit. However, comparing the predicted orbit to further observations, at points along the orbit different from the ones he had used in his calculations, he realized that there was a discrepancy of approximately 8 minutes of arc. This would have been entirely satisfactory for Copernicus, since the observations he had access to were barely accurate to that degree. Tycho's careful and meticulous observations were probably accurate to one minute of arc and Kepler insisted that the excellence of this observational record had to be matched by an equally excellent and accurate theory. He was reluctant to employ epicycles because, in his view, the geometry had its own foundation in the physics. He saw the Sun as the source of the power that drove the planets along their orbits, causing them to move faster as they came closer and releasing them to a slower pace as they moved afar. The epicycles amounted to an artificial device without a physical grounding. Starting anew in his enterprise, Kepler set out to study first the orbit of the Earth for it was essential to understand the motion of one's point of observation, realizing that if the models for the orbit of Mars were erroneous, so were probably the ones that had been used for the Earth. It was now that he noticed that the Earth, at the extreme points of its orbit, that is, at the points where the distance to the Sun is shortest and longest, had a velocity inversely proportional to the distance. Kepler took this relation to be particularly meaningful and was willing not only to apply it to Mars as well, but to pretend that the it held at all points along the orbit, despite being aware that this was not quite true. Nevertheless, he investigated the type of curve that would satisfy that relation, getting entangled with very complicated mathematics, experimenting with various oval shapes for the orbit without making things much clearer, all the while lamenting that, if these orbits had been ellipses, the calculations would have been more straightforward. At one time, he modified slightly his simplifying assumption requiring that the radius of the planet's orbit should sweep equal areas in equal times, a condition that is indeed satisfied by an ellipse and that, years later, Kepler would recognize as one of the laws of planetary motion. At this time, however, he was driven back to study Tycho's observations until he decided to try to match them with an elliptical orbit and realized that it provided the best fit. Exhilarated, he now realized that the orbit of Mars was most accurately represented by an ellipse and that the Sun was placed at one of its foci. As you recall, an ellipse can be generated tying the ends of a string to two foci rather loosely, then circling around these foci while you hold the string always stretched with your pencil; in this way the points on the ellipse are such that the sum of the distances to the foci remains constant. For the first time it was being proposed that the planets moved on orbits which were constructed of anything other than circles. Each step of his investigation had taken Kepler long months of work and countless calculations. We know the details of the path he followed because he recounts them in the most vivid way in his book Astronomia Nova, which is for this reason one of the most valuable and endearing documents in the history of science. In it he explains to the reader, not just how things are, but how he came to discover them, with all the detours, the subterfuges and compromises, the illusions and the deceptions, as well as the errors that led him astray and the lucky coincidences that put him back on track. These were the most fruitful years of Kepler's life. He never ceased to be dedicated or industrious but in those years he completed the work that would give him lasting fame. In addition to the Astronomia Nova he also finished a work on optics that he had started some years before coming to Prague. This work had been motivated in part by Tycho's observations, as well as his own, of some intriguing phenomena associated with light during eclipses, the peculiar coloration of the Moon during lunar eclipses or its apparent change in diameter during a solar eclipse. But Kepler went on to study other matters related to the geometry of light rays; he analyzed the process of vision and he also addressed the problem of atmospheric refraction, of such importance in astronomy, although he was unable to arrive at its correct mathematical formulation. These two books, the Astronomia Nova and the one on optics, Kepler dedicated to his benefactor, the Emperor Rudolph II. The office of Imperial Mathematician afforded Kepler relatively good working conditions, but the position was not as alluring as its pompous title suggested. The Emperor's good will and simple-minded largesse extended well beyond what the treasury allowed, so that the Imperial Mathematician spent a good deal of his time and energy trying to secure his salary. In addition, Rudolph II was particularly fond of astrology and Kepler was asked to provide advice on these matters as well. For instance, when a new star appeared in 1604, Kepler wrote a report to the Emperor, very much in the style of the times, speculating on what this ominous event could portend for mankind, whether it would lead to the fall of Islam or to a massive migration from Europe to the newly discovered lands in America, while at the same time he was cautioning that one should value those celestial occurrences for what they are worth and in accordance to their effects, that we would be acting foolishly if they mean nothing and we still seek to interpret them in arcane ways. Kepler's attitude toward astrology was complex and not easy to understand from a modern viewpoint. He drew up dozens upon dozens of horoscopes and it is impossible that he would not have taken his task seriously to some degree, but he also decried the abuses of many astrologers and pointed to their mistaken predictions. We can say with certainty that he had an expansive nature and a lively imagination. He believed that order and meaning are ubiquitous, that man was made in the likeness of the world and has therefore the capacity to understand it. And even when this belief may have been born out of some sentimentality or childish wonder, he was mindful, through his own studies, of all the delusions that we fall prey to and the hard work that is required to extricate and understand the simplest secrets of Nature. He did not abdicate his conviction that he who does not seek will not find; hence, he refused to dismiss astrology altogether. He could say with some irony that even out of a foul smelling pile of dung, perhaps a good little grain, a golden corn could be found and scraped for by an industrious hen. Kepler opposed the aspect of astrology that sought to find omens in the Heavens, signs that would enable us to guide our daily affairs, but he clung to the notion that the configuration of the planets does influence the sublunar world and, ultimately, our own constitution. The type of astrology he sought was of a physical nature and he kept records of weather observations for many years in the hope that they might provide the appropriate clues to unravel how the planets affected events on Earth. Regarding his years as Imperial Mathematician, it must also be said that the rank and prominence he occupied helped to provide him with the assurance and self-confidence that he lacked by his own nature. He was independent intellectually and very much enjoyed a good controversy, but in more mundane matters, perhaps because of the recurring periods of bad health or depression he endured, he was not always sure of his own station in society. Moreover, he was free of conceit and arrogance and he was prone to praise others or their work, eager to be on good terms with those on whom he depended. Despite his differences with Tycho Brahe, he showed great deference to him and the same can be said of his relation to his former teacher Michael Maestlin, a well known astronomer of Tycho's generation, who outlived Kepler, in fact, and with whom he always kept a very warm relation, sometimes seeking his advice and assistance, often without success. Kepler, a native of Swabia, had studied at the University of Tübingen, well recognized then for its teaching of the sciences and where Maestlin was a professor. It was under his guidance that Kepler's mathematical talent flourished and it was also Maestlin who introduced him to the theory of Copernicus, of which he became an ardent supporter. Kepler had intended all along to become a Lutheran priest. However, in the last years of his training a vacancy arose for a professorship of mathematics in the city of Graz, then in the district of Styria, and Kepler was encouraged to fill it, thereby interrupting his theological studies and starting his career as a scientist. It was in Graz that Kepler's first original ideas came to fruition. Starting from Copernicus' model of the Heavens, he had puzzled about the arrangement of the planets' orbits around the Sun, seeking to explain, in particular, why the radii of these orbits stand in the proportions they do. He then realized, and he took this idea to be a most felicitous inspiration, that there are six planets and there are five regular solids. These are the convex solids that can be constructed joining regular polygons along their edges. There is the tetrahedron or pyramid, one of the three that can be built with regular triangles, then the cube, of course, and the icosahedron built with twenty pentagons. Kepler thought of nesting these solids one inside the other, each containing a maximal sphere within it and, in turn, being contained by another minimal sphere, which is again contained in another solid and so forth. As it turned out, Kepler found an appropriate order to carry out this procedure, so that the ratios amongst the radii of the resulting spheres were in rather satisfactory agreement with the ratios amongst the orbits of the planets as given by Copernicus. Kepler was elated for having made such a discovery, giving further proof of the rational and harmonious architecture of the Cosmos. He felt that the remaining task was to explain the discrepancies, rather than justifying the model itself; there were indeed some residual ambiguities since the sphere for each planet, centered at the Sun, was supposed to have a certain thickness. A planet's orbit was eccentric and its distance to the Sun, therefore, did not remain constant. Kepler presented these ideas in his first book, titled Mysterium Cosmographicum, that was published in Tübingen under Maestlin's auspices. Although Kepler was very proud of his work, his fame in Graz was due instead to some events he had forecast in one of the first calendars he was bound to assemble as district mathematician. In it he had announced extreme cold weather for the coming winter and an invasion by the Turks, both of which events came to pass, thus securing his reputation. Of his first book Kepler sent a copy to Tycho Brahe, who read it indeed, sending back a warm letter with some cautious encouragement. It was in this way that he came to hear for the first time of the young German mathematician who would later become his assistant. Even before joining Tycho in Prague, Kepler was already making plans to write a series of books that would extend and perfect his initial work on the architecture of the Heavens. One of them would be strictly astronomical, later he would explore mathematical harmonies and their relations to music and he also intended to write on the constitution of the Earth, on the origin of mountains and the nature of the tides and then, finally, on the links between meteorology and astrology. During his stay in Prague he was unable to undertake these projects in any serious way. Most of his time was taken up by his work on optics, on the orbit of Mars and not least by events in his own private life. The year 1611, in particular, proved to be quite eventful. The Emperor Rudolph II, who had already lost the territories of Austria and Hungary to his competing brother, the Archduke Mathias, was finally forced to abdicate the crown of Bohemia in his favour. To defend himself the Emperor had summoned friendly Catholic troops from Germany; this in turn aroused the Protestant population of Prague, who found common cause with Mathias. For the first time after a long period of calm, the city saw now considerable turmoil and bloodshed, so that Kepler, seeing also the patronage of Rudolph imperiled, decided to move elsewhere. He attempted to obtain a position in his native Germany, but his suspected Calvinist inclinations made him undesirable in the eyes of his former colleagues in Tübingen. In the end he chose to return to Austria, this time to the city of Linz, where a position was created for him, also in part because it was more favourable to his Austrian wife, whom he had married in Graz. However, after Kepler had returned from an initial trip to Linz and before they could move, she contracted one of the diseases brought to Prague by the invading troops and very soon she was dead. Earlier in the year the family had already grieved the loss of one their children, a six-year-old boy. Kepler's marriage had been an unhappy one. From what we know through his letters, disaffection, dissimilar interests in life and constant financial worries, as is so often the case, had each played its role. Now, at age forty-one, he found himself starting a new life in Linz under somewhat uncertain circumstances. In any event, one year before these accidents, a more momentous occurrence from the point of view of astronomy excited Kepler greatly. In April of 1610 Kepler received from the Tuscan ambassador in Prague a copy of the Message from the Stars, the brief tract in which Galileo Galilei described his first observations with the telescope and where he announced the discovery of the four Moons of Jupiter. In fact, the rumour regarding the discovery of some new planets had reached Kepler a few weeks earlier but, despite his excitement, he had been unable to confirm whether the discovery concerned planets around another star, planets in our solar system, in which case his scheme of placing the five regular solids between the six existing planets would have been shaken in its foundation, or it could also be that these so called planets were moons orbiting another planet. The first rather crude telescopes or spy-glasses had been built only a couple of years earlier in the Netherlands by an artisan whose request for a patent at The Hague was denied, probably on account of the apparent simplicity of the instrument and the ease with which it could be replicated. In retrospect, considering that spectacles had already become quite common at the time, it seems surprising that such a device had not been built earlier, perhaps even by accident. Nevertheless, given its usefulness in military campaigns, the news of the invention traveled fast throughout Europe and, despite the secrecy surrounding it, a number of artisans began experimenting, seeking by trial and error the most appropriate combination of lenses to replicate the instrument.

Plebeius: It must have occurred to a number of individuals that such an instrument was possible, without their trying to build it. Paolo Sarpi, the theologian from Venice and long a friend of Galileo while he was a professor at the University of Padua, says he had thought of the possibility himself. Apparently, some Dutch emissaries had offered to sell the device to the Venetian government, but the authorities decided to consult with Sarpi first. Sarpi was quite knowledgeable in scientific matters, kept a vigorous correspondence with other experts in Europe and his loyalty to the Venetian state could hardly be put in doubt. Just two years earlier, when Pope Paul V had issued a bull of interdiction against Venice, he refused to obey a summons from the Church in Rome and was summarily excommunicated. Sarpi advised the government against the purchase of the Dutch telescope; he was not ready to look into the matter himself but he was confident that Galileo, with whom he immediately conferred, would be able to produce an instrument equal or superior in quality. Writing to a friend, Sarpi expressed that he felt his mind, either through age or habit, was now a bit dense for such complications. You would hardly believe, he wrote, how much I have lost both in health and composure through attention to politics. There must be some truth in his contention, not the least because the year before he had barely escaped an assassination attempt, which he believed had been orchestrated by the Curia in Rome. But it is also true that during those years he discontinued the extraordinary books of notes he had been keeping on all manner of scientific ideas and speculations and began to concentrate more on his work as a historian, gathering information for what would be his massive chronicle of the Council of Trent. His earlier notebooks are quite remarkable in that they reveal his vast erudition and the scope of his speculations, on matters such as the tides, magnetism, the corpuscular nature of light or the nature of the vacuum and on the circulation of the blood too, which he is said to have discovered independently. Within the Church, the only other contemporary of Sarpi, comparable to him by the range of his interests and his influence, was the Frenchman Marin Mersenne, who also maintained a voluminous correspondence with the most learned scientists of his time and could consider himself one of them. He is likely to have gathered fewer enemies too. Sarpi also acquired a reputation as a theologian and wielded unusual power in Venice on matters of doctrine. Especially for those whose sympathies were with the Church in Rome, his decisions often seemed motivated by a sly or sinister intent. Galileo, on the other hand, held him in great esteem and thought few men in Europe were as versed in scientific matters as he was. In any event, after the two men had conferred on the construction of a telescope, as you well know, Galileo worked with such haste and diligence that, in a few weeks, he was able to go to Venice with his first instrument and show to the authorities, from atop the Campanile in Piazza San Marco, how easy it was to spot the distant ships entering into the lagoon with his device.

Albertus: His first instruments probably offered a magnification of only a few. Nevertheless, the service he was rendering to the republic allowed him to negotiate a hefty increase in his salary as professor at the University of Padua. Galileo continued to grind lenses for several months and it was not until the next year that he achieved magnifications by a factor of twenty and could begin to observe the skies with profit. It is curious that, until this time, his interest in astronomy had been only secondary and never the subject of his researches. Now, at age forty-five, he was launching into a new pursuit that would take a good portion of his time and energies. His early discoveries proved to be nothing less than sensational. Observing the Moon, he noticed the variations in the contours of light and shade when its position relative to the Sun varied, deducing that its surface was dotted with mountains and craters and was very far from smooth. He was also able to resolve some white patches in the Milky Way into thousands of stars that had been invisible until then to the naked eye. Then came the discovery of the four moons orbiting Jupiter, which he rushed to put into print in his Message from the Stars, suspecting that other observers might anticipate him and claim priority. In this regard he had little to fear, for his telescopes were of better quality than those of his competitors and no one, as it turned out, had his dexterity and perspicacity as an observer. I mentioned earlier that, within days, the Tuscan ambassador to Prague had brought a copy of Galileo's exposition to Kepler's attention. It is not clear whether he did this at the request of Galileo or by order of his government, for the simple reason that Galileo, no sooner had he obtained a promotion in Venice, began negotiations with the Duke of Tuscany to be appointed at his court and, thus, be able to return to Florence. It is possible that the Tuscan authorities decided independently to solicit a report from the Imperial Mathematician in Prague regarding the accomplishments of his Italian colleague. Kepler was very enthusiastic about Galileo's results and wrote a brief commentary on them, even before he had the chance to look through a telescope himself. He was soon speculating on the number of moons that Mars should have, considering that the Earth had only one and Jupiter had four. Galileo had also pointed to the clear difference between the images of the planets, which were seen as steady, round sources of light, and the stars, whose appearance did not vary from their naked eye view. Kepler was hopeful that the new observations would help solve the riddle of the apparent sizes of the planets, to which he attached great importance. In the meantime, he was prompted to return to the subject of optics and in a few months hastily finished his book titled Dioptrice, which presented, amongst other things, the design of a new telescope using two convex lenses, capable of providing a wider field of view. Shortly thereafter Kepler moved to Linz. Meanwhile, Galileo had already left Padua. After spending several months in Rome successfully demonstrating his findings, he settled in Florence. During the decade that followed his earliest observations Galileo continued to make a number of remarkable discoveries. He was the first to observe the phases of Venus, which gave him conclusive proof that the planet rotates around the Sun. He also turned his telescope to the Sun, projecting its image onto a screen, and studied the behaviour of the sunspots. These had long been noticed with the naked eye as occasional dark specks on the surface of a Sun dimmed by fog or clouds. Galileo was now able to follow them as they formed and dissolved on the Sun's surface, like clouds above the Earth. With their help he was also able to prove that the Sun rotates with a period that he first estimated as fifteen days, although it later turned out to be closer to twenty-six days. With the use of the telescope Galileo was able to provide for the first time accurate estimates of the angular diameters of all the planets. Ptolemy had stated that Venus' diameter, for instance, at its mean distance, was one tenth that of the Sun or, approximately, 3 minutes of arc. Galileo came to the conclusion that the diameter was considerably smaller, not even the 200th part of the Sun's diameter. He obtained similar results for the other planets. Jupiter's diameter was barely less than a minute of arc, some 50 arcseconds, when measured at its closest approach. Because Galileo was not concerned with determining absolute distances, he preserved the old values which gave roughly 1500 Earth radii for the distance to the Sun; therefore, with such small angular diameters for the planets and distances that were one tenth of their real values, he came to the conclusion that the planets were all smaller than the Earth, only the Sun had a diameter eleven times larger than the Earth. Regarding the stars, he was persuaded that their angular diameters could not exceed 5 arc seconds. Assuming then that a typical star was similar to the Sun, it was simple to estimate its distance, namely, one would have to remove the Sun to a distance a thousand times its present value in order for its angular diameter to decrease from 30 arc minutes to less than 5 arc seconds. This was consistent with the fact that no annual parallax had been detected for any star. We speak of annual parallax to refer to the angular displacement of a star every six months caused by the rotation of the Earth around the Sun. Its displacement is measured in relation to the background stars assumed at infinite distance and that for all practical purposes remain fixed. Earlier we spoke of the diurnal parallax caused by the observer changing position as the Earth rotates around its axis. By way of comparison, the diurnal parallax for the Sun had to be less than a minute of arc to have escaped observation; this was an immediate consequence of the fact that the radius of the Earth is so much smaller than the distance to the Sun. It requires indeed that the distance to the Sun be greater than a thousand Earth radii. Likewise, if one assumed a similar annual parallax for the stars of less than a minute of arc, below the capability of the instruments of the time, one concluded that the distance to the stars had to be at least a thousand times the distance to the Sun. Turning once again to Kepler, he was to remain fourteen years in the city of Linz, where he remarried and had several children. This time his family life was considerably happier. In all he had twelve children, the last of them in the last year of his life, but only four or five survived childhood. During this period he wrote a great deal, in particular, two of his more comprehensive books, The Harmony of the World and the Epitome of Copernican Astronomy. Nonetheless, his years in Linz were not free of anxiety and turmoil, either in his own private life or in relation to the affairs of the world at large. Shortly after his arrival he was denied communion by the local Lutheran pastor, supposedly because of his lack of orthodoxy, which meant in practical terms that he was excluded from his own community. Confessional problems grieved Kepler throughout his life. At least, in the city of Linz this was still a time when both congregations coexisted rather peacefully. Six year later the first confrontations between Protestants and Catholics exploded in Prague over the succession to the Bohemian throne, disputes that would soon engulf a good part of Germany and Austria in a war of thirty years, complicated by the most bizarre entanglement of international alliances. In Linz Kepler was under the jurisdiction of the same ruler he had served during his stay in Graz, namely, the Archduke Ferdinand, now Emperor Ferdinand II, whose favour he occasionally courted traveling to Vienna. It was only in 1625 when the fervour of the counter-reformation swept through all of Austria and Protestants were expelled from all position of responsibility that Kepler's situation became uncertain. The next year a great peasants' uprising lay siege to the city of Linz, causing much devastation and leading in the end to his departure. Furthermore, a few years before this turn of events, a further ordeal struck Kepler's family, which caused him great vexation and took a great deal of his time. His mother, still living in Swabia, was accused of witchcraft and forced to undergo a long trial. These were not frenzied persecutions and there was still some civility in the proceedings, but things were taken very seriously and a few women had been sentenced to die in the past. Kepler traveled to aid in his mother's defense, which was eventually successful, although she did not survive for a very long time afterwards. Amidst all these troubles, Kepler continued with his scientific work and with the writing of his books. The Harmony of the World was perhaps Kepler's own favourite work. In it he gave vent to his far reaching speculations on the presence and meaning of harmonies as found in the fields of geometry, music and astronomy. He favoured the view that just as our senses respond to certain musical harmonies, so too our soul is affected through the perception of certain symmetries and relations in the configuration of the planets. This was perhaps more elaborate, less crass than the astrological doctrines customary at the time, but it was nonetheless very conjectural and ambiguous. He was also inclined toward the opinion that the Earth possesses a soul that gives life to it. The rising and ebbing of the tides was akin to breathing or similar to the flapping of a fish's gills. In this regard, he had previously ventured the hypothesis that the tides were related to the influence of the Moon's motion around the Earth. Now he did not discuss one idea against the other, thinking perhaps that it is not contradictory to suppose that the Earth's soul would comply with the motions of the Heavens as our souls do. This animistic view applied likewise to the Sun, whose soul exerted the force guiding the planets along their orbits. In Kepler, these ideas do not appear as the endpoints of a detailed and sober study, but are instead the emotional sources that spur him to those studies. The sympathies and affinities he seeks in the world give expression to the ones he feels within himself. If sometimes we find him somewhat gullible and led astray by his own enthusiasm, other times he could also be ironic and self-deprecating, as when he edited the second edition of his Mysterium Cosmographicum late in his life and did not fail to point to his earlier naiveté and inexperience. But we must also remember that it was in the Harmony of the World that, amidst much speculating, Kepler presents the discovery of two new stellated regular icosahedra, two polyhedra that would only be rediscovered more than two hundred years later, and it was in this book, in his discussion of the planets, that he presented the third of his famous planetary laws, making the cube of their periods proportional to the square of their orbits' radii. The Epitome of Copernican Astronomy was in fact a thorough textbook of astronomy, written in the form of a series of questions and answers, beginning with spherical geometry and continuing with the theory of elliptical planetary orbits, the theory of the Moon's motion and the theory of the Earth's rotation and precession. In this book Kepler came back to the problem that had occupied him many times before, namely, the task of understanding the distances to the planets, their sizes and the periods of their orbits, although now he was armed with more accurate information on their apparent sizes, revealed by the observations of Galileo. He considered, for instance, whether the mean distance from the Sun to a planet was in proportion to its diameter or to the surface area of its cross section, or to its volume. This was not merely an argument of geometric harmony but it involved some physical ideas as well. The solar force sweeping the planet along its orbit decreased as the distance to the planet increased, but the outer planets were larger and intercepted a greater number of the solar lines of force carrying it along, so that this effect compensated for the decline in force with distance. Furthermore, it was possible that the velocity of the planet, and therefore its period, were related to its weight or density. Of course, Kepler would have been astonished to hear that the planets were wandering bodies that had arbitrarily fallen into their orbits as a consequence of some cataclysmic random event in the Cosmos. The true scale of the solar system obviously depended on the Earth's distance to the Sun. Kepler had long been preoccupied with determining the true distances to the Moon and Sun. One of the books he had intended to write was precisely on this subject and would have carried as title the name Hipparchus, in honor of the great astronomer, but at his death the book still existed only in fragments. Kepler, like Hipparchus, had laboured with the method of eclipses to determine the solar distance without much success. Later, the arrival of the telescope made him hopeful that the method of lunar dichotomy might be more reliable. Unfortunately, this method remained as impracticable as ever. Tycho Brahe had estimated the diurnal parallax of the Sun to be 3 arc minutes. He had been the first to take into account the all important effect of atmospheric refraction. Indeed, this had led him to revise Copernicus' own measurement of the inclination of the ecliptic. Kepler maintained that the undetected solar parallax had to be smaller than the level of accuracy of the best measurements, or approximately 1 arc minute. This led him to an estimate of the Sun's distance of approximately 3500 Earth radii. In the Epitome he also used arguments from harmony to advocate this value. The conjecture that the volumes of the Sun and Earth were in the same ratio as the Solar distance was to the Earth's radius led him to that value as well. It was also pleasing that with a value of 59 Earth radii for the Moon's distance, it turned out that the radius of the Earth was to the Moon's distance as this distance was to the solar distance. In spite of these arguments and no matter how suggestive they were, Kepler understood that they were no substitute for a true measurement. The main obstacle to this task remained the problem of atmospheric refraction, still poorly understood. The two phenomena of parallax and refraction, we might say, compete against each other. The parallax is the displacement of a light source relative to the background sky caused by the fact that the observer is not located at the center of the Earth. For a source at the zenith the parallax is zero and it increases toward the horizon. Refraction displaces all sources from their true position and it is also nil at the zenith, increasing toward the horizon. But, whereas the parallax tends to push down a source toward the horizon, refraction makes it appear higher than it truly is.

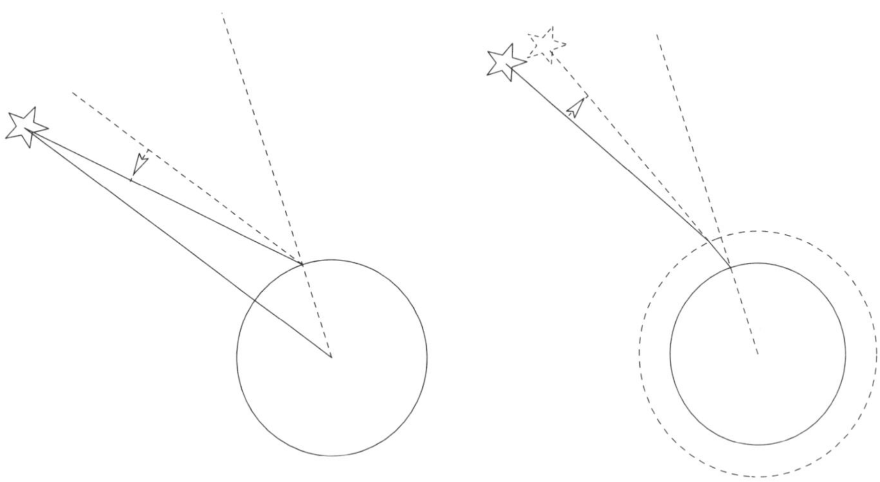

Tycho Brahe had believed that refraction was negligible above 45 degrees. He also had used three different tables of refraction, for the Moon, the Sun and the stars, under the mistaken belief that refraction depended on the distance to the source, being larger the closer it was to the observer. Kepler understood that Tycho's large parallax for the Sun was probably due to his incorrect tables of refraction and he also concluded that refraction affected all sources in equal measure. Later in life he began to suspect that the solar parallax could be as small as half an arc minute. In this regard he probably reflected the view of Johannes Remus, with whom he had corresponded. Remus had also monitored the progression of eclipses closely and had come to believe that they were better understood with a more distant Sun. Likewise, he realized that the different tables for refraction that Tycho had used for the Sun could be made to match those used for the stars under the assumption that the Solar parallax was as low as 17 arc seconds. This, of course, was a theoretical inference and not a measurement but, as we know now, he had come a good deal closer to the true value. However, this entailed a tenfold enlargement of the dimensions of the Solar system from what had become acceptable to Tycho Brahe's generation. Kepler was a convinced Copernican from his student days. At the same time he was a very pious man. It is ironic that his scientific studies and his Christian faith could coexist in him so peacefully and harmoniously, in an age of so many prolonged and unresolved clashes between the defenders of the doctrine and the advocates of the new astronomy. Kepler left Linz at age fifty-five. During the last five years he was to live he led a rather nomadic life, beset by its own set of adversities, not least of which were the great political upheaval and the degree of religious intolerance reigning at the time. He devoted a good deal of energy to finishing the Rudolphine Tables, the task with which Tycho Brahe had entrusted him, and most of all, to securing the funds to publish them and then seeing them through the press. The lack of any desire to exchange ideas between Kepler and Galileo seems to us very peculiar in retrospect. The blame, if there was one, weighed more heavily on Galileo's shoulders. It is true that neither of the two ever felt inclined to travel to distant countries, but this is hardly an excuse, considering that they both kept a lively correspondence. Kepler had sent a copy of his Mysterium Cosmographicum to Galileo in Padua, at a time when he barely had any information regarding his scientific interests. Galileo does not seem to have read the book but he promptly sent a note of gratitude, expressing that he also considered himself a follower of Copernicus and insinuated further that he had found in the movement of the Earth an explanation for some natural phenomena. In later correspondence Kepler encouraged Galileo to be more outspoken in his advocacy of Copernicus' ideas, at the same time that he requested his opinion and information on other scientific matters, but Galileo failed to respond. Kepler was able to guess that the natural phenomena Galileo had referred to in his letter concerned the tides, something on which he had also speculated. Galileo came to the conclusion that the tides were caused by the combined effects of the rotation of the Earth around its axis and its rotation around the Sun. He seems to have subscribed to this opinion to the end of his life. Galileo was seldom given to formulate theories without having solid grounds for them, so it is ironic that, on this matter, Kepler's intuition, generally more inclined toward fanciful and speculative ideas, proved to be much closer to the truth, for he thought the tides were caused entirely by the Moon's rotation around the Earth. Since Galileo did not respond to Kepler's entreaties in the following years, some have speculated that the use of the Tuscan ambassador in Prague to take a copy of the Message from the Stars to Kepler was merely a ruse of Galileo, who wished to have his opinion and support but felt embarrassed to write to him directly. In any event, Kepler's comments were nothing but laudatory, pointing out also in passing that he praised the work because of its merit and not because he owed anything to Galileo. After this rather indirect exchange there was no further communication between the two. Kepler sent to Galileo a copy of his book on optics wishing to have his comments but never received a response. Nevertheless, when Galileo moved to Florence he had recommended that Kepler replaced him at the University of Padua. It is unlikely that Kepler would have accepted the offer given that, years later, when he was offered the chair of astronomy at the University of Bologna, he declined, mindful of the dogmatic attitude of the Catholic Church and its influence. In a letter to a friend, Galileo once wrote that he had Kepler in the highest regard but that their ways of philosophizing were most diverse. This is quite evident to anyone who looks at their works; it is not easy, however, to draw a distinction between the objectives they pursued. Kepler belonged more properly in the tradition of mathematical astronomy of Ptolemy and Copernicus, enriched now by his mathematical skills and by his intuition about physics. Galileo's primary interest and greatest accomplishment was in mechanics and we could say he was a direct descendant of Archimedes. To the extent that both attempted to study the physical world with mathematical tools, we might say that in Galileo's case, the emphasis was always in the physics that the laws should describe, whereas Kepler perhaps put more emphasis in the mathematical form of the laws themselves. It is not just that their ways of philosophizing were different; their personal attitudes and outlook in life also differed greatly. Galileo was more of a worldly man, who cultivated friendships amongst men from all walks of life, artists and artisans, courtiers or priests, ignoring only perhaps his own colleagues at the university. He was pugnacious, never refraining from a good controversy and he did not mind being somewhat irreverent. The witty and nonchalant style that we find sometimes in his writings belie the resolve and determination with which he often pursued his goals. He was clearly ambitious and knew how to court his patrons to advance his career. In scientific matters, most often he was cautious, neglecting to publish his ideas until he was certain they had a solid foundation. Kepler, in contrast, presents himself as more spontaneous and impulsive. He wrote profusely and did not hesitate to expose his train of thought or to report on his tribulations and past mistakes. He did not lack humour but, more often than not, he seems to be driven by a great intensity of purpose. He strikes us also as being more humble, not in regard to his own skills but in the face of the enormous task he wished to accomplish. Kepler belongs in the pantheon of the revolutionaries in science, if for no other reason, because he abandoned, without hesitation, the age-old belief in the circular motions of the Heavens. In some other respects, he held conservative views that would soon disappear in the generations that followed him. He always held on to the opinion that the sphere of the stars circumscribed a bounded Universe and he vehemently opposed Bruno's conception of space as infinitely extended and populated by many worlds. It is probably fair to say that, of all of Kepler's contributions, the ones he cherished the most were the ones that came to mean the least in the end. Amongst his works, the Rudolphine Tables, the most accurate to date, had the greatest immediate impact in making astronomical predictions, in the elaboration of ephemerides and calendars or for astrological purposes. His Epitome of Astronomy was certainly amongst the astronomical textbooks most widely used in the following decades, his work on optics laid the foundations for the ensuing progress in this field and the configuration of telescope he designed was adopted by most astronomers. His most remarkable discovery, however, the elliptical orbit of Mars, was the last contribution of his to be recognized. The influence of Galileo on his contemporaries and his successors was of a different nature. After some initial disbelief, his astronomical discoveries were duly celebrated and they brought him increased fame. It is because of the reputation he acquired then that his later trial and banishment commanded so much more notoriety than many other similar petty prosecutions by the Church. Regarding his telescopic observations, it did not take so many years before they were replicated by others and, in due course, superseded, when further improvements in the design of telescopes were introduced. His more lasting and original contributions in mechanics were recognized much more slowly. The ideas and innovations he introduced in this field were understood by a group of disciples in Italy who developed them further, but his influence in the rest of Europe was indirect and more difficult to ascertain. The revival of an interest in mechanics in England seems to have followed an independent course. It is surprising that Newton read Galileo's works only toward the end of his career. More immediately, on the continent, Galileo had no influence on René Descartes, who spurned his ideas and preferred to be guided by his own whim. The case of Descartes is noteworthy because his brand of philosophy acquired wide currency and dominated the scene for at least half a century. Central to his thesis was the notion that matter and everything else in the Universe is formed of minute corpuscles, constantly in motion, circulating around vortices and conveying the effects of forces from one object to another. It is not too much of an exaggeration to say that, from this premise, Descartes attempted to explain all of the sciences, from the tides to magnetism, from gravitation to the reactions of chemistry and much else. The phenomenon of the tides, for example, which had preoccupied also Galileo and Kepler, Descartes explained by the fact that the celestial matter, circulating in a vortex around the Earth, is squeezed as it passes between the Earth and the Moon and compresses the waters of the oceans in the process. In retrospect, it is easy to understand how Descartes could have scorned Galileo's results, which concentrated on studying the motions of projectiles in unreal circumstances, such as a vacuum, where the effects of friction can be neglected. Quite to the contrary, maintained Descartes, the Universe is a plenum and one must seek the laws of physics in the interactions between contiguous vortices. Today we can deride these ideas of his, which became so fashionable for a time but proved to be totally unprofitable in the end. On the other hand, we should not neglect the instances where his intuition led him to the correct destination. Nowadays we associate the concept of inertia and the equivalence of uniform motion with rest to the investigations of Galileo, but Galileo was content with accepting uniform circular motion as inertial, whereas Descartes realized that inertial motion could only be rectilinear. Similarly, in the field of optics,

Descartes was able to formulate the correct law relating the angles of incidence and of refraction, a law that had eluded Kepler in his studies. We have already discussed his contributions in mathematics and how, through the introduction of algebraic methods in geometry, it became possible to study the curves known as conic sections with greater ease. It is ironic that only a few years earlier Galileo had shown that flying projectiles describe parabolas and Kepler had demonstrated that the orbits of planets consisted of ellipses; in spite of this, Descartes seems to have remained oblivious to the possibility of applying his new acquired tools outside the field of mathematics and into the study of Nature. Thus, his lasting scientific accomplishments appear to us somewhat scattered and at odds with one another. His goal was to construct a comprehensive and coherent system of thought, but in the attempt he became too speculative and relied perhaps too much on his own imagination, which led him astray into barren territory.

Plebeius: In his grand vision, he contemplated explaining not only the simple interactions between corpuscles of matter, but he wished to advance in steps and conquer also all of physiology, which formed a substantial part of his interests. It may be a little unfair to say that he became too speculative; he may not have conducted a great number of experiments, but he spent countless hours dissecting the brains of many animals in an attempt to understand what are the constituent elements of our perceptions and emotions. It was a foolhardy attempt, but brave by any measure. Furthermore, he may have been the first to conceive of a mechanistic interpretation of life, dismissing the need for an immaterial vital impulse of spiritual origin. He fashioned himself in the mold of Aristotle and he failed perhaps in the same way, that he pushed his method beyond its limits and attempted to explain too many things in too simple a way. It may well have been his followers too, who adopted his doctrines too rigidly, too unconditionally and formed a school similar to the old Aristotelians, not heeding Descartes' own advice of casting a doubt on all received knowledge. Voltaire, who had little sympathy toward the Cartesians, thought that those who hold the truth seldom adopt labels; it is only the ones who are in error that tend to group themselves in schools and sects.

Albertus: To those who accused him of contradicting Aristotle, Galileo used to respond saying that he was the true Aristotelian, for Aristotle was not himself dogmatic and was willing to change his views and yield to the force of evidence. I would liken Descartes to Aristotle perhaps in his excessive wordiness, that is, in his belief that something had to exist merely because he had defined it properly in words. I suspect that Descartes went even further, laying too much emphasis on his capacity to construct the truth predominantly by the activity of the mind, without due observation and experimentation. His desire to reach great generality and to explain all the effects by their causes through logical reasoning put him at the opposite pole of Galileo, who always began studying the simple and the particular, attempting to reach the causes by observing the effects. Nevertheless, it is true that in his ambition to reach great generality, Descartes conceived ideas that were quite novel and daring in his time. He thought that the Earth might be an old, cold, extinguished star, so that in this way he introduced the concept of the physical history of the Universe, a topic on which very few had speculated before him from a scientific point of view. It is also worthwhile to compare the outlook of Descartes with that of Kepler. Descartes had studied Kepler's Optics but took no interest in mathematical astronomy. Their ways and methods, one rational and cool-tempered, the other inspired and passionate, seem as different as the goals they pursued. If Descartes sought a mechanistic account of the processes of Nature, Kepler can be said to have favoured an animistic explanation. In his writings, Descartes gives the impression that he wishes to dismantle the great construction of Nature and show that its incomprehensibility can be reduced to simple and understandable laws. Kepler, to the contrary, seems to have searched for laws of Nature that could only add to its mystery. One wished to explicate things and the final goal was all that mattered, the other did not mind to complicate things while the path followed was as important as the goal pursued. It is also striking how they differed in the way they perceived the use of mathematics. Descartes thought that philosophy should adopt mathematics as a model and imitate its methods, stating its postulates and proceeding by deduction. Kepler would certainly have objected that it was not a matter of imitation, that all the necessary philosophy was already in the mathematics itself, in form as much as in content.

Plebeius: They probably differed also a great deal in their attitudes as Christians. Kepler, as you stated earlier, saw himself as a student of theology diverted from studying the book of Scripture and turned to the book of Nature, but applied himself to both with the same spirit. In Descartes, who saw himself as a devout Catholic, we find only a rather perfunctory attachment to the Church; his philosophical writings, however, are never influenced by the experience of faith or by the doctrinal writings of the Church Fathers. He is only concerned with the intellectual task of proving the existence of God, as an artificer or engineer of the Creation, and he applies his greatest gift, the use of reason, to know his deeds and his power. You mentioned that Galileo had no influence on Descartes' scientific views, but he influenced him indirectly when Descartes, upon learning of Galileo's condemnation by the Church, decided to withdraw the book he had intended to publish with the title The World. As he wrote to his friend and frequent correspondent Marin Mersenne, he felt his essay was also too explicitly Copernican to win any favour and he preferred to conform with the orthodoxy. One may suspect that he was proceeding with excessive caution, given that Mersenne himself had written to Galileo offering to publish his book in France, where it would have found less opposition than in Italy. Galileo's attitude, in turn, ought to have been altogether different. Judging from his letter to the Princess Christina, one of the few writings of his I have read, I confess, and where he seems to discuss these matters most candidly, he grappled with the question of how to reconcile his faith with his scientific work. He quotes Tertullian's assertion that God is known first by Nature and then again, more particularly, by doctrine, by Nature in his works and by doctrine in his revealed words. Galileo seems to imply that on matters of salvation, of moral judgment and the general life of the soul, one should follow the teachings of the faith, but when it comes to the material world, direct observation is much more reliable than the Scriptures. This distinction between the worlds of faith and of knowledge, coexisting in the spirit of the same individual, became commonplace in a later age, but no one seems to have formulated it so clearly before him.

Albertus: I suspect that Kepler, given his temperament, would have found the distinction somewhat incomprehensible. For him the sphere of the stars bounded a finite Universe. The Sun, the realm of the planets and the sphere of the stars reflected the Trinity of the Father, the Son and the Holy Spirit. This was not just an analogy but a manifest expression of the unity of the Universe as he conceived it. But let us return to our subject and consider, once again, the state of astronomy and the age-old problem of the various dimensions of the Heavens as they stood at the end of Kepler's life. The adoption of Copernicus' point of view had made it clear that it was necessary to determine only one absolute distance, to the Sun or to one of the planets, in order to determine all other distances from it. The problem of the relative sizes of the planets also attracted much attention, in the belief that they might be related to the planets' positions in the sky. The arrival of the telescope slowly transformed the task of estimating these sizes from mere guesswork into actual measurements. At the time, the relative distances and the relations between distances, diameters, volumes and densities of the planets all seemed equally significant. Kepler's legacy was grandiose in conception but little headway had been made in confirming his speculations. The ultimate goal envisioned then amounted to understanding the rules according to which the master plan of the Universe had been designed and the perfect numbers or ratios that were involved. It would not have crossed their minds to postulate that the planets were random boulders cast away negligently around the Sun after some accident in the history of the Universe. Part of Kepler's legacy were his Rudolphine Tables, the best available for the purpose of charting the course of the planets. Kepler himself was able to anticipate the transit of Mercury in front of the Sun for the year 1631, followed a month later by a transit of Venus. These occurrences represented the best opportunities to measure the sizes of the planets observing the silhouettes they cut out on the surface of the Sun and, with some luck, to measure their parallaxes too. This amounted to observing not only the planet orbiting the Sun but also detecting the effect caused by the observer being displaced on the rotating Earth. The transits of Venus are quite rare and the following one would not occur, Kepler estimated, until a century later, in 1761. These transits had been discussed since the time of Ptolemy in relation to the order of their orbits around the Earth, above or below the Sun, but their observation had always remained somewhat doubtful. Kepler explained the manner in which the image of the Sun should be projected through a telescope onto a surface or, alternatively, through a pinhole in the ceiling onto a surface in a dark chamber. The latter method would have fooled any one who attempted it, for the shadows of the planets are far too small to be seen in this way. Kepler had been misled in his estimate of their sizes, believing that the diameter of Venus would be one fifth that of the Sun and Mercury's approximately one fifteenth. As far as the transit of Venus was concerned, it was estimated that it would be visible only in America and, although there was an attempt, so it was said at the time, to encourage any men of learning in the high seas or across the Atlantic to pay attention to this event, there is no record that the transit was observed anywhere. The transit of Mercury was indeed detected by a handful of observers, amongst them Pierre Gassend in France and Johannes Remus in Germany, who were careful enough to project the Sun's image through a telescope. Gassend probably devoted a good deal more time to his philosophical studies than to astronomy, in particular to his long controversy with Descartes, but he was also a careful observer and he was familiar with the slow motion of sunspots on the surface of the Sun, following its twenty-six days rotation period. When he first searched for Mercury's image on the Sun he found a small speck which he thought to be a sunspot; yet, he soon noticed its fast motion and realized that it would leave the surface of the Sun in a few hours, confirming that it had to be Mercury's shadow. Its size, contrary to Kepler's estimate, did not amount to more than one percent of the Sun's diameter, almost ten times smaller than expected. Johannes Remus, a German astronomer, obtained similar results. He had corresponded with Kepler on these matters and he was not so much surprised to find such a small size because it fitted well within his own thesis that the planets' diameters were proportional to their distances from the Sun. Others, however, found these observations suspect. It was argued that the Sun, being larger than the planet, would illuminate from behind more than one hemisphere, reducing the size of its shadow. Moreover, it was well known that light tends to wrap an object as it goes by it, giving the appearance of a smaller size. Lastly, it was possible that Mercury's outer layers were made of a translucent material, so that we would be deceived as to its real size. Man is never short of explanations and, in order to hold on to an old belief, he seems ready and willing to believe almost anything. Regarding Venus, Kepler had missed in fact a transit that was to occur in 1639, nine years after the one that remained unobserved. In northern England, during these years, despite the turmoil caused by the civil war, there was a group of young, enthusiastic, brilliant amateur astronomers who, working entirely on their own, had an extraordinary success in their endeavours, a success that was unfortunately very short lived. One of them, William Gascoigne, was already in possession of a telescope of the type designed by Kepler, which had the advantage that it offered a larger field of view. More importantly, despite the fact that it produced an inverted image, it had a focal plane within the telescope, so that a real image was produced within it. According to legend, Gascoigne once noticed the web of a spider inside his telescope, coincidentally along the focal plane, which could therefore be seen sharply against the background sky. This gave him the idea to place cross hairs in his telescope and use such a device to measure angular diameters with greater precision. Thus, it is said that in this way the first micrometer was born. The story of the spider web seems doubtful, but we are attracted by its moral, because it reinforces our belief that we must always observe everything and discard nothing. It so happens that many decades later, when telescopic observations were gaining in precision, it was always very difficult to find wires or silk threads thin enough to adapt into the telescopes and some observers resorted to using spider webs. Be that as it may, Gascoigne was the first to use this device and he shared his invention with some of his astronomer friends, before his death in battle at the age of thirty-two, defending the royalist cause. His invention did not reach the wider world of astronomers and the micrometer was rediscovered years later in the continent. Amongst those whom Gascoigne had shown his device was the younger Jeremiah Horrocks, who never got to use the micrometer since he died even earlier and quite suddenly at age twenty-two. Horrocks' career is one of the most remarkable in the history of astronomy. Although he was sent to study at Cambridge when he was fourteen, he remained there for only two years. He returned to his native Liverpool without having obtained a degree and was for the most part self-taught. At age seventeen he began studying the orbit of Venus and, having found most of the published tables inaccurate, he became a devoted follower of Kepler after using his Rudolphine Tables, convinced that they provided the best representation of the orbit. Quite surprisingly, Kepler had failed to use his own discoveries at the time he was readying the tables for publication in order to improve on some of the old Ptolemaic parameters for the planets' orbits. Horrocks noticed that such improvements forced a modification of the eccentricity of the Sun, which in turn affected the orbits of the planets. This led him to compute the orbit of Venus afresh and to the discovery of its imminent transit across the surface of the Sun, a transit that had eluded Kepler. Indeed, no other astronomer was aware of its approach and he was the only one to observe it. As he had expected, the diameter of Venus, at its maximum, was much smaller than Kepler's estimate of 7 arc minutes. By his account, it was merely 1 arc minute. Still influenced by Kepler, he was led to speculate on a different set of proportions that might adequately explain the relative distances between the planets. In the few months remaining before his death, he wrote about his belief that the apparent diameter of all planets might be the same when seen from the Sun, approximately 28 arc seconds, an idea that had already occurred to Johannes Remus. Horrocks concluded that this should apply to the Earth as well, in which case the solar parallax, which is precisely one half that angular diameter, should be 14 arc seconds, hence, much less than previously estimated. The distance to the Sun would be accordingly much larger, as high as 14,000 Earth radii. In his all too brief career, besides these singular achievements, Horrocks also found time to work on a theory of the lunar motion. He was able to account for that inequality that had come to be known as the evection, combining successfully a slow rotation of the lines of apsides with elliptical orbits of varying eccentricity. It was this representation of the Moon's motion that was used many years later by Newton and quite a few others, and for several decades thereafter.

Plebeius: But, surely, that description had not made use of the law of gravitation, which was discovered later. Do you mean to say that Horrock's theory remained in use even after the true equations of motion had been formulated by Newton?

Albertus: Indeed, it is one thing to formulate the equations of motion and a different matter to solve them. They are utterly complicated when several gravitating bodies are present and simplifications or shortcuts were required, thus affecting the predictions in uncertain ways. Horrocks, on the other hand, had started from the observations and matched his orbits to them. In the beginning there was little advantage in replacing a proven method by a tortuous and unsettled one. Horrock's work was slow to have an impact. His writings did make their way to London, but they were scattered during the confusion of the civil war and the great fire that ravaged London in 1666. Thus, they appear to have suffered quite a few vicissitudes before they were brought to the attention of some members of the Royal Society, which had been founded only a few years earlier, and could then begin to reach a wider audience.

Plebeius: It seems remarkable that until the emergence of these scientific societies, for a period of some two hundred years, ever since the teachings of the scholastic philosophers ceased to have their earlier appeal, most of the innovations and original ideas in science were the contribution of individuals acting on their own, away from the universities. Both Copernicus and Tycho Brahe had spent several years in their youth visiting various universities, but then they pursued their investigations without any association with them, the one exploiting his wealth and social standing, the other working as a church administrator. Kepler never returned to the university since his student days, although he perhaps regretted not having the security that such a position would have provided him. Galileo, on the other hand, wished nothing but to free himself from his association to the university, which seems to have been only a burden to him. The universities, of course, would have offered much less of a home to those who were more interested in applied disciplines, ballistics, fortifications or irrigation. Within the more affluent classes, there were many amateur scientists whose interests did not draw them toward academic circles. I am thinking, in particular, of those who were called virtuosi with derision, even though sometimes they used that label themselves. These were well educated men with plenty of leisure time, who had an interest not only in science but in the arts and letters as well. It is sometimes said that their motivation was not really scientific, that they were only collectors of exotic things, curiosity seekers, more interested in recording monstrosities, birth defects, the unnatural in Nature, than studying Nature itself, but I wonder whether these distinctions are meaningful at all. Probably the majority of the members who came to found the Royal Society broadly belonged to this class. These men were also very prominent in the Netherlands and certainly in Italy, where they drew on a long tradition. In France too there were several informal groups of cultivated men who gathered periodically to discuss scientific matters. I mentioned earlier Marlin Mersenne and his relationship with Descartes and Galileo. Just as in the case of Paolo Sarpi, it is enough to peruse a little their vast correspondence to realize the broad range of interests these men had, their industriousness and dedication, and also how well they kept themselves informed about the activities of their peers. The French Académie des Sciences, which Louis XIV established at the instigation of his minister Colbert only a few years after the Royal Society was founded, differed from it in that it was not a private institution and its funding, as much as its objectives, were controlled by the State. But in either case these institutions offer the first signs that the pursuit of science was beginning to change from an individual's adventure to a more institutional enterprise. In the past, those who had not had the private wealth that would have been necessary to support their investigations, such as was the case with Descartes or Tycho Brahe, often found refuge in the Church, as in the cases of Mersenne and Pierre Gassend.

Albertus: I suspect that, amongst English scientists, it would be more difficult to distinguish the social classes to which they belonged, given that it was often the more affluent members of society who also often took orders in the Church and sometimes occupied the coveted chairs at Cambridge or Oxford. Jeremiah Horrocks, incidentally, had dedicated himself fully to astronomy, but he had also taken holy orders and, had he lived, was probably destined to a career in the Church of England. He and Gassend are examples of those who, like Kepler or Copernicus himself, had accepted the Copernican outlook in astronomy without this interfering with their faith. Gassend is an interesting case because his scientific judgment was sound and quite modern, his proposed experiments to elucidate the concept of inertia or to measure the speed of sound were cogent and sensible; in his philosophic writings, however, he attempted to reconcile his view of the material world with Christian doctrine. At all events, he always expressed himself very cautiously, perhaps because he had acquired some prominence and respectability in the Church hierarchy, but also because of his lack of dogmatism, the quality he had most criticized in Descartes. I wish to mention the case of Johannes Hevelius, a Polish astronomer from Danzig, who had followed in the footsteps of Tycho Brahe. He studied law and made a living by administering a brewery he had inherited from his family, but in his spare time he built a private observatory that was perhaps the most advanced of his time. It also included a printing press. Some of his efforts were oriented toward obtaining a detailed topographical map of the face of the Moon, an activity that had become fashionable then, and we owe some of the designations in the Moon's landscape still in use today to him. At one time, a great fire laid waste to his observatory, destroying all of his instruments. It is a proof of his energy and determination that he set out to rebuild it very much as it had been, despite the fact that he was already close to seventy years of age. Hevelius may have been one of the last astronomers of the first rank who worked entirely independently, although he was by no means the last. You spoke of the rather poor record of the universities in providing a suitable environment for scientific work; it is ironic that the Church, through benign neglect, often offered more comfort and greater opportunities. The Jesuits, then a relatively young order, were rather loosely associated to the Church but, from the very beginning, they showed an interest in education and had a partnership with several universities. In the city of Bologna there was a group of successful scientists who belonged to the order, one of the more prominent being the astronomer Giovanni Battista Riccioli. He had been teaching theology for many years, but was finally allowed to devote himself entirely to astronomy, so that he could finish writing a lengthy book of his, entitled Almagestum Novum, a rather ambitious title that revealed his attempt to emulate Ptolemy, providing a detailed summary of the astronomical knowledge that was available then at mid-century. Still bound to the Church and quite conservative in his views, he never fully subscribed to Copernicus' theory. He also devoted some time to make maps of the Moon and a few other names in use today are Riccioli's names. We may see his devotion to the planetary model of Tycho Brahe in his naming one of the more prominent of the Moon's craters after him. One of Riccioli's students at the university was Gian Domenico Cassini, whose name in Bologna is associated to the great gnomon in the Cathedral of San Petronio. Very much like the one in Florence, it was essentially a hole pierced high in the ceiling, in such a way that the position of the Sun, projected onto the floor, could be traced throughout the year, making thus of the Church a gigantic astronomical instrument. The devise had been introduced early on, but the young Cassini refurbished it after it had fallen in disrepair, and made then some very precise measurements that contributed to his reputation. Cassini had been brought to Bologna from northern Italy by the whim of an extravagant nobleman, who needed help in his astrological researches, a subject for which Cassini had shown some early inclination. Later, however, he was won over to the study of astronomy by his Jesuit mentors at the university, where he did not take much time to become a professor. With the gnomon at San Petronio, by following the altitude of the Sun throughout the year and marking the occurrence of the equinoxes, Cassini could determine the inclination of the ecliptic and the eccentricity of the Earth's orbit with great precision. To make these determinations, it is clearly necessary to correct for the effects of refraction and parallax. As we have already seen, this practice had been already began by Tycho, who had composed various refraction tables, depending on whether the object observed was the Sun, the planets or the fixed stars. Riccioli had followed suit, authoring new tables and adding the complication that there were now different tables for each of the seasons throughout the year. While making use of these tables, Cassini was perhaps the first to emphasize the fact that there were no accurate observations regarding the solar parallax and that its value had always been taken more or less on faith. He further observed that, when it was reduced from the value of 1 arc minute, which had been adopted by Kepler, by a factor of 4, to 15 arc seconds, then the corrections for refraction would be uniform throughout the seasons and also necessary at all latitudes, contrary to Tycho Brahe's belief that they were relevant only at low altitudes, far away from the zenith. Cassini was also the first to employ Snell's law of refraction. If his suspicion was true, it required the Sun to be at a much greater distance than was then accepted. It was the effort to disentangle these various effects and the desire to understand the phenomenon of atmospheric refraction properly that eventually opened the way to solving the long-standing problem of the distance to the Sun. This by itself justifies that we include Cassini in our account of the evolution of astronomy. But he is also an excellent representative of this time of transition you referred to a moment ago, in the manner that scientific studies were supported and promoted. Amongst other occupations, Cassini was brought to mediate between Italian cities in a number of disputes that concerned the distribution of waters, hydraulic projects and changes in the course of rivers. Eventually, he was made superintendent of water for the Papal territories and the Pope requested that he take priestly vows. Cassini refused, perhaps on the grounds that it might interfere with his astronomical interests. Indeed, when the Académie des Sciences was inaugurated, he was invited to join it and he at once obtained permission for a temporary visit to Paris, which turned out to be a permanent move. Cassini was received by the King and promised a comfortable residence at the Paris Observatory. The Observatory was then under construction and Cassini was going to be its director. This was a time of splendour and great architectural projects in France, when the palace of Versailles was also being built. In fact, the buildings of the Observatory were designed in such an elegant style that later there were not enough funds to equip it properly. At any rate, Cassini had followed the improbable path of starting as an assistant to an amateur astrologer, then joining a group of Jesuit astronomers in Bologna and finally becoming a rather conspicuous member at the court of Louis XIV. The presence of the Cassinis in France proved to be long lasting, given that Gian Domenico started a dynasty of four generations of astronomers who remained at the Paris Observatory until after the time of the Revolution. When he first arrived in Paris, the other prominent foreign member of the Académie was the Dutch polymath Christiaan Huygens, who had been received with equal distinction. They were joined by Ole Römer, who was Danish, and a group of French astronomers, amongst them Jean Picard. Huygens had had much success polishing lenses and building telescopes of excellent quality; a decade earlier he had been led to the rediscovery of the micrometer, the device that was placed at the focal plane of the telescope for the purpose of measuring angular diameters. His version of it was technically superior, in that the separation of the measuring wires was adjustable by means of a screw and lent itself to greater precision. Huygens had also gotten into the habit of coating his lenses lightly with soot from candles, in order to reduce the glare in the planets' light, so that all these tricks enabled him to improve on the measurements of their diameters. Following the same path that Remus and Horrocks had already traveled, he was led to speculate that these diameters were proportional to the planets' distances to the Sun. The accuracy that was now attainable in these observations had also led Jean Picard, amongst others, to carry out a new systematic investigation into the apparent sizes of the planets, so that this problem continued to be in the focus of attention of many astronomers. Cassini, on the other hand, had brought with him from Italy the problem regarding the yet undetermined effects of refraction and parallax. This problem affected not only the solar parameters, the obliquity of the ecliptic in particular, but the elements of the planetary orbits as well. The task of measuring the compensating effects of parallax and refraction was now probably within reach, given the precision obtained with micrometers. One obvious possibility was to conduct observations near the equator. Beyond the tropics, the sun crosses the zenith in its yearly motion and at such a time the effects of refraction are nil. Succinctly stated, Cassini expected that, at any location away from the zenith, the effects of parallax and refraction are both small but present everywhere, whereas, according to Tycho's theory, down to relatively low altitudes, the effects of refraction would be absent while those of parallax would grow steadily. A comparison between observations carried out at Paris and at the equator would surely help to solve this dilemma. In 1672, three years after Cassini's arrival in Paris, an opposition of the planet Mars was expected, a propitious opportunity to measure its parallax, given that the planet is then at least three times closer than the Sun. These circumstances more than justified the organization of an expedition to the equatorial region in order to complement the observations that would be made in France itself. By this time, transoceanic trips had become almost routine, although not precisely for scientific missions, but ships from the French Navy were then actively exploring the St. Lawrence and Mississipi valleys or they were carrying goods to and from China; the French were also trading Caribbean islands with the Dutch or with the Spanish crown, attempting also to set foot in Senegal, to keep control of the island of Madagascar and to establish a colony in Pondichery in Western India, all of these things more or less simultaneously. The chosen destination for the astronomical mission was Cayenne, in French Guiana, on the northern Atlantic coast of South America. Its main objectives were to determine the parallax of Mars, to measure the obliquity of the ecliptic, to monitor the orbit of Mercury and to obtain the positions of as many southern stars as possible. The measurement of parallax can be done essentially in two ways, either by determining the position of the planet relative to the background stars at the same time from two widely separated locations or, taking advantage of the rotation of the Earth, from one single location at two different times. During this expedition both methods were attempted. For the first method it is obviously necessary to be in possession of synchronous clocks at the two locations. The keeping of time with acceptable precision had always been a cherished goal and a recurring frustration for astronomers as much as for navigators. One of Christiaan Huygens' technical contributions was the improvement and development of the pendulum clock to the extent that, for the first time, mechanical clocks were now beginning to form part of the arsenal of instruments in any observatory. Some of these clocks had been taken on sea voyages, equipped with all sort of devices to keep them stable, but the results had been most disappointing. It was still impossible at the time to contemplate the transport of a clock across the ocean and preserve any degree of accuracy. Alternatively, the orbits of Jupiter's moons had been monitored for some time, so that having a chart of their course as they go about the planet while being observable from various locations around the Earth simultaneously, they could provide a common clock for all observers. Cassini, in particular, had been involved in providing tables for the motions of these satellites. It was then Ole Römer who, studying these tables, noticed a discrepancy that seemed to depend on the distance from the Earth to the planet, leading to the suspicion that it was probably caused by the difference in the travel time for the light to make its trip from Jupiter to the observer. In this manner he was the first to provide a reasonable argument for the finite speed of light. Of course, he also attempted to estimate this speed but, without a reasonable notion of the distance scale within the solar system, his estimate was quite inaccurate. By this time, Huygens had already written his treatise on light and he supported Römer's conclusions without reserve. Cassini, who had been in fact the first to notice the discrepancy in his tables and conceived the same explanation, remained skeptical and was not persuaded. The second method to determine the parallax, by repeating the observations at two different times from the same location, has to take into account the fact that neither the planet nor the observer are at rest in absolute space. Therefore, it becomes necessary for the observer to determine the position of the planet at the same hour, on successive days, so as to be able to trace its uniform motion across the sky and only then, once he is able to predict its position at any other time, compare the predicted position to the one actually observed after the Earth has rotated around its axis a certain angle. It is this discrepancy, namely, its periodic displacement every twenty-four hours, which is superimposed on the uniform motion across the sky as recorded over several days, that must be ascribed to the effect of parallax. The desire to measure the parallax at opposition is obviously due to the fact that the planet is then so much closer. However, you remember that at opposition the planet's motion is not always uniform, it is precisely at this time that it performs its retrogressions. Therefore, it was necessary to choose very carefully the moment to carry out the observations. More importantly, it was convenient and conducive to greater precision to observe the planet when it was passing in the neighbourhood of bright stars which could provide a frame of reference easily monitored. The expedition to Cayenne lasted approximately one year and was awaited with great anticipation in Paris. As it turned out, it had had a stupendous success. Regarding the problem of refraction and parallax, as Cassini had estimated, the use of Tycho's prescriptions would have meant that the obliquity of the ecliptic, or the inclination of the Earth relative to the plane of its orbit, had a different value at the equator, which was certainly absurd. At the same time, a small solar parallax not greater than 15 arc seconds with refraction corrections up to the zenith gave an obliquity that coincided with what Cassini had measured in Bologna. The answer seemed now clear. This was not quite the same thing as a measurement of the solar parallax but it provided a lower bound to the Sun's distance. As to the parallax of Mars, the results were less conclusive. Cassini announced the initial estimates stating that it amounted to approximately 25 arc seconds. This value was consistent with a solar parallax close to 10 arc seconds and, as it turned out, it was a very good estimate. Nevertheless, it appears that those early estimates were done hastily and without taking proper care of all the sources of error, so much so that the publication of the final results was delayed for a decade. One interesting dividend of this expedition was the verification that the period of the pendulum clocks they had carried with them lengthened relative to the measurements made in France. In a few years time, this was interpreted, in the light of the new theory of gravitation, as a proof of the bulging of the Earth at the equator, which causes the gravitational pull to diminish as the distance to the center of the Earth increases, thus easing the oscillations of the pendulums and lengthening their periods. Once again, Cassini remained skeptical of this interpretation. He was even distrustful of the general theory of gravitation and the elliptical nature of the planets' orbits. All in all, even if the excellent estimates of the solar and Mars' parallaxes to which Cassini contributed were less exact than they appeared to be, he may still be considered one of the more influential figures in astronomy of that era. It is clear that he pointed the way toward the resolution of the problem regarding refraction and parallax. It is also clear that he erred in his refusal to accept the finite speed of light or in his rejection of the theory of gravitation. It is all too tempting to contrast Cassini with his colleague at the Académie, Christiaan Huygens. Astronomical observations had been only one of Huygens' activities and in this regard he had been fortunate enough to discover a Moon around the planet Saturn. He had also found a way to demonstrate his technical inventiveness polishing lenses, building telescopes, developing the screw micrometer and perfecting the pendulum clock. But he also had a genius for theoretical ideas and he devoted much of his time to exploring the new sciences of mechanics and optics, the underlying mathematics as well as their broader philosophical significance. Huygens belonged to a cultivated and prosperous Dutch family, in whose home Descartes had been an occasional guest during his childhood. He benefited not only from having an excellent education but also from the fact that he was able to devote himself entirely to his scientific pursuits and to travel until well into his thirties, when he became a member of the Académie des Sciences in Paris. The versatility and breadth of his genius allowed him to apply himself to almost every subject that was then at the frontier of science, from the practical design of instruments to the abstract theories of probability in mathematics or gravitation in mechanics. Sometimes it is tempting to peek through the distortions of history and consider the attitude one might have taken toward a scientist, had one been on of his contemporaries. That is, would we have shared his opinion regarding the fruitfulness of some of the theories that were then in the making? Would we have approved of the lines of research he pursued or the problems he chose as significant and those he discarded as unimportant? These questions are difficult to answer. I wonder whether, in his own time, one might not have sided with the more cautious Cassini. Would we have been justified in accusing Huygens of being a bit of a dilettante? In retrospect, this seems more preposterous than unfair, but some of the theories he embarked on were at the time tentative and he did not pursue them to leave a conclusive and polished result. His theory of light contained insights that eventually proved to be more fundamental that those presented by Newton, against whose theory Huygens was writing. Regarding the theory of gravitation, Newton's work appeared when Huygens was in his sixties. To his credit, he appreciated the power and effectiveness of the new mathematical tools, as opposed to the more geometrical approach he had favoured. But regarding the origin of the gravitational force itself, he did not accept the notion of action at a distance and clung to ideas based on Descartes' vortices, on which he had worked for years and that eventually led to no results. From the early trips in his career, Huygens had remained in contact with the major scientists of his time. During his visit to London in 1661, precisely on the day that Charles II was restored to the throne after his years of exile, he attempted a measurement of a transit of Mercury in collaboration with Thomas Streete, one of the more noted English astronomers of his time, but they did not succeed. It was also in London that he obtained a copy of the works of Jeremiah Horrocks from the Oxford mathematician John Wallis, the first to have begun collecting them. Huygens forwarded them to Johannes Hevelius, in Danzig, who printed them for the first time. The restoration and the establishment of the Royal Society created a climate in England that was propitious to pursuing scientific studies, but the astronomer whose work proved to be the most influential in the next generation emerged at the margins of these developments. James Flamsteed had not had a university education, either because of his poor health that continued to bedevil him for the rest of his life, or because of the opposition of his family, whom he helped support. Surreptitiously, he persisted in his astronomical endeavours and at one time, still in his youth, he submitted a list of future lunar occultations to the Royal Society, the sort of tables that were valued then for navigational purposes. This brought him some official recognition and the possibility of obtaining better lenses for his telescopes and a useful micrometer. Still acting entirely on his own, he set out to determine the parallax of Mars during the same opposition of the planet that the French were observing in France and in Cayenne. He followed very much the same method and, after finishing his observations and carrying out the necessary computations, he arrived at a very small value, which implied that the solar parallax could not exceed 10 arc seconds. He communicated his results to Cassini and was delighted to learn that the French measurements had led to a similar conclusion. In 1675 the King appointed Flamsteed Astronomical Observator. It is not certain that Charles II had a great interest in science; at one time he had ridiculed the members of the Royal Society because they did little else besides attempting to weigh the air. But he seems to have had a genuine interest in the development of his Navy and, by this time, the problem of determining the longitude at sea was a pressing one, a problem to which the astronomical observator or Royal Astronomer, as the position was later called, had to dedicate himself. Even with his move to Greenwich a year later, where the rudiments of an observatory were to be established, the position Flamsteed had acquired did not amount to much. Initially, he had to supply his own instruments and, as far as income was concerned, he was only relieved somewhat after taking orders in the Church but, even then, he was often pressed to take private students in astronomy. His great achievement was a catalogue of 3000 stars, for which he obtained positions that were more precise than those in Tycho's catalogue, still in use at the time. The motion of the Moon was a subject to which he returned again and again. After making careful observations and reducing his data, he compared the results with some of the theoretical models proposed, which convinced him that Horrocks' theory was the most accurate. In later years he got involved in a dispute with Isaac

Newton and Edmond Halley, the latter once an assistant to Flamsteed. It is said that Newton was impatient to obtain some data on the observed lunar orbit for the verification of his own theory, which Flamsteed was reluctant to deliver until he had satisfied his own standard of accuracy. Newton and Halley eventually obtained and published the data under the auspices of the Royal Society, without Flamsteed's permission, on the grounds that he was a public official and ought not to have restricted access to that material. This may have been true, but it is also true that Flamsteed had acquired his instruments himself and had even invested his inheritance in the observatory. In reprisal he seems to have found a way to set fire to many of the copies Newton and Halley had printed. Edmond Halley began his career as an astronomer at age twenty by traveling to the island of St. Helen to conduct a survey of the southern stars. Thereafter he spent a few years as an assistant to Flamsteed at the Greenwich Observatory until they eventually parted company in disagreement. In an era that produced many brilliant minds and when many new theories were being debated, Halley was very much at the center of this agitation and excitement. He traveled to Danzig to intercede in the dispute that had arisen between Hevelius and Robert Hooke, of the Royal Society, concerning the accuracy and reliability of Hevelius observations. Many years later, he attempted to mediate again in the disputes between Newton and Leibnitz. His association with Newton had begun quite early; he contributed some of his own observations to Newton's work on optics and most likely, being a man of considerable means, helped finance the first printing of Newton's Principia. His interests were wide, from mathematics to archeology, but he concentrated primarily on astronomy and it was in this discipline that he made several valuable contributions. Perhaps owing to his cold relationship with Flamsteed, he occupied himself at first with other tasks which were not strictly astronomical, such as the fortification of ports or surveying the English coastline; only close to the age of fifty did he become professor of astronomy at Oxford and finally, fifteen years later, he succeeded Flamsteed as Astronomer Royal. It was during these latter years that he began to publicize the importance of the next transit of Venus, still some fifty years away, as the most reliable method to determine the distance to the Sun. He attempted to persuade other astronomers of the importance of equipping several expeditions to observe the transit from various stations around the world and of the need to strive in earnest for the greatest accuracy then possible. This campaign of his assured him of some renown after his death in 1743, when he was already eighty-seven years old, but his fame grew still larger in 1758, when the comet of 1682 returned as he had predicted. The study of the orbits of comets according to Newton's methods was one of the activities he pursued with greater interest. He deduced from the observations that those orbits were not parabolic but, in fact, very elongated ellipses. To be more precise, he proved that the supposed parabolic orbits of some comets were identical, which led him to interpret them as being the same comet making repeated appearances. Halley was probably right in insisting that the transit of Venus would afford the best opportunity to improve on the solar parallax, given the observational tools available at the time. It was admitted then that, although the most recent measurements indicated that this parallax was probably very close to 10 arc seconds, it had not been ruled out that it was nil, so that the distance to the Sun might be infinite, for all practical purposes. Clearly, on the next opportunity the measurements would have to become more refined. The need for more accurate observations and for more reliable and effective instruments was felt for other reasons as well. The desire to determine the longitude at sea using the occasional occultation of stars by the Moon required the compilation of precise tables providing the positions of the stars. Moreover, the need to determine accurate coordinates for the stars was also felt for purely astronomical purposes. The old wish to detect the annual parallax of a single star, to which Tycho had given some attention, had never come true. But there was also now the intriguing possibility of observing the intrinsic motion of stars, of which Halley seemed to have already found conclusive proof. In France, the astronomers of the Académie des Sciences had turned their attention toward improving the measurements of the size of the Earth and revising what was then known of its shape. For more than a century, geodesy remained one of their major enterprises. Already in 1662, Jean Picard had measured the length of one degree of latitude some distance north of Paris, that is, the length of meridian one must travel to see the zenith shift by one degree with reference to the background stars. It is interesting to note that, when Newton was speculating on the mathematical form of the law of gravitation and the hypothesis that the acceleration of objects falling to the Earth is inversely proportional to the square of its distance, he wished to verify it by comparing how the Moon falls toward the Earth, so to speak, as it departs from an inertial straight path, to the acceleration of objects on the surface of the Earth. Now, one can assume in this comparison that the entire attracting mass of the Earth is concentrated at its center, so it all reduces to comparing the distance to this center from its surface to the distance from the Moon. Ironically, Newton had a fairly good estimate of the distance to the Moon, in terms of the Earth's radius, but as to the radius of the Earth itself, or the circumference, he had used an estimate of 60 miles per degree and the expected law of gravitation didn't seem to be verified. As it turned out, this estimate was erroneous by more than 10 percent and it was only when he learned of the correct value of 69.5 miles per degree obtained by the French astronomers that things fell into place. I mentioned that a few years earlier, during the expedition to Cayenne, it had been discovered that the periods of pendulum clocks lengthened as one approached the equator. This was interpreted to mean that the Earth is a sphere only approximately, flattened at the poles and more bulgy along the equator. Newton, in turn, interpreted this effect as having been caused by the rotation of the Earth and was soon led to deduce that the precession of the equinoxes was caused precisely by this lack of sphericity. Indeed, given that the axis of the Earth is tilted somewhat toward or away from the Sun, from its vantage point, the Sun cannot exert an even pull on all sectors of the planet; one of the hemispheres, being nearer, is attracted more forcefully, disrupting its equilibrium and making the axis of the Earth go in circles parallel to the plane of the ecliptic, in the same manner that the axis of a tumbling top sweeps a cone as it rotates parallel to the floor. Now, given the peculiar shape of the Earth, a degree of latitude cannot correspond to the same distance at all latitudes, being longer near the poles where the Earth is flatter. Partly to unravel its precise shape, two teams of French astronomers launched expeditions to Lapland and to Peru in 1735 to measure the length of a degree at those latitudes, expeditions that confirmed and refined the expected results. The fundamental application of the theory of gravitation was surely to the orbits of the planets and comets and, very prominently, to the motion of the Moon, which is affected by the Earth and the Sun, a problem that already presents the theory in all its complexity. We have seen that the essential characteristics of the Moon's orbit were known since the time of Hipparchus. As described by Horrocks, for instance, the Moon's orbit traces an ellipse around the Earth, but this ellipse does not remain fixed in its orientation. In fact, its elongation, the line of apsides properly speaking, rotates ever so slowly around the orbit, lazily following the Moon. Similarly, because this elliptical orbit is slightly tilted relative to the plane of the ecliptic determined by the Earth and Sun, the Moon alternatively dives below the ecliptic and reemerges half an orbit later; the points of intersection, the so-called nodes of the orbit, determine a line that also changes positions, rotating along the orbit but this time in the opposite sense. Well, when the orbit of the Moon was computed according to the general law of gravitation in the presence of the Earth and Sun, precisely these effects were all seen to arise from the laws governing the motion and this was very satisfying indeed. Yet, the predicted motion of the Moon did not accord entirely with the observations and the desire to confirm the validity of the theory in its finer details provided further incentive to obtain more accurate observations of the Moon's motion. These investigations were taking place during the initial years of James Flamsteed's tenure as Astronomer Royal, when he was observing with less than perfect instruments and under difficult circumstances. Nevertheless, he managed to supply Newton with the data he needed to validate his theoretical findings, a contribution that Newton, if anything, grudgingly acknowledged, even when it had been crucial to show that he was following the right path. This may have contributed to increase the ill feeling between the two that years later ended in open hostility. Flamsteed spent some twenty years trying to refine the quality of his observations, only to discover each time that his instruments were defective or insufficiently calibrated. To calibrate an instrument one has to see how it performs against a well understood and stable scale. In those days there was no such measuring tool as precise and stable as the positions of the stars on the surface of the sky. Therefore, it was sensible to test one's instrument by examining how well it reproduced a result of spherical geometry, for example, when one attempted to reproduce it doing measurements on the sky. In this way, one was using the stars to observe the instruments, before using the instrument to observe the stars. Allow me now to say a few words regarding the accuracy with which observations were made during those years, because this was a time when the requirements for further precision, as well as the sources of error in measurements, became more clearly focused and better understood. To take a rather simple illustration, if we wished to mark the arc of a quadrant with subdivisions, so that arc seconds should be clearly distinguishable to the eye, by being, let us say, 1 millimeter apart, then the quadrant must have a radius of 200 meters. Obviously, it would be impossible to work with an instrument of that size. As we decide to reduce its size ten times, if you wish, we would still be faced with a sizeable instrument that most likely would deform under its own weight and would cause insurmountable difficulties when trying to obtain stable and precise measurements. If we then reduce the instrument further, you can well imagine that the markings of one arc second we had wished to be able to distinguish on our quadrant get awfully tight and we must devise some very fine art of engraving; furthermore, we must be able to accomplish the subdivisions of the quadrant in 90 equal degrees and each degree in 60 minutes and each minute in 60 seconds, with some confidence that all subdivisions are equal to one another. For instance, Tycho Brahe had been able to achieve a resolution of 1 arc minute in his measurements, not only because of the large size of his instruments, but also because of his systematic checking of one instrument against another and the overall care with which he carried out his observations. The fact that these were in error due to his misunderstanding of the effects of refraction, for example, does not alter the assurance that he was measuring what he was seeing with consistent precision. Now, the ingenious and prolific Robert Hooke, who seemed to have more ideas than he was willing or capable of pursuing to fruition, conducted some of the first experiments to test the visual acuity of human beings. This certainly depends on the type of objects one is trying to observe, but in the case of two point-like light sources, such as two stars at night, Hooke noticed that the human eye can separate them up to one arc minute, beyond which they coalesce to become indistinguishable. You can see then that, if you were aligning your instrument with the naked eye, however accurate your instrument is, you would not be able to achieve greater precision in your measurements beyond what Tycho Brahe had achieved. In this sense, the arrival of the telescope and the incorporation of telescopic sights into quadrants, sextants and other instruments, as well as the use of microscopes to read the scales, it all amounted to a substantial progress in the art of measurement. It represented a challenge to the engravers and instrument makers to achieve the precision that matched the artificially enhanced resolution with which the observer could now distinguish an object. In a general sense one could argue, at least from a historical point of view, that for more than a century, the true value of the telescope resided not in its ability to magnify objects, although we saw that making detailed maps of the Moon became quite fashionable for a time, but in its contribution to the task of assigning precise coordinates to the stars in the sky. There is another obvious difficulty we ought to mention regarding the practical aspects of astronomical observations. When an astronomer attempted to determine the angular separation between two stars, for example, he had to contend with the fact that the rotation of the Earth always proceeds apace, amounting to 15 arc seconds every second of time, and it was impossible for him to fix his telescope or sextant into position for anything more than an instant. The meticulous observer Johannes Hevelius in Danzig was the first to produce a star tracking device that rotated his sextant and was guided manually with a screw, but the attempt did not prove entirely successful. Much of the delicate craftsmanship that was required to make these instruments successful was borrowed from the workshops of watchmakers. Robert Hooke attempted to design a quadrant endowed with worm wheels, so that the telescope adapted to it would not rotate freely, but would do so driven by a screw, in such a way that so many turns would correspond to so many arc seconds; he even persuaded Flamsteed to use it at the Royal Observatory, but it proved to be very slow and clumsy to use and, in addition, the gear wore off so fast that the instrument became unreliable almost as soon as it had been calibrated. Nonetheless, some of these new ideas were ultimately adopted in one way or another when the manufacturing techniques became more refined. In a a rather crude sense but without exaggeration, the quality of one's astronomy was in direct proportion to the quality of the screws one was able to manufacture. In the beginning, the grooves on the screws were produced by wrapping a paper around the bolt with a right triangle inscribed on it, so that the hypotenuse would trace a helix on the surface of the bolt that the artisan could follow with his file, but the exactness of this method for precision astronomy left something to be desired. The pendulum clock was unquestionably one of the more remarkable tools that came into use in astronomy toward the end of the XVIIth century. James Flamsteed was one of the first to use it to great advantage in obtaining a degree of precision in the coordinates of his catalogue of stars that had been inaccessible until that time. The procedure by which Tycho Brahe had measured the position of a star relative to the Sun might include, firstly, the measurement of Venus relative to the Sun and then, after sunset, he would measure the star's position relative to Venus. Flamsteed was able to use an excellent and reliable mural that he had ordered built at the Royal Observatory after a great many vicissitudes, so that he could track precisely the rise and fall of the Sun throughout the year to determine precisely the instant of the vernal equinox. Then, instead of using the planet as an intermediate step, he followed the rotation of the Earth with his clocks and read off the angular separation of a few lead stars relative to the Sun, directly from the clocks at the time of observation. Only later did he used these lead stars to measure the positions of others by triangulations. This method was later extended and used in the form of a transit instrument, as it came to be known. It was a telescope held fixed by an axis oriented directly east-west so that the telescope could only follow the meridian in the direction north-south. It appears to have been used first by Ole Römer to observe the passage of a star through the meridian and in this way be able to check the accuracy of his clocks, but at a later time it was used in conjunction with a clock to record the coordinates of the stars obtained from their ascension and their time of passage across the meridian. Given that the rotation of the Earth corresponds to 15 arc seconds per second of time, clearly the accuracy of your clock would have to be better than a tenth of a second in order to record positions with arc second precision. By the time James Bradley became Astronomer Royal in 1742, pendulum clocks were reliable to within one second a day. The changes in the length of the pendulum itself due to variations in temperature remained the main source of error, but a great deal of effort was invested to study the expansion coefficients of different metals and the use of appropriate alloys gradually helped diminish the influence of temperature. James Bradley, the third Astronomer Royal, succeeded Edmond Halley who had held the post for some 20 years since 1721 until his death. Bradley's interest in astronomy was fostered by his uncle, a rector at the parish of Wanstead, just outside London, who was also an accomplished amateur astronomer and had one of the best private observatories in England. In addition, he was very helpful in providing for his nephew's education at Oxford. The young Bradley followed in his footsteps, not only through his interest in astronomy, but in that he was also ordained in the Church, although he eventually abandoned it altogether in order to devote himself entirely to astronomy. When Halley had relinquished his professorship at Oxford to become Astronomer Royal, Bradley was appointed to succeed him, while at the same time he was continuing his observations from the observatory at Wanstead. It was during these years that he made some of his more remarkable discoveries. After Flamsteed's death, his associates had removed all the instruments from the Royal Observatory, so that Halley had been forced to order new ones. Although he enlisted the help of some of the best instrument makers working at the time, Halley at this stage in his life did not have the dedication to monitor their accuracy and preserve them in good order, nor did he have the zeal to pursue systematic and meticulous observations over extended periods of time. These are precisely the qualities that Bradley had aplenty and that, many years later, he would bring to Greenwich, along with his accumulated experience, transforming the observatory, during the twenty years of his tenure as director, into what could be called the preeminent such institution in the world. In 1727 Bradley had ordered a new instrument for his observatory at Wanstead, a zenith sector, as it was called, by the same manufacturer that had made the instruments for Halley. He had been observing with a zenith sector built by this outstanding craftsman, by the name of George Graham, and he had been very impressed with its performance. The zenith sector was essentially a refined and more specific version of a transit instrument. It was a telescope of long focal length that was held by a pivot at the top, so that its tube was left to hang down seeking the vertical by the action of gravity, thus aligning itself with the zenith. The telescope had a screw micrometer and a delicate mounting that made it stable and very precise to observe the positions of stars near the zenith. It was the type of instrument that, in a less elaborate form, Jean Picard had used many years earlier to measure the length of one degree of latitude and, perhaps, had first been used by the ingenious Robert Hooke, who thought of this device in his attempt to discover the parallax of some zenith stars. It was with the intent of verifying some of these early claims regarding the detection of stellar parallaxes that Bradley began monitoring the positions of a few stars, when he noticed that they were not culminating at the same altitude on successive days, but were all displaced toward the south. After a few months, the stars began to climb again, proceeding further to the north and returning to their original positions after one year. The total oscillation amounted to some 20 arc seconds and it had clearly been below the discriminating power of earlier observations, that is to say, the random oscillations in the errors of those observations were at least of that same magnitude and were disguising this small systematic oscillation. After pondering about the cause of this effect, Bradley came to the right conclusion, declaring that he had detected the aberration of light caused by the motion of the Earth around the Sun. The effect is explained in entirely the same manner as our own instinctive reaction to tilt our umbrellas forward when we walk in the rain, even when the rain is pouring straight downward. Likewise, as the Earth proceeds along its orbit we must tilt the telescope in the direction of our motion, which does not coincide with the true direction toward the star we are observing. For a star directly overhead, the downward motion of the light wave, combined with the forward motion of the Earth, from the perspective of the observer, has the effect of producing a slightly tilted wave front. The amount of aberration depends on the speed of light and on the speed of the observer. Unfortunately, the speed of the Earth could only be determined from the size of its orbit, so that it was required to know the distance to the Sun. On the other hand, the speed of light had been deduced from the delays observed in the periods of Jupiter's moons and this demanded, once again, that one knew the scale of distances within the solar system, something that was still somewhat uncertain. Nevertheless, this was the first time that one had a physical proof that the Earth was in fact in motion and that the theory of Copernicus was not merely a mathematical device or a convenient way of arranging the celestial orbits. One year after making this discovery, when Bradley had already installed his own zenith sector at the observatory in Wanstead, he began to notice another discrepancy in the position of the stars passing through the zenith. For the stars in two opposite directions in the sky, their displacements relative to their expected positions were greatest and in opposite directions, whereas for those stars directly at right angles the effect was minimal. This is precisely the effect that would be produced by tilting the axis of rotation of the Earth or, if you wish, by rotating ever so slightly the Earth's globe around an axis that lies on its equator, so that looking outward, the stars in the direction along that axis remain fixed, but for those at directions perpendicular to it their displacements are greatest. In any case, the displacements observed did not amount to more than a couple of arc seconds, but it was their systematic pattern that made them interesting and that led Bradley onto the right path toward their interpretation. In his Principia, Newton had already explained the precession of the Earth's axis of rotation as caused by the distorted shape of our planet and the fact that the Sun's attraction exerts a torque on it. Bradley realized that the gravitational pull from the Moon on the uneven Earth would have a similar effect. The Moon is much less massive than the Sun but compensates somewhat by being so much closer. The orbit of the Moon intersects the ecliptic at two points, the nodes, which slowly rotate backwards along the orbit. This is tantamount to saying that the plane of the Earth–Moon system, slightly inclined with respect to the ecliptic, wobbles like a fallen coin before coming to rest. It is this cycle, already identified by the Babylonians, that takes eighteen and a half years to complete itself. The Moon attempts to drag the Earth's axis of rotation parallel to this shifting plane, just like the Sun drags it parallel to the ecliptic. Of course, on average, the plane of the Moon's orbit coincides with the ecliptic. Bradley continued to monitor the displacement of the stars and waited for nineteen years to verify that the stars would return to their original positions, as indeed they did. Only then did he published with great satisfaction his discovery of this already anticipated phenomenon, to which he gave the name of nutation. The total displacement of the stars in this case did not amount to more than 9 arc seconds. Along the way, Bradley did not forget to lavish praise on George Graham, as he had already done previously, the superb craftsman whose instruments he had the privilege of using. In addition to these two great discoveries, Bradley accumulated throughout the years thousands of precise observations for an eventual star catalogue. When he moved to Greenwich to take the position of Astronomer Royal, he refurbished the instruments left by Halley and introduced some of his own, raising their level of accuracy to an unprecedented level. The mural quadrant that was formerly affixed unto a wall, was now adjustable and monitored constantly. So too were the other instruments, the transit telescope and the zenith sector, as well as the various clocks. After years of observing, Bradley realized that many of the star positions he had determined were probably erroneous, mainly due to the effects of temperature and pressure in the atmosphere and he began the practice of keeping records of the weather at Greenwich, alongside the records of his astronomical observations. He had reached such a level of precision that he began to notice that the instruments themselves were highly sensitive to temperature and that their alignment and accuracy could be affected when the came into contact with the observer.

Plebeius: It is perhaps not very illuminating to try to discern what advances in the sciences are brought about as windfall from attempts to solve practical problems and in what cases it is a purely scientific argument that spurs the development of a new technique, which may then find multiple practical applications. I am thinking of the efforts to develop a clock that was stable and seaworthy, a tool badly needed by the navies of that era and that must have stimulated the ingenuity of many would-be inventors. But, at the same time, you mentioned the desire to confirm the theory of gravitation in its finer details and this required a precision in the observations and in the construction of instruments that was a challenge to the watchmakers and artisans of the time. It is probably fair to say that for the larger part of the history of astronomy, the tools used were relatively simple and the ability to conceive theories always outstripped the capacity to conceive tools that would confirm them. Now, for the first time, the instruments in use were acquiring a sophistication and complexity that required as much attention to themselves as the objects for which they were used. It is interesting that a tool, designed to accomplish a certain task, may pose its own problems or raise unanticipated questions, diverting our attention and leading us onto a discovery different from the one we intended, as if we were the tool's instrument and not the other way around. We should also point out that these scientific advances, in theories as well as in techniques, were taking place in a social climate that had changed quite drastically from decades past. The early burst of interest in the sciences that had accompanied the first telescopic observations or the invention of the microscope was restricted to a rather small and affluent segment of society. As we discussed earlier, the men that came to form the Royal Society and other such institutions in Europe belonged to an enterprising class, cosmopolitan and well educated. The aristocratic Edmond Halley may be a good representative example. Particularly in England, the disruptions of the Civil War fifty years earlier, or perhaps the plagues that ravaged the country in the 1660's, blunted any motivation or inclination toward self-improvement in the majority of the population. Then, within a few decades, the circumstances were completely different. The Royal Society was struggling for lack of funds, but the citizenry at large was motivated with a new spirit of enterprise and initiative. There was a growing class of small business owners, professionals and craftsmen, animated by a desire to learn and to innovate. This was a fairly widespread phenomenon in England and in Scotland. It may not have had any immediate influence on the evolution of the sciences;

nevertheless, the degree of participation and enthusiasm on the part of the public is attested by the establishment of public libraries in towns large and small, by the countless public lectures, as well as by the large number of clubs and coffee-houses where people used to gather for discussions and debates. Should one consider this awakening as the expression of a democratic society eager to ameliorate its own condition? Or was it instead that all this curiosity and inquisitiveness were rather superficial and ephemeral as some contend, stimulated perhaps by the many reports and goods brought from the four corners of the world, from those distant and exotic lands that these diminutive European countries, so it seems by comparison, had now set out to roam and conquer? It is also said that one should look for the true underlying causes that were transforming this society, even without it being aware, either the rapid growth in population or the introduction of the steam engine in coal mines, which brought with it plentiful fuel and a machine that revolutionized industry, or perhaps it was some other change in the material conditions of life, such as the expansion of trade or the spread of banks and the easing of credit. Of course, many of these things must have acted in concert, but it seems to me undeniable that there was also a new attitude in the society as a whole and the image it had of itself, perhaps as a consequence of comparing its own predicament to that of other societies with which it had not had contact until then. By all measures, it was now more optimistic and confident and it did not see itself in the hands of superior historical forces; quite the contrary, it considered the human experiment as a progressive undertaking and it felt it had the capacity to steer it. The knowledge it sought was of a more practical nature. Nothing illustrates these changes more eloquently than the monumental Encyclopaedia compiled in France in mid-century. When you consult any of the encyclopaedias that had been published until then, you find that the vast majority of them was devoted to present the genealogy of kings or their military campaigns, or perhaps to introduce the Biblical texts or to recount the myths of the Greek gods. Now there was a new encyclopaedia that, as it title expressly indicated, was a reasoned dictionary of the sciences, arts and crafts. Its entries ranged over a large number of subjects, from philosophical ones, such as a discussion on human nature, to very practical ones, such as the art of masonry, with profuse illustrations of all the tools in the trade. The change in the mental landscape of that age is truly remarkable. Clearly, the fact that there was now an optimistic climate and that a practical and resolute bent prevailed in society does not imply that the conditions of life changed dramatically overnight. Most towns were still relatively small, agriculture still took the largest share of economic activity and the burgeoning industry was still centered in small workshops, not having yet undergone the transition from hand to machine production. Then, a few short decades later, this world would transform itself again and the changes, the commotion that came about could not have been foreseen, in particular in those northern English cities that were soon turned into little infernos, greyed by rows of smokestack industries employing a new, poor population in bleak and tedious jobs. In the end there were larger forces at play that proved to be overwhelming. We can no longer speak of a society in control of its own fate and what had been a promising project in social organization was torn asunder; however, examining the conditions that gave rise to the original optimism, it does not seem to me that it had been misplaced.

Albertus: But we should also remind ourselves, Plebeius, that this era full of promise, as you call it, was also a bellicose one, plagued by those senseless wars of succession, like the war in Spain that entangled almost every other country in Europe or, a little later, the quarrels over the Austrian crown. Hardly a year seems to have gone by without some major conflict raging in Europe. The dispute over Saxony, the contention about Silesia, plus the incessant skirmishes between England and France over their colonies, not to mention the seven year war with Prussia that must have been so costly and disruptive to all. The optimism you spoke about was no doubt tempered by much turmoil and insecurity.

Plebeius: These were rather belligerent times, without question, but it is also true that many of these conflicts were wars of maneuvers, in which strategy and the adoption of defensive positions counted more than fighting cruel battles. In addition, the dynastic wars were carried out for the most part by relatively small professional armies with little disruption for the civilian populations. Even in the major confrontations, a code of war did prevail, the cases of pillaging were rare and the treatment of prisoners quite humane. This was not true of the wars fought in the east of Europe; the war between Russia and the Ottoman Empire, for example, was brutal and without rules. Plunder, wholesale killing and everything else in the inventory seemed to be allowed. But the situation was different in the West. Of course, one cannot speak of civility in war, it is always about destruction and carnage, but I should say that in those times it was largely confined to the warriors and there were boundaries that determined what conduct was admissible and what was not, something that we have come to see less often in our own times.

Albertus: It must be that spirit of civility that prevailed in the end and did not disrupt, except in a few instances, one of the first examples of extended international cooperation in the pursuit of scientific goals. Allow me to come back to the still pending problem of the distance to the Sun and the transits of Venus, so much advertised by Halley as the best opportunity to solve that riddle, transits that were now approaching, the first one in 1761, in the midst of the Seven Year War. During the first half of the century astronomers had not really abandoned this problem. They attempted to measure Mars' parallax during one of its opposition and they also took advantage of the more frequent transits of Mercury. The French astronomer Nicholas de Lacaille had traveled to the Cape of Good Hope to observe the transit of 1753 after having sent his assistant Joseph Jerome Lalande to Berlin, approximately at the same longitude, to make simultaneous observations, in addition to those that would be conducted from Paris. The results, despite their efforts, proved to be disappointing. They managed instead to determine the parallax of the Moon with increased accuracy. Thus, as Halley had anticipated, renewed hope was placed in the upcoming transits of Venus. In the years leading to it, another French astronomer, Joseph-Nicholas Delisle, proved to be the main promoter and organizer of expeditions, even during the years that he spent in St. Petersburg, where he had been lured by Peter the Great, then eager to assemble a court of scientists and scholars. In those years, numerous studies were conducted to determine which were the best locations to observe the transit, where the ingress of the planet onto the surface of the Sun would be observable sufficiently high in the sky, where its exit would be clearly visible, which locations would afford a complete view of the transit from beginning to end and, of course, much effort was devoted also toward studying the climate of each of those locations, to determine their latitude and longitude and to consider finally the difficulties of travel or the possibilities of setting up temporary observatories. The French were certainly the best organized; they dispatched three expeditions, to the island of Rodriguez in the Indian Ocean, to Siberia and to India. They also strove to convince their colleagues abroad, in the United Provinces, England, Spain and the American Colonies, to make every effort to observe the transit too. Cassini of Thury, from the Paris Observatory, the grandson of Gian Domenico, traveled to Vienna to assist in the observations from that city. The British, despite Halley's admonition, were late in their preparations and had difficulty collecting the large number of instruments necessary to disperse in several expeditions. Nevertheless, they organized one to the island of Sumatra, which was later cut short and sent to the Cape of Good Hope. Nevil Maskelyne, on the other hand, the Astronomer Royal who had succeeded Bradley, traveled to the island of St. Helen. All in all, the transit attracted enormous attention and it was the subject of much commentary and debate in the popular press. The lengthy expeditions required considerable investment of resources and, if this was not enough, both the French and British expeditions, given that the two countries were in a state of war, in several instances suffered at the hands of each other's navies despite the warnings and recommendations for safe passage. One imagines that those captains, roaming the coasts in far flung places in search of a good booty, would not be persuaded too easily by the assurances that these expeditions were devoted to worthy causes that would benefit all mankind, as their documents stated. In the end, the observations were very much a disappointment. Even from those places that had enjoyed excellent weather, much confusion was created by the fact that the planet had not been seen to enter onto the Sun cleanly. Different observers had different views as to when the moment of first contact had occurred, or when complete immersion had taken place. Obviously, the Sun did not have a well defined edge and the irradiation from its very bright surface around the contour of the small planet had caused greater difficulties than anticipated. The timings of ingress and exit differed by up to several seconds, which resulted in a rather ambiguous solar parallax. From the multitude of observations and after much study, the values obtained oscillated between 8.28 and 10.60 arc seconds. The second transit, to occur in 1769, aroused as much interest as the previous one and numerous expeditions were organized. Just as on the previous occasion, approximately sixty stations reported having observed it. The Russians traveled to Siberia, the British went to the recently discovered island of Tahiti and the French set up observing posts in the island of St. Domingo, in California and in India. From California they did manage to obtain valuable results although several members of the expedition died of pestilence, while Guillaume Le Gentil, the French astronomer stationed in Pondicherry, India, could not see the transit because of cloudy weather. Le Gentil had been in charge of the expedition to India in 1761 but, on that occasion, he had been unable to record the transit because Pondicherry had fallen in the hands of the British. Indeed, he saw the transit from his ship offshore, but could not make any use of his instruments. Afterwards, he decided to stay for some time in India to study its culture and the history of Indian astronomy. Eventually, as the transit of 1769 was only a few years ahead, he desisted from returning to France and made plans to observe the transit in Asia. He traveled to the Philippines and found that it would be an excellent location to set up an observation post. However, he received word from France that he should return to Pondicherry and set up his instruments there. So he did, only to be frustrated by bad weather, while in Manila, as he learned later to his chagrin, there had been an unobstructed view of the entire transit. Thus, Le Gentil spent more than a decade away from France without ever observing a transit with his instruments. To add insult to injury, by the time he returned home, it appears that his family had already divided his inheritance and the Académie des Sciences no longer counted him as an active member. Nevertheless, he spent many years thereafter writing several volumes on his experiences, on the natural sciences in India and on its astronomical tradition. This was part of the return from all these expeditions, a bounty of information on the flora and fauna of distant lands and the customs and history of other peoples. Such was the case also of Alexandre Gui Pingré, who had traveled to the island of Rodriguez and to St. Domingo for the second transit. Pingré was an exceedingly prolific writer and left detailed accounts of his travels. This was not unusual at the time; Joseph Jerome Lalande, another astronomer I have already mentioned, also filled lengthy volumes with accounts of his trips, in addition to his very well received textbooks on astronomy. But Pingré wrote on a variety of subjects, even more disparate. A volume of his retraces all the comets that had been seen throughout history; he also wrote on the major astronomical achievements of the century, on literary criticism, on liturgy and some Latin poetry as well. Most of his writings remained in manuscript form, I believe, and are still hidden in some library awaiting a curious archaeologist of a future age. In any event, regarding the measurements conducted during the second transit of Venus, the observers were better prepared this time and the value they obtained for the solar parallax was also better constrained, not differing by more than two tenths of a second of arc, centered at 8.7 arc seconds. Thus, the distance from the Earth to the Sun was finally set at approximately 23,000 Earth radii. This was certainly not the end of the attempts to determine the distance to the Sun and, throughout the centuries, every new advance in instrumentation prompted new efforts to determine this distance more precisely. However, since we are not so much interested in precision but in estimating the order of magnitude of the distance, let us say, to within an error of a few percent, it can be stated that the transit of Venus more or less settled the problem of the scale of the solar system. Now we must turn our attention further afield and consider how astronomers strove to determine the distance to the stars. As we do so, we shall set aside and ignore one of the glorious chapters of astronomy, for the second half of the XVIIIth century saw the culmination of what came to be known as physical astronomy or celestial mechanics, namely, the application of the law of gravitation to the motions within the solar system. Even before these advances, the theory could already count other successes. It is interesting to note, regarding Bradley's discovery of the phenomenon of nutation, that he anticipated its period of 18 years; while his interpretation was not entirely correct, nevertheless, he made a justified use of the predictive power of the theory of gravitation, which by this time was fairly well established. Fifty years earlier, it could have been argued that the law of gravitation merely provided a suitable synthesis for the interpretation of facts that were already known, such as the main features of the Moon's orbit or the precession of the equinoxes or the outline of a comet's orbit. Furthermore, it also seemed as if the theory still had to reach the level of precision in practical computations that the astronomer had already achieved through his observations. Now, however, the theory had been used to anticipate something that was yet to be observed. A similar instance occurred some years later, in 1774, when a similar prediction was confirmed experimentally. According to the theory, the gravitational attraction is exerted between any two arbitrary masses, so that under the appropriate circumstances, one should be able to detect the side attraction of a mountain on the surface of the Earth. This is what Nevil Maskelyne, the Astronomer Royal who succeeded Bradley, set out to prove. The experiment was simple in conception but required a delicate execution. Maskelyne performed it next to Mount Schahallion in Scotland, which has the peculiarity of being sufficiently symmetric and largely isolated. The task involved determining the position of the vertical with a zenith sector at opposite sides of the mountain and measuring the corresponding displacement caused by the attraction of the mountain on the hanging telescope. In this instance, the total displacement amounted to 11 arc seconds, while the instrumental error tolerated was below 1 arc second. The experiment also allowed for a fairly accurate estimate of the density of the Earth, by comparing the relative volumes of the mountain and the Earth as a whole. Here again we have an example of the predictive power of a theory that leads to some new knowledge, not accessible by other means. After the fact, the experiment appears quite straightforward, but without the guidance of an underlying theory, the idea of measuring the zenith's position at two sides of a mountain would have seemed extravagant and without purpose. The problems posed by the orbits of planets, comets and the Moon within the solar system were of great mathematical complexity and they engaged some of the more gifted minds of the time, such as Leonhard Euler, Joseph Louis Lagrange and Pierre Simon Laplace. There was, for instance, the problem of reconstructing the orbit of a comet in a three dimensional space from a few positions on the surface of the sky. The trajectory followed by the Moon in its orbit is complicated enough to have kept busy generations of astronomers. In fact, not everything had been in accordance with the predictions of theory from the beginning. Newton had calculated the the rotation of the line of apsides of the Moon's orbit had to occur three times as fast as it was observed and it took decades before the miscalculations were explained. Similarly, Halley had noticed, by studying old records of the motions of the Moon, that our satellite had been accelerating in its orbit throughout the ages, something that seemed to contradict the predictions from the theory of gravitation. The evidence for this acceleration was so conclusive that it led temporarily to seek a replacement to the inverse square rule for the law of gravitation. Some speculated that a diffuse aether permeated space, which in turn had forced the Moon closer to the Earth, speeding up its pace. All these things we shall set aside; however, I wish to mention only one small detail as a curiosity. By the time Laplace came to study these problems in celestial mechanics, the methods used to reconstruct orbits from the observations and the elucidation of the perturbations by one planet on the orbit of another were fairly advanced. Studying one perturbation of the Moon's orbit that depended on the angular separation between the Sun and the Moon, Laplace found that the best agreement with the observations could be found assuming a solar parallax of 8.6 arc seconds. Undoubtedly, he felt very proud stating that he could determine the distance to the Sun from the comfort of his office in Paris, without having to travel to the tropics to observe the transit of Venus. Turning now to the daunting problem of determining the distance to the stars, you recall that Tycho Brahe had attempted to measure the annual parallax of a star in the hope of establishing whether the Earth orbited the Sun or not. Unable to do so and believing that his instruments would have detected a displacement of the order of a minute of arc, he could only conclude that, if Copernicus' theory was correct, the stars had to be at a distance at least 10,000 times the distance to the Sun. Johannes Kepler had estimated distances only using arguments derived from harmony and, as far as measurements was concerned, he thought the angular diameters of stars were as large as 2 arc minutes. However, Jeremiah Horrocks had used the telescope to observe the occultation by the Moon of some bright stars in the Pleiades and, in his view, their disappearance had been instantaneous. The stars were point-like and not resolved, he concluded. Years later, Robert Hooke had attempted to measure a parallax without success and so too had Flamsteed who, with greater means, had seen the seasonal displacement of a star, not realizing that he was observing the effects of aberration and not parallax. Christiaan Huygens had thought of a very ingenious and simple method to estimate the distance to the stars, on the assumption the they were comparable to the Sun in absolute luminosity. He let the Sun's light into a dark tube through a small aperture, which he then reduced in size until the light he could observe from the other end of the tube was analogous to a bright star like Sirius. He then estimated that the aperture was letting in only approximately 1/30,000 of the diameter of the Sun. Therefore, he concluded, the Sun would have to be removed to 30,000 times its present distance to resemble the other stars. The difficulty with this strategy is that Huygens could not have made a measurement of such a small aperture. Even if his tube was one hundred meters long, he would have required an aperture of less than a tenth of a millimeter in diameter. Isaac Newton, some years later, described a similarly ingenious method, published posthumously in his work entitled System of the World, a method that had been conceived several decades earlier by to the Scottish astronomer James Gregory. From the angular diameter of Saturn, as seen from the Sun, he could deduce the amount of solar light that was intercepted by the planet at any one time. Then, allowing for some absorption of the light and assuming that three quarters of it was reflected, he could estimate, using the distance that separates us from Saturn, the fraction of that reflected light that was intercepted by an observer on the Earth. This, Newton concluded, was approximately the same as would be observed by removing the Sun to a distance 100,000 times further away. Since the apparent luminosity of Saturn is only slightly brighter than a star of first magnitude, he deduced that this was probably the distance to the nearest stars. These numbers were indeed quite reasonable, even if they fall short of the true values. Nevertheless, the assumption that the Sun was a typical star was at the time merely a bold hypothesis and, needless to say, an estimate derived from theory could not replace a direct measurement. Amongst those who set out to measure the parallax of a star I must mention William Herschel, a self-taught astronomer who had turned to science rather late in his life, when he was well into his thirties, although he then dedicated to it the larger share of his energies for the remaining forty years he was to live. In the end, he never managed to measure a stellar parallax. Nevertheless, his extraordinary vigour and originality and his devotion to astronomy brought about nothing short of a revolution in this discipline. We cannot say what would have been the course of events had he not entered the scene, but it is probably no exaggeration to say that he advanced the subject by as much as the rest of mankind would have accomplished in a century. He was born in Hanover, in Germany, in 1738 but he moved permanently to England at the age of nineteen. He had visited the country earlier and having already learnt the language, when Hanover was taken over by the French during the Seven Year War, he decided to avoid being drafted in the army and moved abroad. He settled down in the city of Bath, where he was offered the position of organist at a local church and, for many years, he earned his living as a musician, composing and teaching. A moment ago you were describing the inventive and enterprising spirit that gave this era its optimistic glow; I do not think one could find a more representative example than William Herschel, for he had an insatiable curiosity and seems to have been always overflowing with activity. It was his studies in musical harmony that apparently led him to learn more mathematics and geometry which, in turn, directed him to optics and finally to astronomy. In any event, Herschel was soon doing observations of his own and, being dissatisfied with the quality of the telescopes he was using or, at least, those he could afford, he decided to build a reflector telescope with his own hands. In the spare time from his music lessons he learnt the art of polishing mirrors and very soon he had some very good quality instruments with which he began surveying the skies. In all, throughout his life, he must have polished several hundred mirrors, which he often sold to supplement his income. It is not clear what strategy he followed in his astronomical observations, but he gives the impression of having had the desire to observe and record everything. His earliest achievement, after some three years of observing, was the discovery of what he at first believed to be a comet, but turned out to be instead the planet Uranus, twice as far removed from the Sun than Saturn and the first planet to be discovered since the most remote antiquity. This accomplishment brought him enormous fame and a number of accolades, amongst these a recognition from King George III, who awarded him a life pension and an invitation to move to Windsor, so that he could entertain the royal family from time to time. The pension allowed him to give up his music lessons and dedicate himself fully to astronomy. By this time, he had brought his younger sister Caroline from Hanover, whom he also steered away from music and into astronomy. They both settled in Slough, near Windsor, where Herschel would remain for the rest of his life. Indeed, he traveled very little; only twenty years later, for example, did he travel to France where he met some of the more distinguished scientists, amongst them Laplace, and where he was also received by the then First Consul, Napoleon Bonaparte who, Herschel found, was not as informed on astronomy as the King and whose general air was something like affecting to know more than he really knew. The attempt to measure the parallax of a star involved the old method, suggested by Galileo, of monitoring the relative positions of a bright star and an adjacent fainter star, under the assumption that the brighter star was much closer and, thus, as the Earth circles around the Sun, it would be seen to perform a circle around the fainter one in a year. With this in mind, Herschel sought stars that were not farther than 2 arc minutes from each other, so that if they were visible to the naked eye at all, they would appear almost as a single star. One year after his discovery of Uranus, he already published a catalogue containing some 260 of these double stars. Another related problem that attracted his attention during his early years as a professional astronomer was the proper motion of stars, of which quite a few were known by that time. Here Herschel did not carry out the original observations, but gathered those of other astronomers, such as Nevil Maskelyne and Joseph Lalande, and applied them to the task of determining the motion of the solar system. This, once again, had already been suggested by Bradley, amongst others, but Herschel was the first to solve the problem satisfactorily. The principle involved is not different from the observation we all make when traveling along a straight road and see everything around us, the road, the trees and the clouds vanish behind us at one point in the horizon. If we then place at each and every star an arrow pointing in the direction of its motion and these arrows tend to converge to one point in the sky, we may suspect that it is not due to their concerted motions. Instead, it must reveal our own displacement, along with the Sun, in the midst of all the stars. Of course, the stars could be moving on their own, which makes the task somewhat more difficult, but Herschel was able to identify the correct direction in the constellation of Herculis toward which our solar system is moving and he also attempted to estimate the magnitude of this motion, something that he could not really do, given that the average distance to the stars was still unknown. Nevertheless, using the best guesses then available and estimating the displacement of the nearest stars to be approximately one arc second a year, he concluded that the solar system traveled in a year a distance comparable to the distance to the Sun from the Earth, a value very close to what came to be accepted later. There were several other projects that Herschel undertook and that he carried out more or less simultaneously over many years, often returning and adding or correcting something that he had done earlier. For example, three years after his first catalog of double stars he published a second one, containing some 400 additional stars, and he still added a few more many years later. Almost half of his writings concern matters related to the solar system and I will not say anything about them because they are the less interesting ones. He was of the opinion that the heat from the Sun comes from its atmosphere, whereas the Sun itself he thought to be colder and habitable. Herschel had not been trained as an astronomer and early in his career, in particular, his style was somewhat exuberant and very idiosyncratic. Later he became more circumspect and not so distinct from the professional astronomers whose approval he sometimes courted. Of his great accomplishments one must first mention the great telescopes he built. Extending the magnification power from 400 up to 6000, they were much better instruments than those available at the Greenwich Observatory or anywhere else. The most successful of his telescopes was a 6 meter (20 foot) long reflector that he used with mirrors of 30 and 48 cm (12 and 19 in) diameter. For the giant 12 meter telescope with a 1.47 meter (57 in) mirror, financed by the King, unable to find a workshop that would cast the mirror according to specifications, he set up a foundry and, after some mishaps, manufactured it himself. But this telescope proved to be too cumbersome, the mirror deformed under its own weight and it never lived up to the expectations placed in it. One of Herschel's early undertakings was a long project to map the three dimensional distribution of the stars or, as he put it, to describe the construction of the Heavens. Ever since the first observations with a telescope, it had become apparent that the Milky Way consisted entirely of stars. It was only the extraordinary condensation of so many stars along these peculiar directions in the sky that gave it its uniform appearance and justified that we call the Milky Way by its name or, as the Chinese would call it, the Silvery River. Already some astronomers and philosophers, such as Immanuel Kant, had considered the distribution of the stars and its lack of uniformity and had speculated whether we were immersed in a rather flattened disc of stars, itself immersed in an infinite void. Herschel set out to map this structure. To simplify a very complicated task, he first assumed that the stars were all of equal brightness and, furthermore, that their spatial distribution was uniform. We might encounter a similar situation when we go for a walk in the woods; how do we size up its dimensions? If we see fewer trees in one direction, it may mean that the forest is coming to an end, or that it continues indefinitely but the trees are becoming more sparse. Similarly, if the trees appear smaller, is it because they are farther away or because they are intrinsically smaller? Both assumptions, the uniformity in the distribution and the uniformity of brightness, were known to be false in the strict sense. From his catalogue of double stars, Herschel was very much aware of the unlikely coincidence of so many pairs of stars that seemed to have fallen on the same spot in the sky. This was simply a matter of statistics. If you are about to place a star in the sky at random, the surface area of those locations that are somewhat removed from other stars is fairly large; the locations that are at most an arcminute from another star is infinitesimal by comparison, despite the large number of stars. Therefore, the chances for a star, any star, to be contiguous to another star ought to be infinitesimal, but indeed this is not what we see. There were hundreds of such pairs. They could not be lucky alignments along the line of sight, they had to be true physical associations. After much observing, some twenty years later, Herschel could produce convincing evidence of six such pairs of stars, where the stars were circling around each other, with periods that he estimated to be of the order of a few hundred years. Hence, the original goal of determining the parallax of a star transformed itself into another unexpected discovery. It was also the case that the stars in many of these pairs were very different in luminosity and, given that they happened to be at the same distance from the Earth, one had to admit that not all stars had the same intrinsic brightness. As for their uniform distribution in space, even disregarding these close associations in doublets and sometimes triplets, it was clear that there were other less tight but nonetheless close groupings of stars, amongst the Pleiades for example, where there are many stars of similar brightness, apparently in greater proximity amongst themselves than the Sun is to any of its closer stars. Despite these irregularities, Herschel proceeded along, counting the number of stars in various sectors of the sky. If you decide, let us say, that the typical distance between two stars is the distance between the Sun and Sirius, then you can scatter roughly a dozen stars, equally distant, on a sphere around the Sun. On a sphere twice the radius, you can scatter four times as many stars. Then, if you triple the radius, there will be nine times as many stars and so on. Looking into your telescope, you sweep through a cone that catches always the same fraction of each sphere, so that you can estimate how many stars you ought to see up to a certain distance, or to the discerning power of the telescope. In this manner Herschel sketched an approximate map which revealed that, in directions that pointed more or less along the Milky Way, there were stars up to 800 times the distance to Sirius, whereas their distribution, in directions directly perpendicular to it, did not extend beyond 150 times that distance. An indirect consequence of these observations was the fact that the solar system was not too far from the center of the star system. Let me now turn to the last of the great contributions of Herschel. Quite early, in fact, he had received a catalogue by the French astronomer Charles Messier, containing the locations of some one hundred nebular objects in the sky. Messier was primarily a comet hunter and he had provided this list for the benefit of those who shared his interest and might confuse these objects of fuzzy appearance with actual comets. Herschel soon realized that, with his superior telescopes, he could observe many more similar objects and he could also begin to resolve some of them into their constituent stars. This launched him on another project, the study and cataloging of the so-called nebulae. Buoyed by his initial discoveries, he stated that all nebulae were condensations of stars which he could resolve, or nearly resolve, with his telescopes. In a few years Herschel presented to the Royal Society his first catalogue of nebulae, which contained approximately a thousand objects. Other catalogues followed throughout the next decade with some fifteen hundred additional nebulae. His analysis of some of these nebulae was essentially correct, particularly in regard to what have come to be known as globular clusters, round concentrations of thousands or hundred of thousands of stars with very dense cores, the outskirts of which Herschel could resolve in their individual stars. On the assumption that the stars were of average luminosity, he concluded that these clusters were located at about six hundred times the distance to Sirius, at the edges of what he thought was the Milky Way. He correctly pointed out that, had he considered the distance between any two stars in the cluster to be typical, he would have had to place the cluster much farther away, at six thousand times the distance to Sirius but, of course, in such a case all its stars would have been unusually bright, a hypothesis he discarded. This was all very sound. He also felt inclined to speculate on the evolution of the Universe, observing that perhaps in the beginning all stars had been widely scattered and that the force of gravitation had gradually created these agglomerations over very long periods. He also stated, like others before him, although with more justification, that some of the unresolved nebulae were probably island universes very much similar to the Milky Way and he did not fail to notice that they were placed in regions of the sky preferentially away from the Milky Way. According to his estimate, they were probably at two thousand times the distance to Sirius from the Earth. Despite the fantastic enlargement of the Universe that his discoveries had brought about, here he fell very much short of its true scale, barely accounting for one percent of the distance he was trying to appraise. As the years progressed and Herschel observed many more nebulae and each in greater detail, his opinion regarding their constitution gradually changed. Like a few other astronomers, he had seen the great nebula of Orion change its appearance in a matter of years, which indicated that it was most certainly gaseous. He was also the first to describe another type of nebula, which contained a bright central star with a small, round, dimmer aura surrounding it. He came to the sensible conclusion that, if these were not gaseous nebulae but unresolved distant clusters of stars, then the central objects were of exceptional brightness, unlike anything else seen nearer in the Milky Way. In any event, the great variety of objects that Herschel described defied any easy classification. He had seen stellar nebulae and nebulous stars, milky nebulosities and planet-like nebulae, nebulae with bright centers and nebulae without them. The acceptance of the existence of true nebulosities also modified his ideas regarding how the Universe had come to be what it is. He thought now that, in its original state, all matter had probably been in a gaseous state out of which, by the effect of gravity, the stars had formed, fast in some places, more slowly in others. This was probably the first time that someone was using scientific arguments to advocate an evolutionary view of creation, contrary to the prevailing view of the time, largely inspired by religion, that saw the Universe for the most part as a finished invention. Herschel could now contemplate the Milky Way as a luxuriant garden, which contains the greatest variety of productions in different flourishing beds, and one advantage we may at least reap from it, he wrote, is that we can, as it were, extend the range of our experience to an immense duration. Continuing this simile from gardening, he asks: is it not almost the same thing, whether we live successively to witness the germination, blooming, foliage, fecundity, withering and corruption of a plant, or whether a vast number of specimens, selected every stage through which the plant passes in the course of its existence, be brought at once to our view? It was in this vein that he spoke of the Milky Way as a mysterious chronometer and the same idea has been flirted with, with greater or lesser success, ever since. Herschel's contribution to astronomy was enormous and long lasting. As an individual, he seems to have been good natured and of a friendly disposition. He married at age fifty, at which time his sister Caroline left the house, although she continued to collaborate with him in his work. From his marriage only a son was born, John Herschel, who continued his father's illustrious work and was a distinguished scientist in his own right. After William Herschel's death, Caroline decided to return to Hanover, where she died, some twenty-five years later, almost a centenarian and not without having received her share of honours and recognition. Herschel's successes gave new impulse to the use of reflector telescopes. Introduced a century earlier and having undergone various modifications and improvements in design, notably by Newton and by Herschel himself, reflectors gradually became an important tool for astronomical research. Originally, their appeal had been based on the fact that refractors presented some shortcomings difficult to overcome. It was difficult to produce large pieces of glass that were sufficiently smooth and uniform, so that object lenses ended up being smaller than was desirable, and the aperture of the telescopes remained always too small. Furthermore, these instruments were also plagued by the aberration of the lenses, chromatic aberration, due to the different index of refraction for each color, and spherical aberration, which tends to focus parallel light rays at different distances depending on their angle of incidence relative to the central axis of the lense. All this tended to produce very blurred images. Astronomers had resorted to increasing the focal length of the telescopes, demanding in this way less bending of their lenses and producing thereby less distortion, to the point of placing the object lense so far removed from the eyepiece that they neglected to enclose their apparatus in a tube, although now it was the air currents in circulation that tended to distort the image in the focus. Reflectors had the additional advantage that their large mirrors could collect more light and, consequently, they detected fainter objects and in greater detail, probing deeper into space. This capacity made them most suitable for the type of study that Herschel had undertaken. By contrast, reflectors were more difficult to align and, for the astronomy of precision, when the task was to place a point-like source of light in the sky, refractors performed their work more effectively. Moreover, by the time Herschel was assembling his own instruments, the fortunes of the refractor telescope were being revived with the development of the achromatic lense. Newton had stated that dispersion is an unavoidable consequence of refraction, implying that the hope for a lense without aberration was probably a futile chimera. Others, amongst them the mathematician Leonhard Euler, taking a cue from the structure of the human eye which is apparently free from aberration, worked toward that goal, experimenting with compound lenses or lenses immersed in water, but seldom successfully. The task at hand seemed to have more to do with experimentation and perspicacity than with any theoretical understanding of optics. John

Dollond, an instrument maker based in London, was the first to obtain a patent for a successful achromatic lense. Although it might be an exaggeration to say that he had stolen the idea, it appears that he hid the fact that an encounter with an English lawyer and amateur astronomer, who had been successful decades earlier, did put him on the right path, using a combination of a convex lense made of regular crown glass and a concave lense made of flint glass. Be that as it may, Dollond and his son soon developed a very profitable business building instruments of excellent quality. Years later, once the patent had expired, other London optician joined in the competition, taking advantage of the great demand, in particular from the British Navy, interested mainly in sextants and portable telescopes. For many years the industry remained almost exclusively British, for the government controlled very carefully the trade of glass and saw to it that the secrets of the production of flint glass were not divulged. George Graham, the craftsman who supplied Halley and Bradley with an assortment of very fine tools, had initiated in England the tradition of producing top quality precision instruments, a tradition that was continued by such men as John Bird, whose great ability was to engrave a scale on a circle to unprecedented accuracy. This expertise of graduating an instrument and of providing it with a stable mounting was now complemented by the quality of the optics, which insured that Britain retained, for the time being, a pre-eminent position in the manufacture of the best quality telescopes. The craftsmen involved acquired all the reputation and prestige that they deserved. One of the more prominent ones was Jesse Ramsden, who had married the daughter of John Dollond. His workshop employed several dozen artisans, but the production of his more elaborate instruments was a very slow process all the same. It took him twenty years to complete a transit telescope for Nevil Maskelyne, the Astronomer Royal at Greenwich. A few other of his excellent instruments, which facilitated a number of discoveries in the next generation, were only finished by his associates after his death. Ramsden was elected Fellow of the Royal Society and awarded its prestigious Copley Medal for his contributions to science. In France, Jacques Cassini had sought the help of Minister Colbert for the development of the glass industry without much success. In the end, the zeal with which Britain kept the secrets of its glass manufacturing and the way it prevented the free trade of glass across international borders turned to its disadvantage, when glass of better quality began to be produced in the continent, leading to the development of an industry of precision instruments, predominantly in Germany, that remained unequaled for most of the next century. At the very start, this had been the enterprise of a single man. A Swiss cabinet-maker by the name of Pierre Louis Guinand, in the town of Les Brenets by the Jura, had specialized and worked for many years in clock cases when he became interested in the manufacture of glass. The woods in the Jura mountains provided him with abundant raw material for his wood work as well as fuel for the glass furnaces. With persistence and patience, he managed to solve the riddle of flint glass manufacture and, in addition, found a way to mix and stir the molten glass so thoroughly that he could then produce lenses twice the size of those common at the time and, at the same time, sufficiently pure and striae free. Guinand's work attracted the attention of a German Army officer who, in 1802, had founded in Munich the Mechanical and Optical Institute with the intention of manufacturing portable telescopes for military use. Soon an agreement was signed so that Guinand would not divulge his newly acquired expertise and would furnish all of his lenses to the Munich Institute. A young apprentice from the Institute was also sent to learn the craft at Guinand's workshop, which had now been moved to Bavaria. The youngster, Joseph Fraunhofer, had started his apprenticeship at another glass manufacturing plant in Munich, where he had been placed at the death of his father, also a glass artisan. That plant, however, had been destroyed by a big fire from which the youngster had barely escaped with his life. In compensation, the elector of Bavaria awarded him a stipend and the very young Fraunhofer soon attempted to start his own business without much success. It was then that he began working at the Optical Institute, where he not only became a superb artisan, grinding lenses up to 26 cm (10 in) in diameter, unheard of at the time, but also, through sheer dedication and a good deal of talent, he taught himself enough mathematics and physics to become a leading expert in theoretical optics and a brilliant example of the new generation of scientists that was emerging then in the German speaking countries. By age twenty-five Fraunhofer was in charge of the glass manufacturing workshop, after Guinand had decided to return to Les Brenets. Later on he would rise to become director of the Munich Institute, until he was also appointed Royal Professor of Bavaria. In the few years before his death he supervised the construction of some of the best and most sought after refractor telescopes of his era. Fraunhofer was also the first to carry out a detailed study of the refractive properties of lenses using the monochromatic light from various gas flames. It was during these experiments that he discovered the hundreds of dark lines that appear in the spectrum of the Sun, when its light is decomposed through a prism. Initially, he had used these lines as benchmarks to describe the characteristics of his lenses, but he later observed that some of the lines had their counterparts in the spectra of the gases he was using and he took the first steps in their study and classification. Because these lines would later be used to understand the physical conditions and the chemical composition of the Sun, we might call Fraunhofer the founder of astrophysics, a branch of astronomy perhaps, but due to its richness and scope, truly a new discipline altogether. The excellence of his work was sufficiently valued during his lifetime to the extent that he was exempted from paying taxes in Bavaria and, at his death, at the age of thirty-nine, he was awarded a state funeral. One of the astronomers who benefited from Fraunhofer's instruments was Friedrich Wilhelm Bessel, a few years his senior, who had the good fortune of being the first to announce, unequivocally, the measurement of a stellar parallax. A native of Westphalia, he had traveled to Bremen when still an adolescent to become an apprentice in the merchant marine but, just as had been the case with Herschel, whose interest in music led him to mathematics, optics and astronomy, so too in Bessel's case, the prospect of long sea voyages moved him to study languages and geography and the methods to determine the latitude and longitude by astronomical means. Then, reading Lalande's textbooks, he was led to determine anew and with greater precision the parameters of the orbit of Halley's comet. His computational ability impressed the Bremen astronomer Heinrich Olbers sufficiently that he soon became his sponsor and recommended him to the position of assistant at the private observatory of Lilienthal, near Bremen. Thus, Bessel passed up the opportunity to join the merchant marine and began to devote all his energies to astronomy. For the next eight years or so, he took upon himself the task of reducing the vast number of observations that Bradley had left behind some fifty years earlier. Reducing the observations entails accounting for such effects as refraction, parallax and the motion of the Earth, so that the raw data of the observer can be given an objective meaning and put to use by other astronomers. In 1813 Bessel published the results of his labour with the ambitious title Fundamenta Astronomia, although his presumption had some justification, for Bradley's observations, being the most accurate until that time, became now available as a reference for the use of the entire community and, being dated in the year 1750, would allow in subsequent decades the determination of stellar motions that had taken place over the course of one century. That same year, not yet thirty years old, Bessel was appointed director of the newly built observatory in Könisberg, where he was to work for the next three decades. During those years, with the help of an assistant, he determined the accurate positions of some 50,000 stars. Initially he used a transit instrument made by the Dollond's in Britain but, after 1819, he obtained a meridian circle from the Munich laboratories and, in 1829, three years after Fraunhofer's death, Bessel finally received a heliometer that Fraunhofer had designed and begun to build. The meridian circle was similar to a transit instrument, but mounted on a circular graduated frame that afforded greater accuracy. The heliometer was a refractor telescope that had received that name because it was used to measure the diameter of the Sun. Its object lense was cut along a diameter and the two halves of the lense were allowed to slide sideways, along their common diameter, by the action of a graduated screw, always keeping the lense on the same plane, so that out of their central position they produced two separate images of one single star on the line of sight. This was used to make the second image coincide with another star and measure in this way the separation of the two with great precision. With these instruments Bessel noticed some irregularities in the behaviour of a few stars, such as Sirius and Procyon, which did not proceed along a straight line as expected, but gently oscillated to one side and the other of their mean direction of motion, just as a sknake advances on its path. He was led to the conjecture that these stars had a dark, invisible companion and that the oscillations were caused by the motions of the stars orbiting each other. He also estimated their period of revolution to be approximately 50 years. In this manner, Bessel inaugurated what was sometimes called the astronomy of the invisible, namely, postulating the existence of objects that are not observed but help explain others that are, in a way that is consonant with the known laws of physics. The history of this tradition is somewhat checkered and it has left along the way a mixed collection of successes and failures. In 1838 Bessel finally settled to monitor the star 61 Cygnus, a rather inconspicuous star of fifth magnitude, barely visible with the naked eye, but suspected of being nearby, for the simple reason that it had one of the largest proper motions then known, some 5.2 arcseconds a year. Bessel used his heliometer to measure the relative distance from this star to those in its neighbourhood in order to see whether he could detect the seasonal effect caused by the Earth's motion around the Sun. After one year of continuous observations, he emerged victorious announcing a parallax of .3136 arc seconds. Most likely, this measurement was not significant beyond the first decimal place but it represented the first time that a stellar parallax was actually measured, as opposed to merely obtaining an upper bound. The next year Bessel dismantled his heliometer, reconditioned it and began monitoring the star again, obtaining this time a parallax of .3483 arc seconds. A few months after Bessel's announcement, two other measurements of parallaxes were reported. One was made by Friedrich Wilhelm Struve, the director of the Dorpat Observatory, who had in his possession one of the finest telescopes available at the time, a 23 cm (9 inch) refractor built by Fraunhofer. Struve followed the star Vega, which also has a large proper motion and had the advantage of being bright and near the pole of the ecliptic. Thirdly, Thomas Henderson, a Scotsman and for a time director of the new observatory established at the Cape of Good Hope, had obtained an estimate of 1 arc second for the parallax of the star alpha Centauri; although it was later determined to be just .76 arc seconds, it remained the largest measured parallax and made alpha Centauri the nearest star at 270,000 times the distance to the Sun. The almost simultaneous determination of these parallaxes reveals the effectiveness of the new generation of instruments, as well as the relative parity in the accomplishments that well trained professionals could attain with them. It does not appear, as far as we can surmise from the records, that these astronomers were competing against each other. There was perhaps a race to get the job done, but not a race to be the first. This rather petty and childish behaviour seems to have come into fashion at a later date. It had been approximately 150 years since Robert Hooke had attempted to measure the parallax of a few zenith stars without success; Flamsteed had been deceived in his belief of having made such a measurement and it was Bradley who, in repeating these attempts, had discovered the aberration of light. Even in more recent years, the astronomer Giuseppe Piazzi with his very fine vertical circle built by Ramsden, believed he had measured several stellar parallaxes, only to discover later that they had been a product of flaws in his observations. Also the Astronomer Royal at the time, John Pond, had been involved in a long dispute with the director of the Dublin Observatory who believed, mistakenly, that he had measured several parallaxes too. The achievements of Bessel and his colleagues were true landmarks in the history of astronomy and were celebrated at the time as such. After their announcements and during the 1840's, the parallaxes of several bright or fast moving stars were measured, amongst them Sirius, with a parallax of .32 arcseconds and Betelgeuse and Riegel, in Orion, with parallaxes of .02 and .008 arcseconds, respectively. Initially, some of these very small values were highly uncertain and varied, upon repeated measurements, by as much as an order of magnitude. Nevertheless, it soon became clear that the apparent brightness of a star was a poor indicator of its distance and that their intrinsic luminosities differed sometimes by factors of thousands. After Bessel's announcement of the first measurement of a stellar parallax he was awarded the gold medal of the Royal Astronomical Society and John Herschel, its president, proclaimed the achievement as the greatest and most glorious triumph which practical astronomy had ever witnessed. Today we may dispute such an assessment; nevertheless, the event was the culmination of an effort that had lasted for centuries and may still be viewed as a turning point in the history of astronomy or, at least, as symptomatic of the changes that were then taking place, with the advances in instrumentation and the general scale of the projects to be undertaken. Sixty years earlier, the French astronomer Le Gentil had found it reasonable to remain in India with all his instruments for almost a decade, awaiting the second transit of Venus after he had failed to observe the first one. In this new era, the pace at which new ideas circulated or new tasks were carried out quickened considerably. Astronomy, like the rest of the sciences, aroused greater interest and attracted a wider audience. Astronomers began to meet periodically at scientific congresses to better organize their work and periodicals were established to disseminate their findings with greater regularity, some of which have survived to this day. This was also the era of the great observatories. Those of Paris and Greenwich had existed for more than a hundred years and a few others had been established since then in Berlin, Bologna, Uppsala or Göttingen, but much work was also done independently of these establishments by devoted amateurs, most notably by William Herschel. As the XIXth century progressed, a greater share of the work was carried out within the observatories, which were now designed and built following the advice and recommendations of astronomers. Old observatories were enlarged, such as those in Berlin and Göttingen, and new ones arose in Munich, Gotha and Cambridge, in Dorpat, then Tallin, in Estonia, and in Pulkova, near St. Petersburg, this last one, since its inception, under the direction of Friedrich Struve, remained for quite some time one of the finest in the world. The awakening of an interest in astronomy in the United States led to the foundation of a large number of institutions, first in the city of Washington and at Harvard College in Massachusetts, culminating fifty years later with the greater installations at the Yerkes Observatory in Chicago and the Lick and Mount Wilson Observatories in California. Likewise, the desire to observe the southern skies led to the establishment of observatories in the southern hemisphere, such as those in Cape Town, Sidney or Cordoba. Amongst the largest new undertakings in the field of stellar astronomy, there was the task of obtaining precise positions for large numbers of stars. One such project was initiated by Friedrich Argelander, a descendant of a wealthy merchant family in Berlin, who was seduced to pursue a career in astronomy while he was studying economics in Könisberg and attending at the same time some of Bessel's lectures. This goal of obtaining precise coordinates for as many stars as possible was motivated in part by the increasing number of stars that had been seen to have a proper motion across the sky. Argelander was continuing the work already initiated by Bessel himself. First working in Turku, Finland, and moving then to Bonn after his observatory was destroyed by a fire, he set out to compile a vast catalogue of positions, an enterprise that he completed with the help of two assistants, counting some 300,000 stars. Years later, a similar survey was carried out recording more than half a million stars in the southern skies. After he had finished his Bonn survey, Argelander was the initiator of another great project to improve the determination of the luminosity of stars. Despite the many advances, astronomers were still using essentially the same system of magnitudes that Ptolemy had used and many bemoaned not having a tool as useful as a thermometer or a barometer to measure the intensity of light. Initially, Argelander merely refined the system of magnitudes by doing step comparisons amongst stars. Meanwhile, gas lamps were already becoming commonplace; they were being used, for example, to illuminate the cities at night, something that would later become a curse for some observatories but, for the time being, gas lamps were a welcome addition to the arsenal of the astronomer. They were adapted to telescopes in such a way that the flux of light from a star could be compared to a standard candle, which was varied then by changing its distance, by partial occultation or by filters and polarizers. Johann Zöllner, a German scientist based in Basel, who excelled as a designer of very fine instruments, produced the first successful photometer that was soon adopted by many observatories around the world. Thus, Argelander promoted a new survey with more accurate photometry, this time through the German Astronomical Society and enlisting the help of seventeen observatories, which busied themselves surveying different zones in the sky. Matching the visual magnitudes to the photometric measurements was not straightforward and astronomers arrived at an agreement on the appropriate scales only after lengthy debates. It was found that the visual perception is graded on a logarithmic scale, so that a change of two and a half visual magnitudes corresponded approximately to a change in the flux by a factor of ten. By the time these new photometric surveys got underway at midcentury, several decades of experiments studying the effects of light as an agent in chemical reactions were beginning to bear fruit, giving birth to the new discipline of photography. Some of the earlier successes belonged to the Frenchman Nicéphore Niepce, who introduced the silver coated copper plates and also experimented with glass plates, which were the forerunners of what we would call negatives. Niepce had started his experiments trying to make improvements in the art of lithography, also relatively new at the time, and he used the term heliograhy to describe his technique of using sunlight to etch his plates. Photography, a word introduced by John Herschel, became popular and commercially successful after the improvements introduced by Louis Jacques-Mande Daguerre, improvements in the sensitivity of the plates along with his method of fixing the images with mercury vapours. The French astronomer François Arago was one of the first to see the promise of this new resource for astronomy and quickly reported it to the Académie des Sciences, of which he was the general secretary. He also persuaded the French government to purchase the rights to the new technique so that it could be made widely available. Arago always kept a fresh, active interest in most developments throughout the sciences and made contributions in several areas, besides being active in French public life. He once expressed the view that the Sun might be a cool planet after all and suitable for habitation. You recall that Herschel had stated a similar view when he speculated that the Sun's radiation was coming from high above in its atmosphere and not from the body itself. Much earlier, the French astronomer Joseph Lalande had maintained the opposite view that the sunspots were cooler mountains rising above an incandescent sea. These ideas, incidentally, were later disproved when the photographic techniques became efficient enough to photograph the Sun and the sunspots, prominences and flares were seen in greater detail. Nevertheless, despite some early successes and despite its enormous popularity, photography remained of little use to astronomers for more than thirty years until the development of the dry silver bromide plate and the incorporation of cameras to the telescopes. By that time, the more resourceful photographers, such as Paul and Prosper Henry in Paris, began to obtain not only excellent reproductions of the stars, useful for photometric studies, but also pictures of nebulae, like the one in Orion, with a wealth of details that were invisible to the human eye. In the Paris Congress of astrophotographers of 1887, a plan was laid out to photograph the entire sky, a project of 22,000 plates that took almost a century to complete, but served to register stars down to magnitude 20. Once again, photographic photometry was a different affair from visual photometry. What was the correspondence between the intensity of light as perceived by the eye and as recorded on a photographic plate? How was the image formed as a function of exposure time? What were the means to best measure the image density? All these questions required lengthy studies and, given that the techniques have not ceased to change, many are still debated today. In the same manner that precise coordinates for the stars had been sought in order to study their motions across the sky, the drive toward more accurate photometry was partly motivated by the desire to understand the variability of certain stars. Were these stars eclipsed periodically by darker stars? Were they rotating stars with enormous sunspots on their surfaces? Already in 1844, in one of his catalogues of stars, Argelander had published what he called an invitation to all the friends of astronomy, encouraging all observers to monitor repeatedly and precisely the luminosities of as many stars as possible. Argelander was not so much interested in providing answers to the questions related to variable stars, but he wished to amass the data that would enable future astronomers to address those questions cogently. By the time of his death, several dozen variable stars had been recorded. Bessel had once stated in one of his lectures that astronomy had no other task than to find rules for the motion of every star. It is in this sense too that the work of Bessel represented a turning point in the evolution of the science. Indeed, the motions of the stars remained a central problem, but several new tasks came now to equal prominence. Amongst these new astronomies, as they were then called, none caused a greater impact than the emergence of spectroscopy or, more generally, to use the expression coined by Johann Zöllner, the new discipline of astrophysics. Since the pioneering discoveries of Joseph Fraunhofer in 1812, it took almost fifty years until the German physicist Gustav Kirchhoff provided some of the earliest explanations regarding the nature of the lines in the spectra of light from gas flames and from sunlight. Very soon, many astronomers and instrument makers were searching for the appropriate combinations of prisms to produce suitable spectrometers that could be adopted to telescopes. Some of the more successful were those designed by Johann Zöllner. The use of gratings instead of prisms to disperse the light came only much later, since such spectrometers were initially very inefficient and could not be applied to faint sources. From the very first studies of the spectra of stars, it was found that they displayed a great variety of features and the task of their classification was soon undertaken. It became clear that the characteristics of the spectrum were somehow related to the temperature and the chemical elements present in the star, but the early classifications were of a descriptive nature, without any attempt at providing appropriate physical interpretations. By the turn of the century, several thousand stars had already been catalogued and a few more thousands were added each year after then. I have mentioned all these things to show the many routes that astronomy took in the course of the XIXth century. It was not just to determine the positions and the motions of the stars as Bessel had stated. Of course, Bessel himself had also been involved in several other tasks himself, from geodetic measurements to the study of how each planet affects the orbits of the other planets directly, as well as indirectly through its perturbations on the Sun. To paint a faithful picture of the progress in astronomy during the rest of the century one should really devote a great deal of time to examine the advances in the manufacture of telescopes and other related instruments, photographic cameras, photometers and spectrometers. The sophistication of the new instrumentation began to rely on such a broad range of activities that few individuals could master but a few specialized details of the whole enterprise. Engineers in the most diverse fields could contribute their expertise, whether it was in the deformation properties of materials used for the construction of telescopes or on the methods to produce good quality glass for lenses and mirrors. There were advances not only in the design of the optical components of the telescopes, but also in all the accessories that make it efficient, such as the mounting or the mechanical drive with its clock. If one were to examine further all the ingenuity that went into the design of the tools used to polish mirrors, as well as the other machine tools necessary for the construction of the telescopes and for testing their reliability once they were built, one could easily get lost in the midst of many fascinating and challenging tasks and forget entirely the original purpose for which these things were being produced, that is, to observe the sky. However, since the purpose of our discussion today is to understand the architecture of the Universe, we must turn now to more theoretical ideas which will eventually help us make some headway in our quest. To begin let us turn to the discoveries that took place around this time in the field of geometry, for it is the interpretation of astronomical observations in the light of some new notions that arose in pure mathematics that will enable us in the end to understand the true architecture of the Universe. From a historical point of view, it is the emergence of what came to be known as non-Euclidean geometries that we must emphasize first. Due perhaps to the rather diminished role that Euclidean geometry has come to play in education in the very recent past, we fail to realize nowadays the impact it has had for more than two thousand years, both as an educational tool and as an achievement of the imagination and of logical reasoning. In his compilation of what was know at the time about geometry, Euclid introduced his fifth postulate in the following manner: If a line, cutting two other lines, makes on one side two interior angles with those lines whose sum is less than two right angles, then the two lines, when extended indefinitely on that side, will eventually intersect. This is equivalent to saying that through any point exterior to a line we can draw one and only one line parallel to that line. The parallel postulate always seemed less natural or intuitive than the others and Euclid proceeded as far as he could without making use of it. The suspicion persisted that it might be a consequence of the other four postulates and numerous attempts were made to demonstrate it through the centuries, including many erroneous arguments that were publicized as conclusive proofs. Later on, some attempts were directed at showing that one could device a geometry in which the fifth postulate was false, that is, the two lines did not need to intersect.

Therefore, through a point exterior to a line one was not constrained to select one single parallel line. After all, motivated largely by astronomy, ancient mathematicians had also studied geometry on the surface of the sphere, where it is clear that any two lines, or great circles, always intersect. On a sphere there are no parallel lines. One could envisage a case in which the parallel lines were many, with the case of Euclid's plane geometry standing in the middle. These efforts first came to fruition around 1830. The German mathematician Carl Friedrich Gauss, one of the most versatile that has ever lived, was at the time fifty-three years old; he had made several attempts in his youth to deduce Euclid's fifth postulate and had come to believe that it was possible to device a geometry in which the postulate would be violated. Indeed, he developed most of the ingredients of what we know now as non-Euclidean geometry but refused to announce any of his results, initially in part because they were not complete, but also because he disliked controversy and knew that if he were to announce he was in possession of a new geometry different from Euclid's, some of his colleagues would descend on him, in his words, like a swarm of bees. A few years later similar results were published by a Hungarian mathematician by the name of Janos Bolyai. His father, a mathematician too, had exchanged ideas with Gauss before and had made some attempts of his own in this field. Furthermore, considering his lack of success, he had entreated his son to devote himself to other more fruitful problems. But the young Bolyai persisted and was rewarded with success. Yet, when Gauss revealed that, over the years, he had arrived at similar results, it led to some acrimonious protestation on the part of Bolyai, as if Gauss had intended to rob him of his merits. Gauss was perhaps rather cold and distant by temperament, so say some of those who met him. Nevertheless, he could not be accused of scientific dishonesty. Indeed, only a few years earlier, the Russian mathematician Nicolai Ivanovich Lobachevskii, working independently, had been the very first to demonstrate the consistency of a geometry in which Euclid's postulate did not hold, an accomplishment that did not bring him particular recognition, given that quite a few of his colleagues in Russia derided and even mocked him for his vain presumption. However, this was not Gauss' reaction. He saw an affinity between Lobachevskii's ideas and his own and began to study the Russian language in order to be able to read his writings. To express these results as they appear in Lobachevskii's geometry, we depart from a given line along a line that is perpendicular to it up to a distance d. Then, through this point you can draw many lines not intersecting the original line but there are two of them, one on each side, that come closest to touching the original line, so to speak, two lines that form the narrowest angle with the perpendicular line along which you were traveling, still without intersecting the original line. Lobachevskii showed that that narrow angle depends on the distance d in a precise way; clearly, it is close to a right angle when the distance you have traveled is very small, approaching the case of Euclidean geometry, but decreases to zero as our points drifts away to infinity. In any event, the important result is not the expression of this law but the fact that such a law existed at all, that one could not produce an infinite variety of geometries by making the fifth postulate less constraining. There is essentially only one such geometry and Lobachevskii, deriving the trigonometric formulae appropriate to solve triangles in this context, was struck by the fact that they resembled the formulae of spherical geometry, if only he could replace the lengths of the sides in the triangles by imaginary numbers, that is, numbers whose squares were negative. It was this unreal aspect of the geometry, together with the fact that one had to proceed by following the logic of its postulates without the guiding aid of visual images of lines and triangles, that made it difficult for many to confer to this theory a status comparable to Euclid's theory. There is one regard, however, in which this new geometry has a more definite character than Euclidean geometry. When we perform measurements of angles and distances, we are used to the fact that on an Euclidean plane the measurement of angles is defined on an absolute scale, such as a fraction of a right angle, but the distance scale is not defined in advance and must be chosen arbitrarily. This is not the case in Lobachevskii's geometry, where both angles and distances are specified uniquely. This is reflected in the fact that one cannot draw similar triangles, in the sense that once the angles are specified, the sides' lengths are too. But let us stay away from any specifics. For our own objective, it will be more convenient to approach the subject of non-Euclidean geometry along a different route. More than a decade earlier Gauss had been commissioned by the electorate of Hannover to carry out extensive geodetic measurements in their territory. The original motivation behind this work was to correct and improve the cartography available at the time for the purpose of collecting taxes, but it also continued on the scientific tradition of obtaining better land measurements to assess the accurate shape of the Earth. Because of the methods he introduced in this task, Gauss is still considered one of the founding fathers of the science of geodesy. Friedrich Bessel had also participated in this work and he and Gauss became frequent and trusted correspondents. During this work, out of pure mathematical curiosity, Gauss was motivated to use the method of the calculus to study the geometry of smooth curved surfaces, and it is in this field that he made one of his more original contributions to geometry. When we imagine a line winding around in space, we have an intuitive understanding of the notion of curvature as it applies to each point along the line. To be mathematically accurate, we might proceed as follows. At each point of the line we can identify its direction, this is tantamount to drawing the straight line that is tangent to the curve at the point. Since our line is curved it will soon move away from the straight line. To a second approximation, we might think of fitting at the very same point a circle instead of a straight line, which might better approach the trajectory of the curve in the neighbourhood of that point. Now, it is clear that if our line is greatly curved the radius of the best fitting circle will be rather small and conversely, a stretched out line would have a best approximating circle whose radius tends to infinity; therefore, we can be very precise and say that the inverse of this radius is a measure of the curvature at the particular point under consideration. A curved line has yet another characteristic called torsion; this notion would not be required if the curving was taking place inside a plane, but we know that, in addition to rising and falling within a plane, the curve may be at the same time swerving from one side to the other and this is not a quality that may be approximated by a best fitting circle since the circle is only a planar figure. In any event, it must be emphasized that these two characteristics pertain to how the curve is immersed in the surrounding space. By this I mean that an individual constrained to live in the one-dimensional world of a line would not have access to measure any of these things and they would indeed make little sense to him. Within his own world, if he were interested in geometry, he would be limited to measuring lengths between different points along the curve and his geometry would only be a geometry of distances. The tools at his disposal do not enable him to deduce any information regarding how the curve is embedded in the ambient space. Similar ideas can be used to study a curved two dimensional surface that stretches out in space. Imagine that we drop a vertical line perpendicular to this surface at any one point and proceed then to intersect the surface along various planes containing this line; it is clear the we would be cutting the surface along various curves, each of which departs from the chosen point and goes out in a different direction. Each of these curves will have its own peculiar curvature. It turns out that we can always single out two principal directions with their own curvatures and determine then the curvature of all other directions in terms of only these two numbers. Geometers call the product of these two curvatures the curvature of the surface at the point in question.

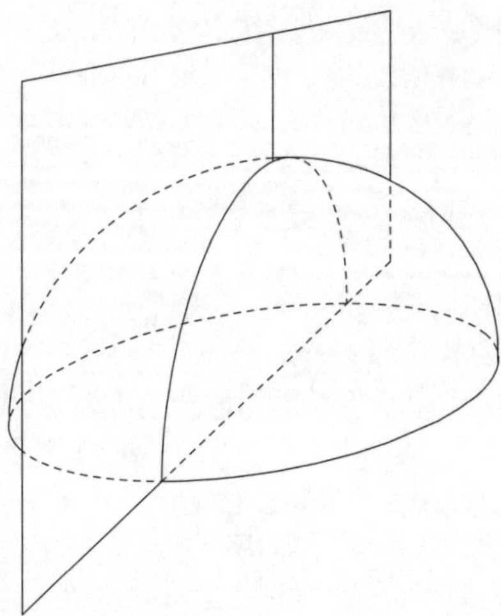

If we think, for instance, of a sphere of unit radius, it is obvious that, at any point we choose and in any direction we move, the best approximating circle to the resulting curve is indeed a circle of unit radius, so the sphere has curvature equal to one everywhere as it should. Of course, in an arbitrary surface the two principal directions of which I spoke need not curve in the same direction and this results in the so-called saddle points. Here the two curvatures have opposite signs and the surface is then said to have negative curvature. Midway between the two directions of maximum and opposite curvature, the saddle point possesses two other directions which are straight lines. Just as in the case of a single curve, we are here talking of the manner in which the surface is immersed in the ambient space.

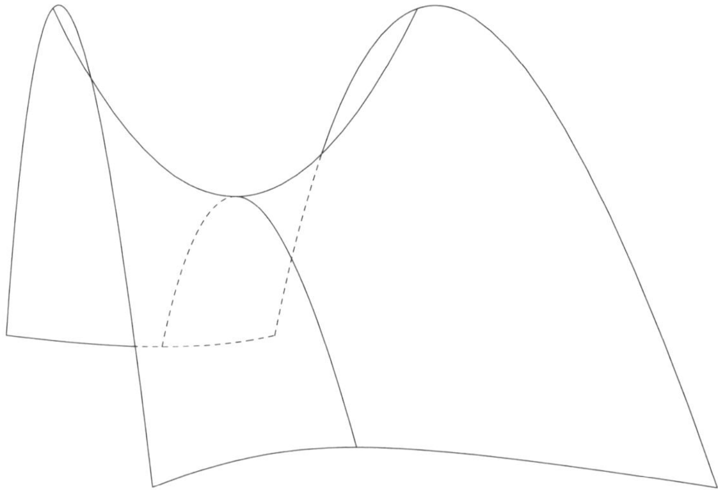

The remarkable contribution of Gauss was to prove that, unlike the case of curved lines, the curvature of the surface at each point is a property intrinsic to the surface and not necessarily determined exclusively by the way it is positioned in space. That is to say, an inhabitant living in such a two dimensional world would not need to venture outside his own world to determine the curvature at any point; of course, unlike his more constrained relative living along just one curved line, this one would have access to measure distances, angles and areas and by so doing could determine curvatures as well. The way to proceed, Gauss said, is to walk away from a given point on the surface in all directions up to a certain distance, in other words, to draw a circle around the given point. Now, on a flat plane the circumference would have a length equal to 2π times the radius traveled and the area enclosed would be equal to π times the radius squared. Meanwhile, the values obtained on a sphere would be somewhat smaller, whereas they would be larger on a surface of negative curvature so that, by measuring the disparity, we can determine the curvature, without ever having to consider the existence of a third dimension in our representation. It is fairly clear to our eyes that the lines departing from a point on a sphere tend to stay close together, while the opposite occurs when we depart from what we called a saddle point; the lines take leave from each other very fast; loosely speaking, we could say that a space of positive curvature is less spacious than a flat plane and that a surface with negative curvature makes for even more roomy conditions. It was clear to Gauss that the very same notions could be applied to a space of three dimensions, that is, an inhabitant of a three dimensional world could, without stepping out of his own world, measure the curvature at each point and determine the peculiar geometry in which he lives. Thus, the abstract speculations of geometry turned out to have an experimental side. It can easily be determined, for instance, that the sum of the angles of a triangle is, in a space of positive curvature, as for a triangle lying on a sphere, in excess of two right angles and is less if the curvature is negative. During one of his extended campaigns making geodetic measurements and with the purpose of establishing in what type of space we live, a considerable effort was spent on measuring a triangle between three prominent peaks, although the results were inconclusive due to the smallness of the expected effect and the uncertainty caused by the errors in the measurements. But we still have one further step to take in our consideration of curved surfaces. The sphere clearly occupies a privileged position amongst those with positive curvature, its curvature remains the same at every point. If we add the plane, which has curvature equal to zero everywhere, could we produce a surface with the same negative curvature at every point? Unlike the previous geometries of Euclid and Lobachevskii, where the starting point was a set of postulates regarding the behaviour of points and lines, here we seek to define the measurement of angles and distances on a surface in such a way that the resulting curvature is constant and negative. The Polish mathematician Ernst Minding, active in Berlin and then in Dorpat, known then as Turku, in Estonia, was the first to explore the consequences of these geometric conditions. Considering the type of trigonometry that would be valid on such a surface, he found that the required formulae resembled almost exactly the ones that applied to triangles on a sphere, except for the fact that one would have to assume an imaginary radius to the sphere, that is, the square of its length would appear as negative in the formulae. It is ironic that Minding published his results in the same publication where three years earlier Lobachevskii had published his, making similar comments regarding his brand of geometry, although the similarity of their observations was not noticed until many years later. Minding restricted himself to proving some of the properties that would have to be satisfied by a surface of constant negative curvature but fell short of providing an explicit realization of it.

Plebeius: On such a surface every point would be a saddle point, but it seems to me that if the tendency of the surface to bend in opposite directions at each point were to persist throughout, these curves will close into circles going in opposite directions and the entire surface would be a complicated maze of circles. On the other hand, you also spoke of the possibility of departing from a saddle point along a straight line, when choosing a direction midway between those of maximum curvature. This must surely be a devilish sort of surface.

Albertus: It is difficult to train our geometric imagination and to dissociate the object our mind is trying to conceive from the demands imposed by our eyes trying to represent them. The closing of curved lines into circles or the straight lines you speak of are only circumstantial properties of the saddle point that manifest themselves when we place a piece of a negatively curved surface into our ordinary three dimensional space; however, as Gauss would have said, these are not qualities intrinsic to the surface itself, where the only truly meaningful concepts are the measurement of lengths, angles and areas. Let me illustrate our situation with a simple example that we have already encountered in our discussion of the astrolabe. At that time we used a stereographic projection to map the celestial sphere onto a plane; in other words, we transported each point on the sphere to its corresponding point in the plane and used the plane as a representation of the sphere. In fact, we could use the same correspondence in the opposite direction to map the plane onto the sphere. You may recall that we had said that this projection maps all the circles on the sphere onto circles and lines on the plane. Thus, if we were interested in a geometric problem on the plane for which we do not want to distinguish between lines and circles, it would be convenient to use the sphere as a representation of the plane, where the distinction disappears and the two objects are of the same type. In such a case we would define the distance between two points on the sphere according to the distance between their corresponding points on the plane; the area of a triangle would be defined in the same way and likewise with all the rest. We would be transporting our very simple planar geometry onto a more complicated representation on the sphere. Our sphere would now be a more peculiar type of space for it would have infinite area; two points in the neighbourhood of the pole from which we carried out the projection would be very far apart and it would take an infinite distance to reach the pole itself. But there is nothing contradictory in this representation. All I wish to point out is that the rules imposed by a given geometry on a specific representation need not always be the rules that appear most evident to our eyes. Eugenio Beltrami, an Italian geometer, was the first to produce an explicit realization of a surface of constant negative curvature. He also pointed to the similarity between the conclusions of Minding and Lobachevskii and, with a concrete representation at hand, had no difficulty in showing that the two objects satisfied the same properties and were indeed identical, that is, a geometry defined in terms of axioms that violated the fifth postulate of Euclid turned out to coincide with a geometry arising out of a curved surface that had constant negative curvature. Soon thereafter other representations followed, each having its own advantage to treat specific problems. One of them that is particularly easy to understand was discovered by the French mathematician Henri Poincaré. It uses the interior of a circle as the representation of our surface. Once again, you could say that a circle lying on a plane is flat and has null curvature, but here we are defining the distance between two points in a peculiar manner and the geometry that results is not the intuitive one you would immediately guess with your eyes. Indeed, in the interior of our circle the shortest path joining any two points, or the straight lines, in Lobachevskii's terms, are the arcs of circles that meet the outer circle at right angles. As in the example of the plane projected onto the sphere, here the distance between two points is not the distance we see with our eyes, but is defined in such a way that distances become larger and larger as we approach the outer circumference, which does not belong to our surface and is, in fact, at infinite distance from any point in the interior. Of course, from this description we do not perceive intuitively that the curvature of this space should be negative. We must use the explicit formulae for distances to compute circumferences and areas to verify it, as Gauss prescribed. However, it is quite easy to show that Euclid's fifth postulate is not satisfied and that a line has infinitely many parallel lines going through a point external to it.

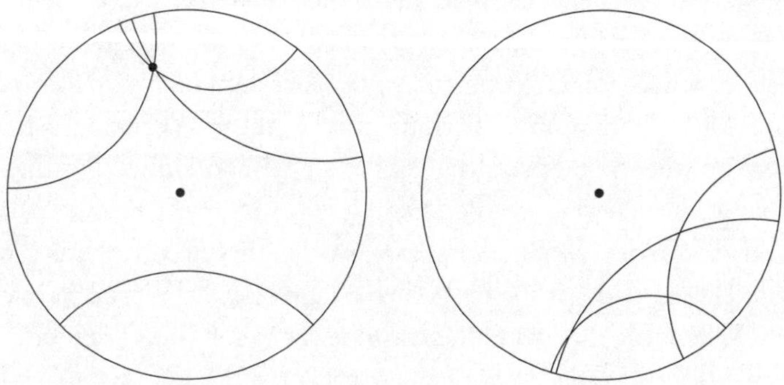

One convenient aspect of this representation is the fact that the circles in this geometry coincide with conventional circles, circles that do not touch the outer boundary. Similarly, the measurement of angles between lines prescribed by the geometry coincides with the measurement as perceived by our eyes. Therefore, it is also easy to conclude that the sum of the angles of a triangle add to less than two right angles. The space we have produced, with constant negative curvature at each of its points has, in all its simplicity, the same standing as the sphere or the plane; despite its being more elusive and resisting our efforts to see it with our eyes, it does not occupy for this reason any less prominent a position from the abstract point of view of geometry. There is one more property we should stress regarding this peculiar space, which is relevant to the geometric ideas we shall be discussing shortly. In the cases of the sphere or the plane, it is immediate to our eyes that no particular point of the space occupies a privileged position in relation to others. The geography of these spaces is identical everywhere, so to speak. However, this is not so apparent in the representation of the negatively curved space we have just found. It is true that the circumference bounding the space is in fact at infinite distance from any point inside; yet, it is not clear to our eyes that the center, for instance, does not occupy a privileged position, since the straight lines defined through this point, having to intersect the outer circumference at right angles, are indeed straight lines in the conventional sense, they are the diameters across the circle. This question I am addressing has to do with the notion of symmetry and the fact that the spaces we are considering, the sphere, the plane and the space of Lobachevskii possess the greatest symmetries. We all have an intuitive, if not very definite notion of the concept of symmetry; even for the mathematician, although he uses this concept in a more precise way, it remains sometimes rather elusive and vague. Nevertheless, the idea of symmetry has come to occupy a central position in many investigations in mathematics, as is the case with the geometries we have been considering, where it may be said that the symmetries themselves characterize the spaces that possess them. Let me be a little more specific on this matter. What do we mean, for example, when we say that a regular tetrahedron has a peculiar type of symmetry? We refer to the fact that if we were to hold the center of the tetrahedron fixed and rotate it around by a certain angle, the solid would come to coincide with itself, although its faces, edges and vertices would have undergone a permutation from their original positions. It is this interplay between something that has changed and something that remains the same that we usually call symmetry and it is the mathematician's task to make it precise and explore its combinatorial aspects. The first to do so was the Frenchman Evariste Galois, perhaps the most charismatic of the men who have cultivated our science. Killed in a duel when he was not yet twenty-one years old, he had a meteoric career as a scientist of the highest caliber, although not fully recognized at first, while at the same time he was an ardent political propagandist during the revolution of 1830 in Paris, activities for which he had to spend some months in jail too. Galois' starting point had been the theory of algebraic equations, which we discussed earlier. Taking one equation, he explored the symmetries exhibited by its roots or, to speak more properly, by the system of numbers created when its roots, occasionally irrational numbers, are added on to the system of rational numbers, or fractions. Thus, studying algebraic equations and the effect that exchanging one of its roots for another had on the systems of numbers produced, Galois was able to abstract some underlying principles that we can now apply to many other mathematical structures. Instead of considering the roots of equations, let us consider, for example, any four distinct objects, such as the first four digits,

1, 2, 3, 4 and the possible permutations amongst them. There are clearly 24 possible arrangements. Instead, when we consider the rotations of the tetrahedron into itself, which effectively produce permutations of its four vertices, there are only 12 possible outcomes. In either case, two permutations can be effected, one after the other, so that the result is a third permutation drawn from the same system we are considering. In this manner, we speak of additions or multiplications, very much as we do with numbers and the systems we deal with are closed, very much like the number systems, in the sense that the results of such operations are always members drawn from within the system. However, these structures often exhibit peculiar properties. For example, when we add or multiply numbers, the outcome does not depend on the order of the summands or factors,

$$a + b = b + a \quad a.b = b.a$$

This is not the case with the rotations of the tetrahedron; the result of combining two of them depends on their order, as you can very easily verify. Let us consider a similar example, namely, the rotations of a sphere. In this case, we can temporarily place an arrow at the north pole pointing leftward, merely for identification purposes. We now rotate the sphere until the north pole reaches some other arbitrary point on the sphere of our choice; then we can rotate the sphere around the axis that passes through this point so as to make the arrow point in any particular direction we wish. This accomplishes a transformation of the sphere into itself and any rotation of the sphere can be obtained in this manner in only one way, for once we have decided on the destination of the north pole and the direction in which our imaginary arrow will point, there is no further freedom to move the sphere onto itself. In the process, a new point occupies the north pole and some particular direction departing from this point is pointing leftward. By placing a new arrow as before, we can repeat the process, that is, we can combine one rotation after another or, as we usually say, we can multiply two rotations, so that the net result is another rotation of the same kind. Thus, the entire set of rotations closes onto itself under these operations. All this is very elementary and simple. Now, in this case too, the result of a multiplication depends very much on the order of the factors. For example, the rotation R_1 of ninety degrees around a horizontal axis followed by a similar rotation R_2 around the vertical axis produces a result that is different from the one we obtain when combining the rotations in the opposite order.

$$R_1 \times R_2 \neq R_2 \times R_1.$$

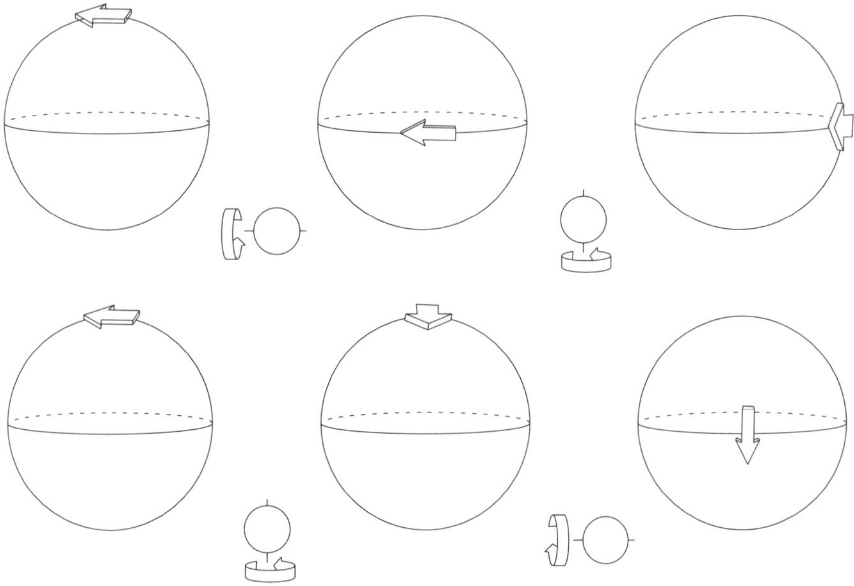

The case of the plane is much simpler. Clearly, by combining a translation and then a rotation we can bring any point on the plane and any chosen direction to any other point and direction of our choice. In this case, we can easily see that when we combine two such transformations, one after the other, the end result does not depend on the order in which we performed the operations. Just as in the case of the sphere, we say that this entire group of transformations realizes the full set of symmetries of our space, showing that every point and direction are equivalent to any other point and direction; in other words, if an inhabitant living on the sphere, or on the plane, had been displaced unawares by one of these transformations from one point to another, he would be unable to discover the change by means of geometric measurements or by exploring the geography of the space around him. This description fully encodes our concept of mathematical symmetry. It is to be noted, however, that despite the fact that the symmetries of the plane appear so very simple, if we had chosen to represent the points of the plane onto a sphere, as we did before, then the motions of translation and rotation that appear so simple when realized on the plane itself, would appear quite complicated in the effects they produce on the points of the sphere, leaving the north pole fixed and rearranging all others, stretching some on one side, crunching others on the other side. This is not as hopeless as it may seem and we could write simple formulae to show how it is done. I mention this because a similar situation arises in the case of the Lobachevskii plane. In our representation of this space inside a circle, we can rearrange the points in such a manner that any point of our choice is moved to the centre of the circle and viceversa, the center can be transported to any other point inside the circle in such a way that all angles and distances are preserved in the process, so that the transformation is a true symmetry of the underlying geometry; once again, an inhabitant of this space cannot distinguish by the tools he has at his disposal, namely, measuring distances, angles or areas, where he is located. It is only our arbitrary representation and our manner of labeling the points by placing them inside the circle, that establishes a convention distinguishing the points amongst themselves, but this does not correspond to any intrinsic property of the geometry under discussion. In summary, we have introduced three distinct and unique surfaces, distinguished by the fact that the curvature is the same at all points in each of them, being unity on the sphere, zero on the Euclidean plane and having the value negative one on the Lobachevskii plane, and further characterized by their complete symmetry, in the sense that their geometric properties look the same when seen from any of their points. More specifically, in each case we have transformations that allow us to transport one point to another in order to make these symmetries explicit. Finally, let me apply the concept of symmetry in another direction that is most important for us. It is a most elementary example and the details are very much similar to the case of the plane we have just considered, except that now we want to incorporate space and time in the same geometric structure. What is most important for us is that in this case the mathematical formalism of symmetry gives expression to an adopted law of physics. Suppose for simplicity we are dealing with time and only one space dimension; thus, we need only consider a two dimensional plane, on which an observer at rest can choose two axes with appropriate units to assign coordinates x and t to each event in his Universe. The first coordinate indicates its location and the second the moment of its occurrence. Now, according to the law of inertia, another observer in uniform motion with velocity v with respect to the first one, could choose another set of coordinates x' and t', with respect to which he is at rest and all the laws of physics would be expressed in the new system of coordinates x' and t' in the same manner as in the old. Although both observers would measure time in the same way, $t' = t$, the spatial coordinates would differ due to their relative motion and we would have $x' = x - vt$, assuming that they had started at the same position at time $t = 0$. The simple class of transformations we are considering here

$$x' = x - vt$$

$$t' = t$$

represent the changes of coordinates between two typical observers in uniform motion relative to each other. The symmetry these transformations express has its root in the fact that the laws of physics can be expressed in exactly the same manner, regardless of the system of coordinates one chooses to use, in other words, the laws of physics are indifferent to inertial motion. However, upon closer scrutiny, it was found out later in the study of electromagnetic phenomena that the true symmetry of the laws of physics is expressed by a different group of transformations than the one we have just considered, although the discovery of this remarkable fact took many years to be understood. The task of redressing those coordinate changes will force us to reconsider our most basic assumptions regarding space and time. The studies in optics and in electricity and magnetism that eventually led to these discoveries were very much advanced during the period that we have now been considering. These days our pupils at school say: Give us the correct formulae and tell us how we should employ them; that will be enough, we shall accept them. This is perhaps a good trait in that they are never encumbered by prejudice, nor are they eager to spare much energy in understanding past mistakes. However, allow me to indulge in telling you how, through the study of the propagation of light and with the emergence of electromagnetism as a scientific discipline, we were pushed into accepting that our common sense was betraying us and that, if we want to do it with precision, we had to modify the way in which we speak of space and time. The history of these studies is exceedingly convoluted and it would be pointless to trace every step since their inception, or the many speculations of the philosophers about the nature of light. Moreover, we are surveying a very large landscape that has taken us away from astronomy into purely mathematical questions and now to problems in electromagnetism. Therefore, I promise to be rather brief, for we shall also have to consider how these things come together in helping us understand the construction of the Universe. Magnetic iron ore had been known since antiquity, as well as the power of amber to be electrified. For a very long time electricity was thought to be one of the many substances that make up the world, along with light itself. Light was thought to consist of corpuscles that are locked inside matter and that can be liberated by heating. The action of electricity was caused by emanations that were similar to light. Matter was permeated by various kinds of effluvia, in the same manner that blood, phlegma or choler affected the human body. At times, even caloric was thought to consist of emanations as well. The notion that heat was not a substance but a state of motion in the constituent parts of matter was one of the earliest to be elucidated. Electricity, in fact, had some analogies with heat, for it could be accumulated in a body by rubbing, it could be transferred from one body to another by contact and, in certain circumstances, could also cause combustion. The fact that there was positive and negative electricity led some to believe that there were two different fluids, the so called vitreous and resinous fluids, depending on the type of materials that held them. Others contended that there was only one such fluid permeating all substances in various degrees; it was its excess or deficit that made electrified objects attract or repel each other, as they tried to restore their balance, acquiring or shedding part of the fluid. Similar ideas were set forth regarding magnetism with the proposal that it was composed of two fluids, boreal and austral, although their behaviour differed substantially from their electric counterparts for one could not, for instance, cut a magnet in two in order to obtain a piece of material that contained just one fluid. These notions were still quite rudimentary; towards the end of the XVIIIth century the study of electricity and magnetism became more quantitative and precise. This was due in part to the development of new tools and instruments and, also, to the introduction of more elaborate mathematical ideas. In 1785 Charles Augustin Coulomb published a memoir in which he established that the forces of attraction and repulsion between electrified bodies as well as the forces between magnetic poles decrease with their separation in proportion to the square of their distance. Coulomb had spent his youth as a construction engineer in the island of Martinique and, upon his return to Paris at age thirty-six, engaged in studies of resistance of materials, constructing mechanical tools such as the torsion balance, which led to his electric experiments. Once the inverse square law was established, another Frenchman, Simeon-Denis Poisson developed a complete mathematical theory to treat electric forces in various configurations. It was reassuring to know that electric and magnetic forces behaved in the same manner as the force of gravitation and that Nature had employed the same mathematical law to rule over them. Moreover, it was now possible to put aside the old notions of effluvia to account for the electric forces; one could accept the direct action between bodies at a distance with the same confidence and success already acquired in the study of gravitation. The second major advance was not the emergence of any new theory but the introduction of a marvelous tool, namely, the electric pile of Alessandro Volta. Originally from a well established family in the city of Como, Volta remained in northern Italy for most of his career, conducting numerous electrical experiments over many years. In 1800 he demonstrated the use of his pile in three lectures in Paris, attended by none other than Napoleon Bonaparte, who was so impressed that he decided to give Volta a pension and make him a count. Before and after that time, the changing fortunes in the territories of northern Italy required that he had to ingratiate himself occasionally with the Austrian authorities, at other times with the French government, although the task of establishing favourable connexions with men of influence must have been eased by the fact that he had five brothers who had become priests in the Catholic Church. With the introduction of the electric pile it was possible for the first time to produce electric currents, as opposed to the sudden discharges that had been possible until that time with the Leyden flask. Amongst those immediately attracted by the use of the pile was the Danish experimenter Hans Christian Oersted, a man of multiple interests, who laboured for many years attempting to secure a professorship at the University of Copenhagen. Nevertheless, after some twenty years of uneventful work, he made a rather sensational discovery which, as often happens, appears to have taken place by accident. During one of his lectures, Oersted observed that an electric current running parallel to a magnetic needle had the power to deflect it, establishing for the first time the connection between electric and magnetic forces. The news of his discovery were soon reported to the French Académie des Sciences by the physicist François Arago and attracted in turn the attention of André-Marie Ampère, who immediately threw himself into the study of electricity. He was then approaching the age of fifty, having spent most of his career devoting himself to mathematics and chemistry as well as to teaching astronomy, physics and philosophy. It is interesting to note that both Oersted and Volta had also made their lasting discoveries close to age fifty. Ampère is said to have been a man of great gifts, who may have misused and scattered them somewhat; however, in his studies of electricity, which he conducted in just a few years, he rose to excellence and this gave him a solid reputation. Most importantly, he provided a quantitative theory to Oersted's observations and made the discovery that two currents had the power of exerting a force between them, which he also formulated in a mathematical law expressing the force between two segments of wire, analogous to Coulomb's law for the attraction between static charges. Thus, Ampére established a new discipline for the study of electric charges in motion, which he called electrodynamics. Many years later, James Clerk Maxwell, writing his treatise on electricity, called Ampère the Newton of electricity. This may have been an exaggeration for, despite the elegant mathematical form that Ampère gave to his theory, it turned out to be not really fundamental and was substantially modified in the course of time. However, one of Ampère's insights that was far ahead of his time concerns the nature of magnetism, which he thought to be electric in origin. His observation that currents flowing through closed loops acted like magnets on external bodies led him to believe that magnets were in fact composed of molecules with small currents within them, all aligned and flowing perpendicular to the axis of the magnet. In any event, the mathematical laws ruling the forces between currents or charges that Ampère and others developed proved so successful that they began thinking about modifying the law of gravitation in a similar manner; that is, the forces amongst electric charges depend not only on their positions but also on their movement and this is something that the theory of Newton had not taken into account. Nothing came out of these efforts, but they are quite fascinating all the same; some of us have had the suspicion that the last word on this subject has not yet been written, but it has proven to be quite difficult to find the correct mathematical language to express these ideas. Meanwhile, in Great Britain, very much unlike the precise mathematical theories of his colleagues in the continent, Michael Faraday was pursuing a very different line of reasoning. Having grown up in a family of few means and with very little schooling, his rise to become one of the preeminent scientists of his age is one of the more remarkable in the history of physics. As an adolescent he had taken employment as a bookbinder, which gave him a chance to read some of the books he was assembling. By a series of accidents, but driven always by his desire to educate himself, he became the secretary to the president of the Royal Institution, a scientific organization recently created, of which Faraday would eventually become the driving force. The Royal Institution became his home for it provided him with a laboratory and with living quarters as well. At the start of his career Faraday occupied himself mostly with chemical analysis; however, in 1820, after the excitement created by Oersted's discovery and the many claims and theories that seemed to be emerging about it, he was asked to write a review of what was known at the time on electricity and this set him on his course, pursuing his own line of research. Like other researchers, he was persuaded that no only currents had magnetic effects but the opposite should also be true and that, through magnetic forces, it should be possible to generate currents. Ten years later, in 1831, after many attempts, he finally discovered the process of magnetic induction. Faraday was the most industrious, resourceful and imaginative experimenter of his time, but his mathematical skills were rather limited. He developed his own ideas on electricity in a more descriptive fashion, introducing what he called lines of force. We are all familiar with the manner in which iron filings, spread on a surface around a magnet, align themselves with the magnetic force. Faraday conceived of these lines of force extending through space between electric charges as well, and attempted to explain electromagnetic phenomena through the interaction of this lines, as one might try to explain the turbulence in a fluid by following the stream lines and their variations in time. Oersted's phenomenon had to do with the circling of magnetic lines of force around currents; similarly, the discovery of magnetic induction he described as occurring when a wire cuts through lines of magnetic force. During his study of electrolysis, Faraday had measured how the molecules in the conducting solution were dissociated in proportion to the current created. Insulators, by contrast, had the ability to retain the integrity of their chemical structure. However, in his incessant experimentation Faraday observed that the capacity to hold charges in the plates of a condenser depended on the type of material he used as insulator between the plates. According to his way of describing things, the lines of electric force piercing through the insulator caused stresses within it and polarized the medium, similar to the alignment of iron filings. Faraday, incidentally, was also the first to build an electric generator. It is true that much of the subsequent research in electricity and magnetism was driven by the prospects of power generation, telegraphy and such applications, but here we are interested only in tracing the theories of electromagnetism and the efforts addressed to understanding it from first principles. The preeminence that Great Britain had acquired in fundamental research in electromagnetism through the work of Faraday was continued for some years thanks to the accomplishments of James Clerk Maxwell. Born to a well-to-do Scottish family, Maxwell showed precocious mathematical skills and was sent to Cambridge University, where he received a first rate education. He read widely, studying early on, in particular, the Researches in Experimental Electricity of Faraday. Given the prevalent climate in Cambridge at the time, when heat flow and fluids were being studied successfully for the first time with appropriate mathematical tools, it is not surprising that Maxwell would be drawn to applying similar ideas to electromagnetism. From his early work he followed Faraday's description that used lines of force, which he compared to fluid flows with sources and sinks accounting for electric charges, these ideas now dressed in precise mathematical language. Maxwell pursued these studies throughout his career, sometimes intermittently, as he was periodically driven to concentrate on other problems in mathematics or physics. After teaching for a few years in Scotland he received an appointment in London, where he stayed for five years and where he had the opportunity to meet Faraday, already retired, but with whom he had corresponded in the past. Before leaving London, in 1864, Maxwell presented a memoir to the Royal Society in which he introduced the final form of his system of equations for electromagnetism, the equations in which, to this day, we find such a beautiful appeal for their succinctness and elegance. In some way, these equations were nothing but a mathematical summary of what was already known, but Maxwell proceeded several steps further in applying Faraday's ideas leading, in particular, to the realization that light was also an electromagnetic phenomenon. It is probably fair to say that the theory of light, on its part, began to take shape with less vicissitudes than the theory of electricity, at least in its early days. More than a hundred years earlier, the lines had been drawn in the battle between those who advocated the corpuscular theory and those who favoured the view that light consisted of waves, the chief exponents of these views being Newton and Huygens, respectively. To some extent, this was a debate on what light consisted of and not on how light behaved. The undulatory theory of light had arisen by analogy with water waves and acoustic waves. The corpuscular theory of light was perhaps more intuitive, but Huygens managed to describe how every point reached by the wave becomes the source of a new wave in such a way that at each stage the total wave front advances along the envelope of these new forming waves. In this manner, he was able to account for the phenomena of reflection and refraction with equal or greater success than were the advocates of the corpuscular theory. For much of the XVIIIth century very little progress was made in advancing either view. During this time, putting aside the case of water waves, which are surface waves, or the vibrations of the head of a drum or a string, the only waves known to propagate through space were acoustic waves, in which case one observes oscillations in the pressure and the density of the medium as the front of the wave passes through, the oscillations occurring along the direction of propagation. Meanwhile, looking at light rays propagating through polarising crystals, it was noticed that rotating the crystal around the direction of propagation affected the intensity of the ray; this was not supposed to take place according to the wave theory. Years later, around 1830, when the theory of the elastic vibrations of solid bodies was developed, it would become clear that one could have oscillations in the two directions perpendicular to the direction of motion and all of them could propagate in waves. Even then, it would have been quite extraordinary to apply these examples to the case of light; indeed, which type of aether would be tenuous enough to allow the heavenly bodies sail through it unimpeded and at the same time have the rigidity to mimic elastic waves? By that time, however, the French physicist Augustin Fresnel had already demonstrated conclusively that light consisted of transversal waves and it was in part his own success with this theory that provided a stimulus to study waves in elastic media. Fresnel developed his ideas almost single-handedly. He had the privilege of studying at the Ecole Polytechnique but it appears that when he started his experiments on light he was barely aware of the work of Huygens. During his youth in Normandie, Fresnel had expressed the ambitious desire to study the propagation of light, heat and electricity and unravel the nature of the medium responsible for all of them. Later, however, unlike some of his contemporaries, he dealt exclusively with the phenomenon of light, although he did so with a degree of perfection that would stand the test of time. His initial experiments on diffraction and interference patterns, which he conducted during his spare time as a road engineer, were rather tentative, since he lacked the appropriate equipment to carry them out with precision. His work caught then the attention of François Arago, who was very diligent as secretary of the Académie des Sciences in that he had a keen eye to spot important new developments in physics and was active promoting and divulging them. Arago collaborated with Fresnel and supplied him with some of the resources available to him in order to improve the quality of his experiments. It must be said that Fresnel was anticipated by a few years in some of his interference experiments by the Englishman Thomas Young, but Fresnel was the first to express his results in precise mathematical language and offer a coherent theory to interpret them. He was a perfectionist and pursued his goals with dedication until he had them within reach. His work, however, remained unfinished. Poor health and his duties as a road engineer at first and as lighthouse supervisor later, kept him away from his researches and, eventually, tuberculosis took his life at age thirty-nine. Thomas Young was a London physician and with a personality altogether different. He scattered the immense power of his genius in many odd directions, not the least significant of which was as a linguist, for he knew a score of languages already as an adolescent. Toward the end of his life he devoted a great deal of energy to decipher the Egyptian hieroglyphs. It was his interest in the physiology of vision that led him to the study of optics. The experiments he conducted on the interference of light rays convinced him of its undulatory nature and set him on the path to interpret them correctly as transverse oscillations. Drawing on an analogy with the vibrations of air columns in organ pipes, which he had also studied, he may have been the first to speculate that the different colours of light corresponded to different frequencies in the oscillations of the aether that carried them. As to the possible relationship between light and electromagnetic phenomena, John Herschel was one of the first to speculate on a link connecting the two. He had observed a certain similarity in the way quartz crystals and electric currents could act to rotate the plane of polarized light. Some years later the German physicist Wilhelm Weber, who had a long association with Carl Friedrich Gauss in Göttingen and a common interest in problems of electromagnetism, discovered an even more interesting link. This had to do with the establishment of electric and magnetic units. Roughly speaking, when we measure electric forces we have to assess the density of electric charges, whereas magnetic forces depend on the flux of currents. Weber made some difficult but quite precise experiments to relate one to the other; now, it is clear that to convert a charge density, or charges per unit volume, into a flux of current, charge per unit area per unit time, the former has to be multiplied by a velocity. Thus, on first principles, we might suspect that any theory of electromagnetism would have to include a fundamental velocity which, as it turned out in Weber's experiment, coincided with the speed of light. Rather surprisingly, the coincidence failed to attract much attention at first. A similar observation was made about a year later by the German physicist Gustav Kirchhoff. He was attempting to elucidate how fast an electric current would move through a wire. As an element of current presses down the wire, it generates a magnetic force around it which, in turn, induces on a renewed current; here again, it was necessary to relate the magnetic to the electric forces; referring to the results from Weber's experiment, Kirchhoff was struck by the association with the speed of light. I should mention that, by this time, the speed of light was known fairly accurately, not only from astronomical observations but also by direct measurements on Earth. Armand Fizeau, a disciple of Arago and a wealthy amateur experimenter, interested in photography from its inception amongst other things, was probably the first to measure the speed of light directly. He let an intense narrow beam go through a toothed wheel that he made to spin very fast, so that the pulses of light, reflected from a distant mirror, upon hitting the wheel again, would sometimes disappear and reappear again in between its teeth, as the speed of the wheel was varied. But let us now return to Maxwell's work and the way in which he furthered Faraday's ideas. According to his reasoning, the polarization of the medium caused by the lines of electric and magnetic forces took place not only within an insulator or conductor but also in empty space; moreover, when we bring in charges or magnets that push in their lines of force causing this polarization, the emergence of these stresses in any medium are tantamount to an electric current and ought to be accounted as such. The idea that these temporary displacement currents could be taking place even where there was no ponderable matter present, let alone electricity itself, was viewed at the time with some skepticism and we can find some physicists, thirty years later, treating it in their writings as some ad hoc assumption that should be superseded by a more fundamental understanding. Nevertheless, Maxwell pressed ahead. If these currents, however fictitious, were present, they would produce magnetic forces to circle around them and the emergence of these magnetic loops would induce further currents and so the chain would continue. Indeed, one could generate a wave in this way and let electromagnetic perturbations propagate through empty space. Realizing now that the velocity of propagation of these waves coincided with the velocity of light led him immediately to the hypothesis that light was an electromagnetic phenomenon, that not only were electromagnetic waves possible, but that we had plenty of evidence of them shining before our eyes. After Maxwell left London, he retired once again to Scotland, away from any academic attachments and devoting much time to writing his famous and lengthy treatise on electromagnetism. However, after a few years, barely at age forty, he was appointed professor at the newly created Cavendish Laboratory in Cambridge, where he spent the last nine years of his life. He proceeded then to draw further conclusions from the electromagnetic theory of light, but he also found time to edit the works of Henry Cavendish, unpublished for almost a century, which had anticipated, in some ways, glimmerings of Faraday's as well as his own work. One of the observations that Maxwell made during this time followed as a consequence of the fact that not all insulators have the power to transmit electric forces in the same manner, which affects the speed with which electromagnetic waves travel through them. Therefore, in the case of light, given that the speed of propagation through a medium affects its index of refraction, one could predict an optical effect such as the index of refraction by following purely electric arguments. Indeed, these conclusions were soon confirmed by experiments. The waves associated with the electric and magnetic fields were transversal waves; this was made apparent in the very structure of Maxwell's equations. It was satisfying to verify, twenty-five years after Fresnel's death, that the theory of electromagnetism also constrained light to propagate along transversal waves. However, the theory still contained quite a few puzzles. Ironically, the ideas introduced by Maxwell were abolishing once again the simplicity of the notion of action at a distance; all electromagnetic information was now required to be conveyed by waves. Despite the successes of the theory and its unification with the entire field of optics, very little progress was made in understanding the nature of the medium that was supposed to convey these waves. Maxwell, along with many of his contemporaries, spent much time and energy trying to develop mechanical models of the elusive aether, sometimes with wheels and gears, other times with molecular vortices or many other fanciful contraptions, and it was always very difficult to choose amongst the various theories or to discern what was really fundamental in the new ideas. Hermann von Helmholtz, one of the more distinguished German scientists and amongst the few who could aspire to have a commanding view of the entire field of physics, referred to the whole state of affairs in electromagnetic theory at the time as a pathless wilderness. Maxwell himself, in his treatise, did not present a unified and consistent theory, but attacked the many difficulties on various fronts, leaving many open possibilities. A few years later another German physicist, Heinrich Hertz, a disciple of Helmholtz in Berlin who had been appointed professor at the University of Karlsruhe, started a series of experiments that led to the construction of an electric oscillator and the production, detection and measurement of electromagnetic waves for the first time. Hertz wrote at the beginning of his book on electromagnetic waves: Maxwell's theory is Maxwell's system of equations. Today we take this view for granted, for we have learnt how to find our way from this most concise set of equations to explain almost any electromagnetic process of interest, but at the time the situation was not as clear as it seems to us now. Moreover, one could hardly say that one had understood electromagnetism without knowing anything about the medium that conveyed its waves. It must be said that not every physicist was blinded by the need to explain the aether mechanically. Fresnel owed part of his success to his willingness to proceed on a rather ad hoc basis, letting experiments point the way and using the most appropriate mathematical tools that could account for what he saw. Maxwell himself stressed from time to time that a scientific theory had to provide an account of the relation amongst objects and not necessarily of the objects themselves. Joseph Larmor, one of the English theoretical physicists who concocted one of the more elaborate descriptions of the workings of the aether, states explicitly that the physicist should proceed like the geometer, who makes precise statements about points and lines, without having to specify what type of material objects he is referring to. This view was eventually vindicated for, when we speak of electromagnetic waves, we comfortably speak of the waving without referring to what is doing the waving. The old dictum that mathematics allows you to speak meaningfully, without saying what it is you are speaking about, although somewhat scornful, shines here in its true merit. In any event, let us now examine at closer range how the study of electromagnetic processes came to modify our conception of space and time. I should like to return to the description I gave earlier of two observers who are in motion with respect to each other, while each casts a system of coordinates upon all the events in the world, in relation to which he remains at rest. Let us say that observer A sees observer B go past him toward the right with velocity v. Observer A now sends a series of light pulses toward the right at regular intervals of time T. Observer B sees the first pulse instantaneously because he happens to be at the same place as A. By the time the second pulse is dispatched, he will have moved to the right and the light ray will have to spend some extra time catching up with him. Observer B receives the light pulses at longer intervals T', that is, we have to add to the interval T the time it takes light, moving at speed c, to travel the distance B has moved during this time T'; thus, $T' = T + vT'/c$, or

$$T' = \frac{T}{1 - v/c}.$$

Since the denominator is smaller than unity, T' is larger than T, which indicates the delay. You remember that we had stated that both observers agree on all measurements of time, so that they would be in agreement when estimating at which time the pulses are dispatched and the intervals at which they are received. This is all very elementary. However, if we examine things in more detail, we must admit that what is really taking place is something quite different. When he sends a light signal, our first observer is in fact exciting a wave in the surrounding aether, a wave that will propagate through the medium until it is detected by the second observer. We should think of a sailor who leans over his boat and creates a wave that is then perceived by a second distant sailor when his boat begins to bob up and down at the arrival of the wave. We had specified the velocity of one observer with respect to the other, but seeing things in this new light, I should probably take into account the motion of both observers on the surface of the sea when I compute the time it takes the wave to propagate from one to the other. We might imagine a situation in which the two observers are not moving with respect to each other but both are drifting through the aether, just as if they were on ships anchored along the banks of a river that is continuously flowing past them. Common sense tells us that the arrival time of the propagating wave should be speeded up when it is traveling downstream and retarded when going upstream. In the case of light, such experiments were carried out by the French physicist Armand Fizeau. He sent light rays along pipes where water was flowing very fast and he verified indeed that its velocity of propagation varied depending on which direction the water was flowing. Now then, we find ourselves within the solar system, drifting through the aether, amidst stars that are also moving in various directions, and we should like to find out what is the preferred frame of reference at rest with respect to the aether itself. The rotation of the Earth around the Sun would allow us to observe various stars at intervals of six months, as we move toward and away from them. The aether permeates the Earth's atmosphere and our telescopes. A variation in the velocity at which the light signal is propagating toward us would also change the index of refraction of the light as it enters the lenses of our telescope, so that the position of its focus plane would suffer small variations depending on our drift through the aether. Such experiments were carried out by François Arago, who was unable to detect any such effects. Incidentally, when we discussed the phenomenon of aberration, you recall that I dispatched it rather summarily, explaining it in terms of the change in the direction of motion of the light corpuscles, just as if they were rain drops. In fact, aberration is a slightly more subtle phenomenon and requires further elaboration when we view light in the form of propagating waves. Fresnel provided an appropriate account of the phenomenon and he was also able to explain Arago's negative results assuming that matter, and the Earth's atmosphere in particular, has a tendency to partially carry along in its motion the aether that permeates it and not, as he had initially believed, that the aether drifts through matter,

228

in his words, like the wind through a grove of trees. Maxwell proposed to use once again the moons of Jupiter as a distant clock in order to detect the changes in the velocity of propagation of light, taking advantage of the fact that our two planets would find themselves approaching each other sometimes, at other times drifting apart, as they proceed along their orbits. This method never gave a conclusive result due to the small effects expected and the inaccuracy inherent in the measurements. Some of the more accurate experiments to detect the drift of the aether were carried out only a few years after Maxwell's death by Albert Michelson, first when he was still an assistant in Helmholtz's laboratories in Berlin and later in the United States where he settled down. Michelson used for this purpose an interferometer of his own design. The instrument split a light ray in two perpendicular directions, allowed them to travel equal distances and reflected them back. In a drifting aether the travel time of the two rays would be different and the phases of the two waves at their arrivals would be displaced, producing interference patterns. The negative result of this very precise experiment led many to believe that, at least within the confines of the earth's atmosphere, the aether must be carried along entirely with matter. After the identification of light with electromagnetic waves and the emergence of one single aether as the carrier of both, it became equally possible to test the various aether hypotheses by purely electromagnetic means. Heinrich Hertz, in particular, applied Maxwell's theory under the assumption that matter in motion carried with it the aether in full. You recall that, according to Maxwell, the displacement current played a fundamental role. An electromagnetic perturbation not only polarizes the molecules in matter and causes a current to surge, but the same takes place in the empty medium of the aether. Therefore, it is clear that, depending on whether the aether is carried along by matter and both participate of the same motion, or one is independent of the other, the ensuing displacement current is predicted to have quite different magnitudes. However, by 1885 the physicist Wilhelm Röntgen demonstrated clearly that the conclusions drawn by Hertz were not realized. His experiments seemed to indicate that the aether is entirely dissociated from the matter it pervades. Without doubt, during all these years when the various experiments gave contradictory results, there was great confusion and uncertainty and a clear understanding of the nature of the aether appeared to be as elusive as it had ever been. If we adopt a theoretical perspective and follow Hertz's dictum that electromagnetism is encoded in Maxwell's equations, the confusion manifests itself in a more clearcut fashion. This viewpoint was further advanced by the prominent Dutch physicist Henrik Lorentz. In Lorentz's theory we already recognize the outlines of electromagnetism as it is understood today. The atomic nature of the electric charge had already been recognized by this time and all electric and magnetic fields were seen as caused by their distribution and motions throughout space. Let us adopt the point of view that all we understand about electromagnetism is in fact contained in Maxwell's equations. An observer at rest proceeds to locate the distribution of charges in his own system of coordinates, identifies their motion, and specifies the electric and magnetic forces present at every point in space and time. He can now account for everything that is taking place, as encoded within the equations themselves. Earlier we described the simple way in which a set of coordinates, denoted (x, t), can be translated into the coordinates (x', t') associated to an observer who is in motion relative to the first one, merely an exchange of labels for each event in our simplified one-dimensional Universe. The second observer could now retrieve all the information regarding the disposition of the charges from the first observer, learn about the orientation of the electric and magnetic fields at each point and dutifully translate every piece of evidence regarding positions and vectors into his own system of coordinates, with respect to which he is at rest. If we understand anything at all about this matter, we should now expect the following: Our second observer could now open the book where he has written Maxwell's equations in his frame of reference and, as he begins substituting in them all the information he has just retrieved, the equations will be satisfied and he should arrive at conclusions that are just as valid as those obtained by the first observer. That is, if the aether is indeed unobservable, all frames of reference should be equally acceptable and the same equations of Maxwell should emerge, once we have done the necessary translations or, as we sometimes say, when we go from one coordinate system to another, the equations should transform into themselves. Unfortunately, this is not what happens. After we complete this process of translation, the equations do not reemerge in their original form and an inevitable conclusion follows. If the theory of Maxwell is correct, there must be only one preferred system of coordinates with respect to which it gives a true account of the physics and this frame must be, undoubtedly, that of the aether. With respect to a moving frame, the equations appear in a different form and predict electromagnetic effects that should be observable. In 1892, the Irish physicist George Fitzgerald made a proposal that seemed to offer a way out of the impasse. In the changes of coordinates that we have used until now for observers in motion, the measurements of lengths remain invariant. The individual coordinates assigned to two locations x_1, x_2 may vary, but distance, measured by their difference, is the same for all observers, $x_2 - x_1 = x_2' - x_1'$. Fitzgerald suggested instead that objects that are in motion with respect to the aether suffer a contraction along the direction of motion. Therefore we would have a formula of the form

$$x_2 - x_1 = f(x_2' - x_1')$$

where the factor f would be larger than one and would depend on the velocity v of the observer. In fact, $f = 1/\sqrt{1 - (v/c)^2}$ but the precise form of this factor need not concern us for the moment. It was as if the pressure against the aether had the power to compress matter, including our rulers and measuring devices. This idea, however improbable, was adopted by Henrik Lorentz, who also defended the notion of an aether at rest in its absolute frame of reference, unperturbed by the motion of matter through it. The fact that the experiments, such as those conducted by Michelson, had failed to reveal the existence of the aether could be explained by the contraction hypothesis. He concluded that the theory was fully consistent with the observations, if one also assumed that there was a contraction in the measurement of time as well, by the same factor. Hence, we should have

$$t_2 - t_1 = f(t'_2 - t'_1),$$

which implies that clocks in motion relative to the aether do slow down, since one unit of time for the observer at rest, because of th factor f, has to be compensated by a smaller time interval on the right hand side, which merely states that for the observer in motion his time coordinate advances more slowly. These ideas appeared somewhat contrived, but they proved to be successful nonetheless. As it turned out, by applying this type of transformations one could see that Maxwell's equations did transform into themselves when going from one coordinate system to another, so that it was hopeless to expect any detection of the aether by electromagnetic means. By some inscrutable conspiracy, the aether caused peculiar effects that conspire against its own detection. The irony of this whole state of affairs is that the success of the aether theory eventually turned into its own demise. However, before this came to pass, a new and more momentous difficulty arose. This had to do with the conflict between electromagnetism and mechanics. Indeed, while the new transformations of coordinates between observers in relative motion appeared to finally explain all of electromagnetism, they did not apply to the laws of classical mechanics. Here, as had been known since the time of Newton, the old formulae were not only satisfactory but they were the only acceptable ones. Of course, no one believed that two observers should exchange information one way when discussing electromagnetism and another way when discussing mechanics. At the turn of the century, some of the more foresighted physicists, amongst them Lorentz himself, as well as Henri Poincaré in France, came to the view that perhaps the time was ripe to revise the solid and time-tested theories of mechanics. This daring thought was not necessarily the product of sheer fantasy. We know, for example, that when an electron, or any charged particle, is accelerated from its position at rest, it not only carries along with it its own electrostatic field but causes a magnetic field to curl around its direction of motion; therefore, some extra work is required to achieve this effect, as compared to the case when the particle carries no electric charge. It is undeniable then that there is something that must be called electromagnetic inertia. It was considerations of this type that led to the suspicion that mass itself, which we associate with inertia, might be entirely of electromagnetic origin and that, eventually, the entire field of mechanics could be subsumed under the laws of electromagnetism. In the end it was through the combined efforts of the best theoretical physicists of the time, notably Lorentz and Poincaré, as well as Albert Einstein and Max Planck that the correct interpretations for the new transformations of coordinates emerged, that the existence of the aether was finally dismissed and that the new mechanics in agreement with electromagnetism was finally established. The path followed by Albert Einstein is quite remarkable because he worked independently and, when he published his investigations in 1905, he had gone farther than most. He confessed later to have been puzzled by the phenomenon of light since the age of sixteen, by the fact that when we receive a light signal of a given colour, we are incapable of knowing, by merely observing the electromagnetic perturbation that reaches us, whether it is a red signal shifted toward the blue due to the source's motion towards us, or a blue light, red shifted because of the source's motion away from us. The signal we receive always manifests itself as a wave moving at the same speed. The same could be said if the source is at rest and we are ignorant of the state of motion of our frame of reference. To put it in Einstein's words, when I dispatch a light signal, I produce a wave train that moves away from me at the speed of light c. If I decide to chase it afterwards, no matter how fast I go after it, it will still be moving away form me at the same speed c. At first, this sounds absurd; nevertheless, it reflects the true state of affairs. Unlike the case of water waves traveling on the surface of the ocean, where we can adjust our motion so that they appear as standing waves, the notion of standing electromagnetic waves is in fact contrary to logic. To see this we must only remember the very description of electromagnetic waves as given by Maxwell. They arise when a change in the configuration of charges creates currents and, consequently, magnetic fields around them which, in turn, induce displacement currents further away, and new loops of magnetic field, and the entire phenomenon that we call a wave cascades down the wave's path; no matter which observer is describing the wave, he must explain its emergence as a truly dynamic process. In other words, the notion of a stationary wave in this case is inconceivable. Therefore, it stands to reason that any effort to chase a light ray and overcome it is completely vain. Einstein saw this clearly and accepted the constancy of the velocity of light as one of its defining qualities. Let me now adopt a frame of reference with respect to which I am at rest, with coordinates x and t. I find myself at the position $x = 0$; if I dispatch two light pulses to my left and to my right at the present time $t = 0$, these pulses will arrive, after time t, at the positions $x = c.t$ and $x = -c.t$. Therefore, I can argue that the trajectories of these light rays are the points whose coordinates satisfy $x - ct = 0$ and $x + ct = 0$ or, taken together,

$$(x - ct)(x + ct) = x^2 - c^2t^2 = 0,$$

given that the product will vanish only when one of the factors is equal to zero. Another observer, who is moving with respect to my reference frame with velocity

v, will adopt his own system of coordinates x' and t', with respect to which he is at rest. For only an instant at $t = 0$ we find ourselves at the same location $x = 0$ or $x' = 0$. This is merely a convention. Now, we state that the laws of physics require that he also sees my light pulses depart with velocity c, despite the fact that we are moving in relation to each other. This condition requires that the points satisfying $x' - ct' = 0$ or $x' + ct' = 0$ coincide with the points that satisfy the similar condition in my system of coordinates. In a way, the condition $x^2 - c^2 t^2 = 0$, or $x'^2 - c^2 t'^2 = 0$, has a universal meaning. The paths traced by the ligth rays are characterized by the same equation, regardless of the observer that is making the observations. That is, we shall require that

$$x^2 - c^2 t^2 = x'^2 - c^2 t'^2.$$

which states that either side of the equality vanishes when the other side does. At first this seems contrary to logic. I see the other observer move with velocity v; yet, we are both seeing the light pulses move with velocity c in our respective coordinate systems. The condition is absurd from the point of view of common sense, but the mathematics says otherwise. Having discarded the transformations $x' = x - vt$ and $t = t'$ that we used earlier, we try something more general, of the form

$$x' = ax + bt$$

$$t' = cx + dt.$$

The computations are truly elementary but I do not wish to torture you with their details. If you replace these expressions for x' and t' in the previous equation where they are squared, so that both the left hand side and right hand side contain only x and t, you could verify that it is impossible to choose a, b, c, and d arbitrarily and still satisfy the condition we imposed, for all values of x and t. There is only one possible family of transformations, which depend on one single parameter. I shall write them in their finished form as

$$x' = \frac{x - vt}{\sqrt{1 - v^2/c^2}}$$

$$t' = \frac{t - (v/c^2)x}{\sqrt{1 - v^2/c^2}},$$

where all the coefficients are fixed once the value of v is chosen. Of course, I wrote the equations in a convenient form, so that I immediately see that the position of rest of the second observer $x' = 0$ requires $x = vt$; indeed, the second observer, at rest from his point of view, is moving with velocity v in my own system of coordinates. The most surprising feature of these changes of coordinates is that both equations mix the coordinates of space and time. It is also immediately apparent that these expressions reproduce the equations of Fitzgerald and Lorentz regarding the contraction of space and time, even though our interpretation is now quite different. I will return to this question momentarily, but concerning this mixing of space and time some persons find these matters too abstract. They say that it may all be true in a mathematical sense, but they wish to retain their ingrained and differentiated concepts of space and time, as being entirely separate entities. I would try to persuade them by insisting that we are not doing anything extravagant; we are assigning labels, coordinates, to events in the world, in a manner that is consistent with itself. I use the analogy of a row of contiguous buildings, many stories high, built along a street that climbs up onto a hill. We should like to arrange matters so that a pedestrian, entering into any of the buildings from the street, would always find himself on the first floor. However, this requires, for the occupants of the buildings, when they move from one building to the next using the corridors in the interior, that they would find themselves on different levels without ever having used any stairs. Of course, we could arrange matters so that every horizontal floor is assigned one level only, but now a pedestrian entering a building up on the hill, would find himself at street level but on a high floor. In other words, here the vertical and horizontal dimensions appear mixed and, no matter which system of coordinates we use, they will still remain entangled. Space and time also remain always entangled, no matter what we attempt to do. In any event, returning to the coordinate changes I have just written, it is now true that if two observers exchange information according to these rules, they will find that Maxwell's equations express the same electromagnetic phenomena in either system. There is no need to distinguish one particular frame of reference that would correspond to the aether at rest. Furthermore, as I already stated, these coordinate changes reproduce the formulae for the contractions of space and time that Fitzgerald and Lorentz had written down, although their interpretation is now slightly different.

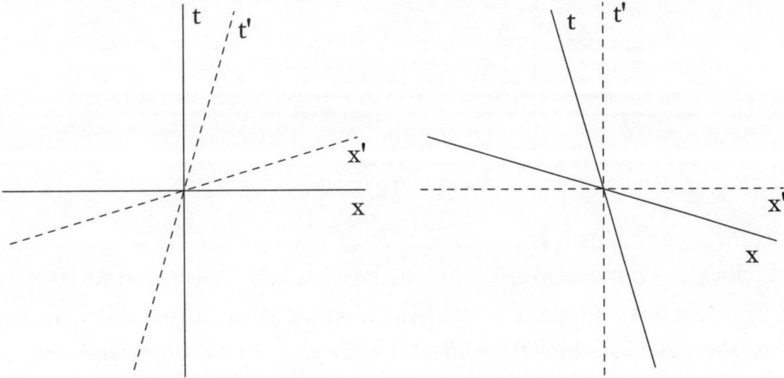

To approach these matters, let me use, once again, a convenient graphic description for the system of coordinates we are employing. I have followed the convention that each observer draws his axes of coordinates perpendicular to each other. In addition, each observer has drawn, in his own representation of the world, the lines that the other observer would identify as his axes of coordinates. In each case, of course, the conditions $x = 0$ or $x' = 0$ denote the positions of the observers at rest in their respective coordinate systems. We also notice that the events that each observer would identify as taking place right now do not happen to coincide, namely, the events singled out by the condition $t = 0$ differ from those that satisfy $t' = 0$. It is difficult to admit it but we must acknowledge that, in a strict sense, the state of the Universe at the present moment is only a convention. On the matter of the contractions, as suggested by Fitzgerald and Lorentz, we now see things in a different light. Let me assume that the first observer holds a measuring rod of length l that remains at rest in his frame of reference. The permanence of this rod throughout time is represented in his coordinate system by a band that stretches out to infinite time. It is now apparent that the second observer will estimate the length l' of this moving rod in a different way, merely because in his coordinate system he will measure the width of the same band differently; this is not more profound than saying that the width of the bread depends on how one slices it. The underlying physics remains unaltered, it is our way of labeling the events that vary.

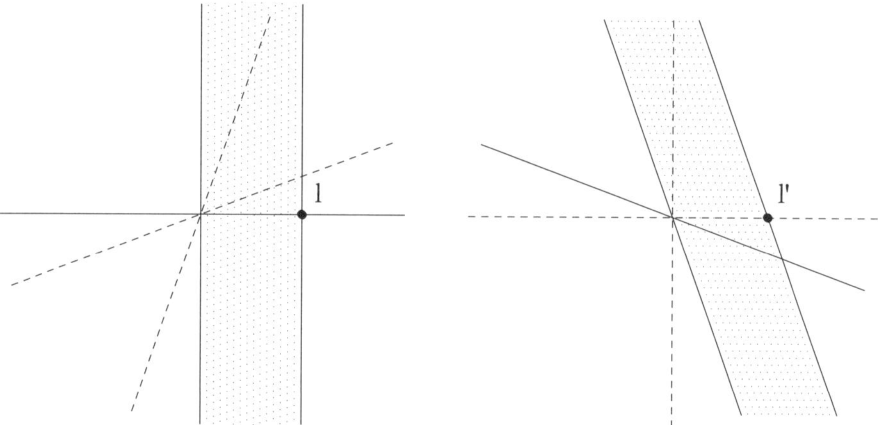

We can now also elucidate how our observers perceive the behaviour of their respective clocks. This has always remained a cause for confusion amongst those who do not understand these transformations in their appropriate context. If I look at my clock at time $t = 0$ and time $t = 1$, I single out two events $(x, t) = (0, 0)$ and $(x, t) = (0, 1)$, whose coordinates in the reference frame of the moving observer are now easy to obtain. For him, the elapsed time amounts to $t'_2 - t'_1 = (1/\sqrt{1 - v^2/c^2})(1 - 0)$, which is nothing but the factor f, greater than unity, proposed by Lorentz. To put it in simple terms, the other observer sees me in motion and concludes that my clock is marching too slowly, since it arrives at $t = 1$ when already a longer time interval has elapsed, according to his clock. Notice, however, that our changes of coordinates are completely symmetrical. If I wish to express the x and t coordinates in terms of x' and t', I only need to replace v by $-v$ in the earlier formulae and all is done. Therefore, I do not need to repeat the argument, for it is clear that, by my account, the clock carried by the second observer is also slow relative to my clock and there is no contradiction in this whatsoever. To make things absolutely plain, let us think of our clocks as consisting of two mirrors, facing each other along the floor and the ceiling, with a light beam bouncing alternatively from one to the other and whose trips we count in order to measure the passage of time. We always agree that light travels at speed c, but, in each case, one observer sees the mirrors of the other one traveling sideways, so that the light is required to travel for a longer time along a diagonal path in order to bounce off the opposite mirror. In this way each observer becomes convinced, quite correctly, that the other observer's clock must run behind his. The trick is accomplished only when we use experiments that are conducted at rest within each reference framework, but are then interpreted by the other observer who is in motion relative to them. Finally, let me return to the first example we discussed, namely, the delay in the reception of light signals by an observer who is in motion away from the source with velocity v. Indeed, we no longer need to mention the oscillations of the aether that will carry the propagating wave. For both observers, the light pulses will travel at speed c. Let us assume that these pulses are emitted, in the framework at rest with the source, at intervals of time T. For the moving observer, those intervals are of length $T' = (1/\sqrt{1 - v^2/c^2})T$; furthermore, given that he sees the source receding away from him, he has to add the time required for each pulse to catch up with the point of origin of the previous pulse. This time amounts to $v.T'/c$, the distance traveled divided b the velocity of light. Thus, he sees the pulses arriving at intervals $T' + vT'/c = (1 + v/c)T'$ or, equivalently,

$$(1 + v/c)T' = (1 + v/c)\frac{1}{\sqrt{1 - v^2/c^2}}T = \sqrt{\frac{1 + v/c}{1 - v/c}}T.$$

This is a different formula from the one we obtained earlier and, as we shall see, it plays an important role in many investigations. If you allow me to indulge in two further curiosities, I wish to illustrate the manner in which the things we have been discussing are related amongst themselves. We have just seen how an observer with coordinates x', t' exchanges information with another observer with coordinates x, t, with respect to whom he is moving at speed v. Suppose now a third observer with coordinates x'', t'' is moving with respect to the second one at speed v'. He uses then an entirely similar pair of equations to translate between one coordinate system and the other. However, in these equations we could replace the appearance of x' and t' by their expressions in terms of x and t, so that the third observer can exchange information with the first one. We expect once again that the set of transformations will be entirely alike and follow the same form given by Fitzgerald and Lorentz. We are using once again our notion of group of transformations which can be combined with one another to produce another of the same kind. But the question arises, what is the relative velocity between the third and the first observers? We are asking how should we add two velocities? Well, I shall let you do the appropriate substitutions in the previous equations; it is all a matter of simple algebra. The result turns out to have the form

$$v'' = \frac{v + v'}{1 + v.v'/c^2}.$$

You might say that this rule introduces a new law of addition that can be applied to any two numbers smaller than c, producing another number of the same kind. We can also perform subtractions and all the rules that apply to the normal addition and subtraction of numbers apply here too. The other curiosity I referred to also illustrates the relationship between these tools that we have introduced for physical reasons and the mathematical ideas we discussed before. Suppose we consider now a space of two dimensions, so that we require a set of coordinates x, y and t. Now, when we refer to another observer in motion with respect to the first one we would also have to specify the direction of motion, in addition to its speed, so that the appropriate transformation of Lorentz will involve one of the two coordinates, x or y, or some combination of the two. Nevertheless, just as in the case of one dimension, all such transformations could be characterized as preserving the validity of an equation of the form $t^2 - x^2 - y^2 = 1$. It is not difficult to represent the points satisfying this condition. Obviously, the coordinate t has to exceed unity in order to be satisfied. As it value increases, we realize that the other two coordinates describe the equation of a circle $t^2 - 1 = x^2 + y^2$, so that we obtain a stack of circles of ever increasing radius. Clearly, all coordinates can be either positive or negative. The resulting surface is called a hyperboloid for the simple reason that, when we cut it along vertical planes, the resulting curve is formed by the two branches of a hyperbola.

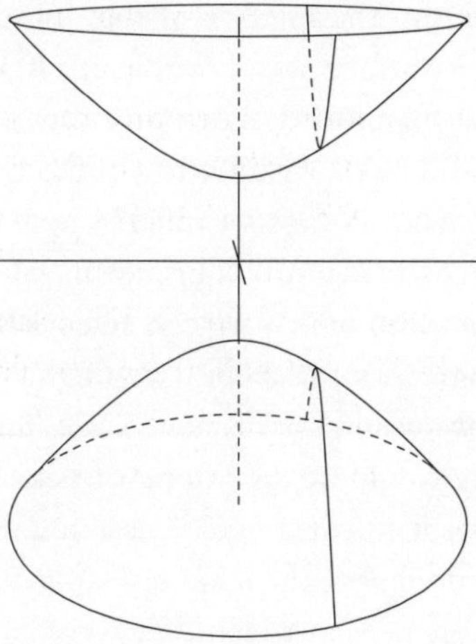

We are now close to the destination I wanted to reach. A transformation of Lorentz, by choosing new coordinates, can also be viewed as placing a new axis along the vertical and horizontal directions; however, if a point satisfied the condition $t^2 - x^2 - y^2 = 1$, it will be assigned new coordinates that still satisfy the same condition, so that our transformation can be interpreted as moving the points on the hyperboloid amongst themselves.

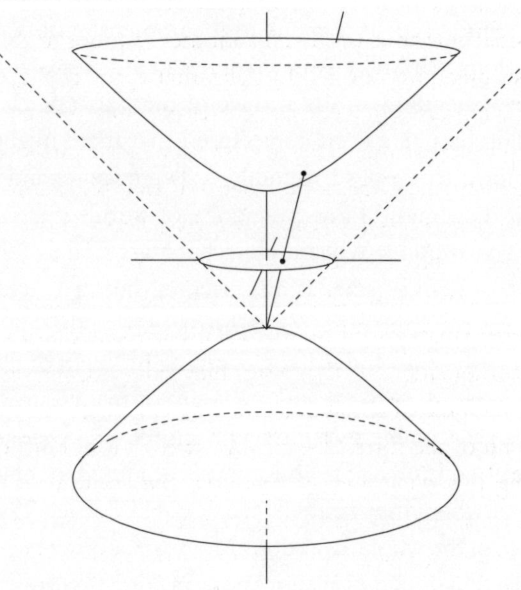

We find now the following surprising result. If we project the points on the upper half of the surface, from the pole of the lower half, onto the plane at $t = 0$, each point is matched to a point in the interior of the unit circle, in such a way that the transformations of the surface above within itself according to the Lorentz transformations I have just mentioned match precisely the transformations of the circle within itself as required by the symmetries of the Lobachevskii geometry we discussed earlier. In particular, it shows explicitly how any point can be moved to the origin by such a transformation. This seems to be a beautiful result and demonstrates the economy with which we can achieve various goals at the same time. After this long digression, Plebeius, we ought to return to the problem of elucidating what we have called the distance scale, the true dimensions of our Universe. Of course, there would be little hope of penetrating into the depths of space without some kind of signposts to guide us along the way. During the epoch we have been considering until now, the main task at hand was to size up the realm of the stars, to learn the distances at which they were located and clarify whether they were distributed according to some order or plan. Bessel had measured the first parallax to an inconspicuous star in Cygnus and the determination of the distances to a few other stars then followed; however, these were just minute steps and the vast array of all the constellations lay awaiting. The earliest attempts to understand the spatial distribution of the stars as a whole had been entirely speculative. One of the more fanciful schemes was suggested by Thomas Wright, a knowledgeable Scotsman who supported himself entertaining the aristocracy with lectures on natural philosophy. Some time before the work of William Herschel, he suggested that the stars were distributed on a thin spherical shell, within which our solar system is immersed. It was due to the immensity of that sphere that we see only the stars along the Milky Way, scattered in a disc around us, in the same way that an observer standing on the surface of the Earth perceives it as being a flat disc surrounding him. He even conjectured that there might be many other such shells unbeknown to us, all surrounding a divine center and whose positions he related to other metaphysical interpretations of worlds possessing various degrees of virtue and corruption. William Herschel was certainly the first to attempt to comprehend the system of stars as an observer, but its extension, or even an outline of its general configuration, the construction of the Heavens, to use his words, eluded him in the end. Indeed, the daunting problem of the architecture of the star system required the efforts and ingenuity of many astronomers before it could be said that an acceptable solution had been found. Could it be, as someone had asked, that the good Lord had promiscuously scattered the stars here and there without attention to where they landed? No, it could not be. There had to be an order and a purpose in the whole invention. Most astronomers would have refused to abandon the idea of a design or plan without making first an attempt to unravel it, were it to exist. Perhaps the solar system was eccentrically situated and the entire array of stars fell into some order when seen from an appropriate vantage point. After all, part of the difficulty in understanding the motions of the planets had derived from the fact that we were not observing them from the central position at the Sun. Some of the early speculations were directed toward finding a central Sun around which all other stars revolved. Was it not Sirius, the brightest star in the sky? Friedrich Argelander objected to this choice because Sirius was known to have a sizeable proper motion. He proposed instead that there was a cluster of stars at the center and his choice was the Perseus cluster. William Herschel thought that it might be the Hercules cluster, perhaps because he had found the Sun gravitating in that direction. The German astronomer Johann Mädler, who succeeded Struve as director of the Dorpat Observatory and was rather opinionated in his views, but also a successful writer of popular accounts of astronomy, found his own reasons to argue that the center resided within the Pleiades and was probably its brightest star Alcyone. The search for the center of the star system continued for some years, until it faded away completely, when the difficulties and intricacies of the problem at hand diverted the attention elsewhere although, like many unsolved problems, it returned decades later in a different guise. In retrospect, we may discern three methods that were used to unravel the distribution of the stars, even when these divisions seem rather arbitrary and inadequate. The first one was the statistical method, already initiated by William Herschel, albeit in a crude and simple manner. He counted the number of stars in various directions in the sky, assuming that their apparent luminosity was an indication of their distance. Without knowing the distance to any star in particular, he also assumed, to a first approximation, that they were all equally spaced, so that, through his counts and observing the disappearance of stars at the faintest magnitudes, he could infer the decrease in density at the farthest distances, at the confines of the starry system. Despite the magnitude of the enterprise, we saw that Herschel came to rather sound conclusions. Taking as unit the distance to a first ranked star, he concluded that the Sun was close to the center of a system that extended up to approximately 800 units on the plane of the Milky Way and 150 units in the direction perpendicular to it. As a consummate observer, however, Herschel was more interested in what he saw in the Heavens than in his past conclusions or pronouncements and, twenty years later, he dismissed much of what he had said earlier. By that time, he stated that the density of stars might not decrease at all in the direction of the Milky Way and that the star system might well be, as he put it, fathomless. This work of Herschel was continued by the German astronomer Friedrich Struve. When John Herschel visited Struve at Pulkova, near St. Petersburg, he took with him a copy of the collected papers of his father, a gift that Struve certainly appreciated, considering that his investigations, even in the search for double stars, had followed closely the path traced by the elder Herschel. In the study of the distribution of stars, Struve assumed that they were arranged in layers along the plane of the Milky Way, like a stack of coins. Each layer, perhaps infinite in extent, had a particular density of stars, which presumably decreased at either side of the central plane. Thus, by looking along an axis in a direction away from this central plane, one would be piercing through the different layers and, by tilting this axis toward the central plane, one would sample, up to any given distance or to the range of one's telescope, more and more of the lower layers. As far as he could determine, the counts of stars did not disprove this simplifying hypothesis. In addition, this layering hypothesis enabled him to introduce in a sensible way the subject of obscuration. He pointed to the fact that well before reaching the limiting power of his telescope, the number of stars that came into view in the lower layers, as one allowed for fainter and fainter stars, did not correspond to the increase in volume that one was presumably sampling, so that there ought to be a loss of starlight by absorption along its path to us. In any event, it was difficult to disentangle this effect from the one produced by a more sparse distribution of stars far away from the Sun. The second method I referred to, instead of looking at the ensemble of stars, studied them individually, attempting to determine their distances or motion one at a time. You recall that Herschel had already used the displacement of the stars across the sky to infer the motion of our solar system amongst them. However, in his initial study he had used only six stars and it may be a stroke of luck that his conclusions proved to be largely correct. His son John, and Bessel too, including many more stars, were of the opinion that their motions were too chaotic to afford any conclusions. Friedrich Argelander, in a later study, supported the initial determination and Otto Struve, who had succeeded his father as director at Pulkova, continuing the tradition of discovering many double stars, further confirmed those results, estimating the velocity of the Sun to be approximately 10 kilometers per second. Thus, in a tentative and not very precise way, it was always assumed that the greater the apparent motion of a star across the sky, the greater its proximity. To determine its actual parallax was a different matter altogether and this process proceeded very slowly. By the end of the century, the proper motions of several thousand stars were already known, but only a few dozen parallaxes had been determined. I should say a word now about the units of distance we are employing. Herschel had used the distance to Sirius as a unit, whatever its true value was. Using parallaxes as a method to determine these distances, we find that only a handful of stars have a parallax exceeding a quarter of a second of arc. It therefore seemed reasonable to use one second of arc as a standard parallax, the distance from which the radius of the Earth's orbit would appear as one arc second across. This distance is approximately 175,000 times our distance to the Sun and it represents a typical distance between stars. It received the name of parsec and has become the standard unit of length on a cosmic scale. It takes light about three years to travel a parsec in free space, so that a light-year, although slightly shorter, is a distance of the same order of magnitude. These distances are so dissimilar to anything that is familiar to us on a human scale, that we usually fail to realize, when we contemplate the aggregate of stars in the Milky

Way, how distant and isolated the stars really are. If I plot two dots at the center of this page, one centimeter apart, barely the width of my pen, and allow them to represent the Sun and the Earth at their true distance, firstly, I have to admit that my dots, however minute, are exceedingly fat in relation to what they represent; the one that stands for the Earth will engulf the Moon's orbit too, and even the dot representing the Sun will be too large, given that one could line up a thousand suns next to each other on the path to the Earth. Nevertheless, I could now envision that the rest of the planets fall within this page and, if I were to place it at the center of a table, much of the dust and debris associated with the solar system would be scattered on its surface. Now, assuming these two dots represent the Sun and the Earth, a nearby star at a distance of one parsec might be identified on this scale with a little dot on the center of a page on a table one kilometer away. Of course, the difference in the distances to the Sun and to a nearby star is evidenced by the difference in their apparent luminosities, but the very fact that this discrepancy is so exorbitant prevents our senses from giving us an accurate representation of its magnitude. During the time that we have now been considering, a new method was introduced to study the motion of the stars, based on the changes in a light signal's wavelength or frequency due to the motion of the source in relation to the observer. We have already discussed how this effect came to be properly understood according to the principle of relativity, but well before this time the effect was discussed following traditional methods, first by the Austrian physicist Johann Christian Doppler and shortly thereafter by the French physicist Armand Fizeau. Doppler drew the comparison between light waves and sound waves and pointed to the rise or the decline in their frequency when the source is moving, respectively, toward or away from the observer. This merely reflects the fact that the succeeding crests of the wave, or the beats, if you wish, seem to quicken when the source is approaching the observer or are delayed when it is receding. Thus, the colour of a light source, which is a reflection of its frequency, becomes bluer when the source is moving toward us and reddens when it moves away. Colours, however, involve many frequencies. For such a displacement of the entire ensemble of frequencies to be noticeable, the velocities required are considerable, amounting to a sizeable fraction of the speed of light. Doppler insisted that the stars did have such velocities and that this method could be applied to determine their motions. This turned out to be nonsense, but not entirely so. As the fellow Austrian physicist Ernst Mach and Armand Fizeau pointed out, one could use, not colours, but the bright lines in the stellar spectra, corresponding to individual frequencies that were well determined and whose displacements could be accurately measured. The method was advocated by Johann Zöllner, the innovator and instrument maker, and it was one his students, Hermann Vogel, who first applied it successfully on a bright source, such as the Sun. By observing one edge of the Sun and then the other, he was able to detect its rotation. Although the initial results were somewhat ambiguous, it was soon verified that the speed of rotation inferred with this method coincided with the speed known from the motion of sunspots. Subsequently, with the improvement in the design of spectrometers, this method was applied to the spectra of stars and the velocities of some bright stars could finally be determined. This procedure, incidentally, proved to be very useful later in discovering double stars, in cases where the two companions were far too close to be distinguished by visual means. Although they appeared as one single star through the telescope, their light spectrum was clearly recognized as the superposition of two individual spectra.

Plebeius: Could these two spectra not come from opposite edges of a rotating star, as in the case of the Sun? It also intrigues me that this method of measuring the velocities of the stars could in fact detect only their motion toward or away from us, but not when they are cruising across the sky, which is what you were speaking of until now.

Albertus: You almost provide the answer to your first question. When the star is rotating, the bulk of it, the central part of the face we see, is moving sideways and does not affect the lines in the spectrum. However, the edges that are approaching or receding contribute their smaller lines at each side, as it were, so that the net effect is a much broader line, as if produced by a smearing of the characteristic frequency, a phenomenon that was later exploited to infer the velocity of rotation of many stars. It is true, as you stated, that this new method accomplishes something very different from what had been done previously registering the movement of the stars on the surface of the sky. Indeed, the two methods were complementary. Nevertheless, one still had to know the distance to a star in order to determine its absolute motion relative to the Sun. Lewis Boss, an able American astronomer working at the Dudley Observatory in Albany, New York, and otherwise busy searching for comets, observing a transit of Venus or contributing to some of the catalogues then in preparation, made an interesting observation while studying the stars in the Hyades cluster. He noticed that their proper motions all pointed toward one common point on the sky. He realized that this was an effect of perspective, that the stars where moving in unison, in the same manner that the trajectories of a flock of geese, when flying overhead in the same direction, all seem to point toward one point on the horizon. Then, using the Doppler effect, he could learn the magnitudes of the stars' velocities along the line of sight. Although he knew only their angular motions across the sky, he could now put together these two pieces of information and, using simple geometry, deduce what was the required distance to the cluster, so that the combined motions made all the stars march indeed in the same direction. This was an unqualified success. The Hyades cluster was found to be at approximately 40 parsec. Soon other loose clusters of stars were studied in the same way. The method not only made it possible to reach greater distances but it also took care of whole assemblies of stars with one stroke. The Doppler effect was also used to study anew the motion of the Sun. This was done very successfully by William Campbell, a well known American astronomer, who rose to be director of the Lick Observatory and later was president of the University of California. He made the study of the radial velocities of stars his specialty. By this time, there was close to unanimity on the direction of motion of the Sun. Campbell verified that the Solar system was moving relative to the ensemble of stars when looking in two opposite directions, whereas the motion relative to the stars along a perpendicular plane was without pattern and random. His observations implied that the Sun had a velocity of 20 kilometers per second, ratifying the remarkable result of Herschel's initial study. Let me now come back to the statistical method of studying the distribution of the stars. We saw that Friedrich Bessel, in 1834, had observed the irregular motion of Sirius and deduced from it the presence of a companion star. Almost thirty years later, the American optician Alvan Clark, stationed in the city of Washington and supplier of superb lenses to many observatories around the world, was testing some new telescope when he discovered the companion star quite by chance. It was subsequently proved to have the orbit that had been predicted from its effects on Sirius, but this star was ten thousand times less luminous than Sirius. Although not a totally unexpected result, it was nonetheless sobering and made it plain that the apparent luminosity of a star was a very poor indicator of its distance. The fact that there was such a wide variety in the candle power of stars did not make the study of their distribution in space any easier. The first astronomer to undertake a comprehensive study of the statistical distribution of the stars was Hugo von Seeliger, a native of Silesia and for many years the director of the Observatory in Munich. Starting in 1884 and for the next twenty years, he wrote a lengthy series of articles addressing this subject. The problem at hand can be stated succinctly in this way: Within an imaginary enclosure of unit volume there is typically a distribution of stars of various intrinsic luminosities, according to some unknown law yet to be determined. On the other hand, each such enclosure may have a different overall density of stars, depending on the region of space from which it was selected. In short, we wish to determine the aggregate density, of stars of all luminosities, a density which presumably varies from region to region, and the relative density of various intrinsic luminosities, which we expect to remain more or less constant everywhere. If we knew these laws, we could surely predict how many stars of each apparent magnitude we would see when we look through a circle cut out on the surface of the sky. It is a different matter to trace our way backward and, from the evidence we see on the sky, to reconstruct the original and more detailed information, since we are now staring into a deep cone where the faint stars of nearby regions are mixed together with the brighter stars of successive shells farther away and we have no a priori way of distinguishing one from the other. von Seeliger sought the mathematical tools that would enable him to recover as much information as possible on the nature of those original distributions, making the minimum number of assumptions necessary. Another astronomer who was attracted to the same problem was Jacobus Kapteyn, from the Netherlands. An extremely enterprising man, when he turned to the problem of the distribution of the stars at the turn of the century, he had just spent ten years measuring the plates of a photographic survey of the southern skies, totaling almost half a million stars. It is said that, at some point, he had enlisted the help of a few convicts lent to him by the Dutch prison system; by all means, it had been a great undertaking. Although he had had a good training in mathematics, his approach to studying the disposition of the stars relied more on empirical formulae than on mathematical theories. He first sought to extract as much information as he could from the stars in the neighbourhood of the solar system. Gathering the thousands for which he knew their proper motions, he assigned to them a parallax or distance, on the assumption that the proper motion was caused primarily by the Sun's movement, so that the magnitude of their motion was a reflexion of their proximity. Of course, it was impossible to guarantee that the star would not have an independent motion of its own which, if it was true for most stars, would spoil the entire exercise. Be that as it may, for each apparent magnitude, he could now collect the stars that fell in that category, determine their average distance, infer from it their typical absolute brightness and, given that he now presumed to know the volume over which they were distributed, evaluate also their density in space. Although the method was not impervious to criticism, it provided a first approximation to the problem of assessing the range of intrinsic luminosities and their relative weight in the star population. Assuming that the same law applied everywhere, Kapteyn could now go about investigating the overall density of stars in various regions of the Milky Way. In the end, he and von Seeliger arrived at similar conclusions regarding the dimensions of the assembly of stars. Along the plane of the Milky Way, von Seeliger estimated that the density of stars declined to 10 percent of the value near the Sun at a distance of 5000 parsec, whereas Kapteyn thought the decline was faster occurring at 3000 parsec. Along directions perpendicular to the Milky Way, von Seeliger expected a similar decline at a distance of 1000 parsec, Kapteyn preferred to put it at 700 parsec. In both models the Sun was assumed to be close to the center of this disc-like configuration. Thus, although both astronomers were cautious to recall their ignorance about the degree of light absorption in interstellar space, they were grossly misled in their conclusions, particularly in regard to the placement of the Sun in this vast architecture. Coincidentally, in case you wish to keep in your mind a simple image of the range of distances we are now discussing, if we were to accept that 5000 parsec is a reasonable scale for the dimensions of the Milky Way, I could extend the previous analogy in which I placed the solar system within the confines of a single page and the nearest star at one kilometer. Having rescaled one parsec to one kilometer in my representation, I can now liken the radius of the Milky Way to the radius of the Earth and conclude that, in order to draw a chart of the entirety of our star system in such a way that our distance to the Sun will be represented by just one centimeter, I would require a map as big as the surface of the Earth. For all its limitations, the information that Kapteyn and von Seeliger retrieved was obtained with sound methods and the hypotheses they adopted were always stated explicitly. To appreciate their value, one should compare their results to the often more fanciful speculations that were common at the time, made with very little reference to the observations. The British astronomer Richard Proctor, never shy to make bold pronouncements, had noted that in the areas of the sky where the density of bright stars was higher, so was the density of faint stars. The bright stars were expected to be nearby for the most part and it was unlikely that the distribution of distant stars could be influenced by the distribution of the nearer ones. Therefore, he concluded, the faint stars are also nearby and the entire Milky Way was a local phenomenon of very many minute stars distributed among normal stars in a ring around the Sun. Of course, he neglected to observe that absorbing clouds near our solar system would affect all stars alike, so that the regions that appear dense or sparse on the sky, do so for all types of stars, bright and faint alike. By the nature of the methods he had chosen to follow, Kapteyn conceived a project, to be carried out as a collaboration, by means of which certain selected areas in the sky would be studied exhaustively, in order to retrieve as much information as possible regarding the distances and proper motions of the stars within them. In the process of carrying it out, he made a rather surprising discovery. Although the proper motions of the stars were random and in all directions, he noted that they had a preference to point toward two opposite directions in the sky. It must be said that, if the understanding of the general arrangement of the stars was very rudimentary during those years, it was all the more so regarding their motions and what these motions could possibly mean. Kapteyn's discovery led to many tentative interpretations. Some tried to revive the thesis of a central Sun, around which all the stars were revolving. If the orbits happened to be elongated, like the comets' orbits within the solar system, their motions would appear to concentrate on two opposite directions in the sky. Others were of the opinion that the effect might be caused by a cluster of stars surrounding us and passing through our neighbourhood that was seen against the background of the rest of the stars. Given that we did not possess a standard of rest, we might perceive this process as two families of stars moving against each other, in opposite directions. There was also speculation as to whether the disc of the Milky Way might be rotating more or less like a rigid wheel, with some stars racing ahead while others where lagging behind. Some even called attention to the spiral nebulae. If these were distant systems of stars similar to the Milky Way, the two streams of stars might be associated to different arms of the local spiral. Even the idea that we might be witnessing the passing of one such nebula through another could not be ruled out. All in all, the task of unraveling these various possibilities was made perhaps more difficult, not by the lack of information, but by the abundance of it. It was observed that belonging to one stream or the other depended on the spectral characteristics of the star, although the meaning of this association could not be explained. It was also argued that there was a dependence on the absolute luminosity of the stars as well. All the while, other phenomena related to the movement of the stars were uncovered. There seemed to be a preference for the nearby stars to be moving away from the Sun, but not in the sense that our solar system might occupy a central position; instead, it was as if the surrounding environment was undergoing some form of expansion. Furthermore, there was the problem of the high velocity stars. These were seen to move with velocities that often exceeded ten times the speed of the Sun, or that of the average star, and the direction of their motions seemed to avoid certain sections of the sky. From this plethora of information it was difficult to extract the defining characteristics, the fundamental features of the motion of the stars. Jacobus Kapteyn was one of the best known astronomers of his era and part of his reputation endures to this day. In his own time, he was celebrated for having discovered the two streams of stars. However, this phenomenon later receded from the focus of attention and was never fully confirmed nor dismissed. Today he is better remembered for his account of the distribution of the stars. As a description of the Milky Way, it was a flawed and rudimentary model, but for a long time it provided a valuable frame of reference, a standard, a prototype that later work came to amend and improve. Kapteyn spent most of his career as a professor at the University of Gröningen, which did not have an observatory. Being more inclined toward the practical aspects of astronomy than to theoretical questions, he relied on the observations carried out by other astronomers and devoted a great deal of energy to organize international projects and promote the cooperations of various observatories. He strenuously tried to keep this cooperation alive even after the outbreak of the great world war. von Seeliger was also one of the more prominent astronomers of his time. He was based in Munich, where there was an observatory, but his inclination led him to address more theoretical problems, on the rings of Saturn or the explosions in stars and he worked many years seeking a modification to the law of gravitation, for he thought that the law, as formulated by Newton, rendered the world unstable and prone to collapse. In his own day, von Seeliger's reputation was probably based on his activities as a teacher and through the influence that he exerted on the astronomers of the following generation. It is always illuminating to compare the activities of scientists who may have been drawn to study the same problem, but went about it in different manners, according to their temperament, their circumstances, or often influenced by unrelated questions and ideas they were addressing in their research.

Plebeius: Before you proceed with your comparison, I recall that you spoke of three methods that were used to study the distribution of the stars but, if my count is correct, you have only mentioned two.

Albertus: The third method I was referring to differs in that it attempts to exploit the peculiar characteristics of each star. The other two methods treated all the stars indiscriminately, merely as beacons of light drifting through a vast ocean. In fact, for the statistical method the vast disparity in their luminosities was more a hindrance than an aid. At any rate, we just saw that Kapteyn combined the method of proper motions with the statistical method in order to better understand the star distribution. Likewise, this third method, which relies on the type of light emitted by the star, on its spectrum, could not be used in an isolated manner to be effective. Joseph Fraunhofer, in his early experiments, had already decomposed the light of a few stars through his prisms and noticed the differences in their spectra, marked by a variety of dark lines at the positions of various colours. Given that these arrays of dark lines often matched the bright lines in the spectra of light emitted by certain gas flames, it was soon understood that the spectrum of a star ought to reflect in some manner its chemical composition. Nevertheless, the spectroscopy of stars remained for many decades a rather marginal discipline, not for lack of interest or dedication, but because it yielded very few useful results. Two pioneers in this field were Angelo Secchi, a Jesuit astronomer at the Vatican Observatory in Rome, and William Huggins, for some time an amateur astronomer in Britain who, in spite of not having a university education, rose to be President of the Royal Society. Secchi, a skillful and crafty observer, specialized on the Sun for most of his career, but turned in the end to observe several thousand stars motivated by the curiosity to compare the variety in their spectra. Noticing some broad similarities, he attempted a first classification in three broad categories, according to their dominant colours, from white and bluish stars to yellow stars like the Sun, to finally red stars, which he later divided in two separate categories to account for the differences in the dark bands of their spectra. Generally speaking, the colour of a star was regarded as an indicator of its surface temperature and, although the details of the bright and dark lines appearing in the spectrum were difficult to understand, this general classification into four categories was soon interpreted as an evolutionary sequence in the life of the star, which started as bright and white, moving progressively toward the red and cool as it aged and exhausted its fuel. Hermann Vogel, the Potsdam astronomer who was using the Doppler shifts of lines in spectra to study stellar motions, was one of the first to seize upon these ideas in an attempt to elucidate the basic physical processes taking place within stars. But his efforts, like those of some of his colleagues, bore little fruit; the fundamental question of how a star shines, the intrinsic nature of these fantastic cauldrons, remained unanswered for several decades. Nonetheless, the study and classification of spectra continued slowly and laboriously, until it became possible to photograph them and the astronomer could compare many spectra simultaneously in one plate. A vast program to photograph and catalogue thousands of them was started at the Harvard Observatory in Massachusetts. It is noteworthy that much of this work was carried out by women astronomers; you must have noticed that, despite a few isolated cases we encountered along the way, such as those of Hypatia or Catherine Herschel, women had been excluded, not perhaps from studying astronomy, but from the chance to participate in its advancement with contributions of their own. Fortunately this is no longer the case. Before the end of the century, more than ten thousand spectra had already been classified and catalogued at Harvard and, twenty five years later, when the program was finished, the catalogue contained more than 200,000 entries. In the meantime, the classification scheme of Secchi was modified to incorporate more categories and subdivisions, taking into account further details and allowing for greater differentiation amongst the spectra. One of the first persons to make practical use of this wealth of information was the Danish astronomer Ejnar Hertzsprung, who had started his career as a chemical engineer but was attracted to the field of astronomy at age thirty by the challenges posed by the new astrophysics. Very soon he was invited to move to Göttingen to collaborate with Karl Schwarzschild, whom he also followed immediately thereafter when the latter was appointed director of the Observatory in Potsdam. Schwarzschild was probably the most brilliant young astronomer and physicist then in Germany and was working at the time, amongst other problems, on some of the fundamental questions regarding the inner workings of the stars and how these monstrous furnaces seem to remain in a steady state. But we need not concern ourselves with these complicated theories, lest we lose our thread. Hertzsprung was interested then in something more immediate. Even before leaving Denmark, he had assembled a large collection of stars for which he could deduce their distances and absolute brightness, either from their parallaxes or their proper motions, using the same principle Kapteyn had employed a few years earlier, namely, that the inferred brightness of a star could be wrong in individual cases but was fairly accurate in the aggregate. He then compared this information with the corresponding spectra from the Harvard catalogue and noticed that these were well correlated with the luminosity of the stars. There was a progressive sequence of decreasing intrinsic brightness, from the white stars, to the yellow, orange and finally the red stars. He also noticed that there was a subclass of red stars that were intrinsically very bright and that could be recognized directly from their spectra; indeed, they had been classified under a separate category because they showed sharp absorption lines, which appeared blurred and less defined in the other, as it turned out, much fainter red stars. Given that the luminous ones were in fact hundreds of times brighter than their dimmer counterparts and, because of their similar colours, were presumed to have the same temperature, he realized that their extraordinary luminosity was a consequence of their size and was thus led to introduce for the first time the distinction between giant and dwarf stars. Of course, the importance of these observations lay in the fact that these giant stars could be observed to very great distances and, once their intrinsic luminosity was well calibrated, could be used to chart the distribution of stars on a scale that had been impossible until then. The publication of these results did not become widely known and similar observations were made independently a few years later by the American astronomer Henry Norris Russell. Herzsprung continued to observe many clusters of stars, where he knew that all stars were at approximately the same distance, so that he could correlate the relative magnitudes of the stars with their colours. To obtain the latter he used a simple method of placing a coarse grating or grid in front of the object lense of the telescope, so that several images of the same star would be produced. Because different colours are dispersed differently, these images, which were in fact minuscule spectra, appeared at varying distances from each other, depending on the dominant colour of the star, and Herzsprung took these distances as an approximate measurement of their colours. Russell, unlike Herzsprung, had had a good training in physics and astronomy in school, completing his studies in England before returning to the United States. At the start of his career he concentrated on problems related to the computation of orbits in celestial mechanics, but he later turned to stellar astronomy and to the riddles posed by the range of luminosities and the differences in the spectra of stars. He was the first to publish a graphic representation of the correlation between intrinsic brightness and colour already noted by Hertzsprung, in which the plotted positions of the stars seemed to align along a a main column with a separate branch associated to the giant stars. It later became known as the Herzsprung-Russell diagram and proved to be fundamental for the understanding of the life of a star. It is found today in all textbooks that discuss this subject. Russell incorporated his observations into a theory of stellar evolution, according to which stars where born as red giants, formed of cool and diffuse matter, then by gravitational contraction became hotter, shifting their colours toward the yellow and white and eventually cooled again when their fuel was exhausted, becoming red dwarfs. As a consequence, he contended, red giants and dwarfs should have similar masses. This prompted him to return to the study of orbits, this time applied to the case of eclipsing binaries, for only when we see the effects of gravitation can we make use of Kepler's laws and infer the masses of the stars. In this task he had the help of one his bright students, Harlow Shapley, who computed many orbits of binary systems. Russell, a professor at Princeton University, became one of the more influential astrophysicists in the United States, giving periodic lectures summarizing the advances in diverse topics or pointing to important problems of interest. He also wrote numerous articles addressed to the general public. Few scientists have this capacity to keep a bird's eye view over an entire discipline, being able to render a service to those who are perhaps a little lost in the petty details of their own endeavours. When Russell began his yearly visits to the Observatory at Mt. Wilson in California, he became interested in the study of the solar atmosphere, at a time when the relations between the chemical composition of a star and its light spectrum were beginning to be understood. However, the underlying theory of atomic structure was then in its infancy and Russell turned then to the theoretical problems associated to radiation from atoms, a subject of a more fundamental nature, whose usefulness went well beyond astronomy and where he was also able to make substantial contributions. His theory of stellar evolution was disproved in due course, but one could say that after so many undertakings he could afford the luxury of being proved entirely wrong in one of them. In the study of the solar spectrum he collaborated with Walter Adams, another American astronomer of the same generation who had followed George Hale from Chicago to California to participate in the establishment of the Mt. Wilson Observatory, to which he always remained very closely associated; he was to become its director for more than two decades and was also very active in the development of the giant 100 and 200-inch Hale telescopes. Before this time, in 1914, Adams elaborated on the observations that Hertzsprung and Russell had made regarding the two populations of red stars and discovered that not only the sharpness of the lines distinguished the two classes, but their intensities were different, so that making the appropriate choices of lines, the ratio of their intensities were an accurate measure of the star's absolute luminosity, distinguishing precisely amongst giants of different intrinsic luminosities. This was perhaps a relatively minor discovery but it had a great impact. Indeed, until that time the largest distances measured with some degree of precision did not extend beyond 150 parsec; now, the red giants could be identified up to a distance of 2000 parsec, an increase by a full order of magnitude in the territory surveyed. After this discovery, the distances to more than a thousand stars were added to the catalogues, whereas at the beginning of this work only the distances to no more than 400 nearer stars were known with certainty. There is another prominent category of giant stars that proved to be invaluable in the determination of large distances. This concerns the so–called Cepheid variables, a type of star named after the first one discovered in this class, delta Cepheus, in 1785, by two amateur astronomers, Edward Pigott and John Goodricke. These two men made variable stars their specialty and, from their own modest observation post in York, managed to discover several of them in the course of a few years, before the death of Goodricke at age twenty-two. The search for variable stars proceeded very slowly during the next one hundred years. This is quite understandable, given that it involved a tedious process of monitoring repeatedly small areas of the sky with visual comparisons of the apparent magnitudes of neighbouring stars. We saw that Argelander knew only of a handful of these stars and in his invitation to all the lovers of astronomy he had asked them to collaborate in this consuming task. The advent of photography provided a substantial boost to the process of discovery and various ingenious methods were devised for this purpose, such as looking at two plates of the same field, taken at different times, in very rapid succession, back and forth, so that the stars that had varied in their luminosity would appear to be blinking. By the end of the century, the catalogues of variable stars contained several hundred entries. Amongst these there were many that were eclipsing variables; they varied only because another orbiting star of different luminosity would periodically come into the line of sight. As we discussed earlier, this had already been inferred from their double light spectrum by Hermann Vogel or, when the second star was too faint, by a single spectrum being shifted back and forth as the stars orbited one another. The spectroscopist William Campbell later discovered many of these stars and it was realized that close associations between stars is a fairly common phenomenon. Some other stars had been seen to become very bright, then fade, never to brighten again. A handful of these appeared to have undergone some kind of explosive events, others varied with irregular patterns and still a few dozen more belonged to the class of Cepheid variables, very luminous stars, which had very regular patterns, rising and falling by approximately one magnitude in a matter of days. The Harvard Observatory had established a station in Arequipa, Peru, in order to extend its photographic catalogues to the southern skies. This had provided the opportunity to observe extensively those two great gatherings of stars known as the Magellanic Clouds. Henrietta Leavitt, one of the women astronomers working at Harvard and in charge of analyzing these plates, discovered in them some two thousand variable stars, almost as many as were known then throughout the skies. Out of this collection she selected a couple dozen Cepheid variables with well determined periods and, realizing that all these stars were at approximately the same distance, she observed that their periods were well correlated with their intrinsic brightness. This entailed, once again, that if one could calibrate the absolute luminosity of the Cepheid variables, they could be used as an indicator of distance. Ejnar Hertzsprung was once again the first to attempt this task. Using just a handful of Cepheid variables for which he could deduce proper motions and tentative distances, he concluded that the Small Magellanic Cloud was at a distance of 10,000 parsec, the largest distance then known to any object in the sky.

Plebeius: To make clear what you have just described, I presume that the red giants as well as the Cepheid variables are exceptional stars so that, if we rummage in the neighbourhood of the Sun amongst the stars for which accurate distances can be determined by trigonometric parallaxes, we would find very few of those. On the other hand, it is thanks to these few stars that we can calibrate their absolute brightness and deduce the distance to the more distant ones.

Albertus: Exactly. Within a radius of one to three parsec you are likely to find just one star of average luminosity, not dissimilar to the Sun. Thus, going tenfold in distance, say, to 20 parsec, you will be able to net about a thousand stars. Amongst these there will be a few of the more modest giants. However, the truly luminous red giants and Cepheids, which can have several thousand times the luminosity of the Sun, are also more infrequent and are not likely to be found in a typical sample of just a thousand stars. Therefore, you must go farther afield to 200 parsec and round up a million stars. Now, of course, they are so much farther away that the accurate determination of their distances is more problematic. I should point out that one gets quite a different impression from looking into the sky at random, selecting stars by their apparent brightness. The average stars fade away once they are outside a very limited volume; the great giants, although very rare, compensate by being visible at immense distances. Thus, at any of the brighter magnitudes these unusual luminous giants will compose a sizeable fraction of all the stars seen. It is only when you reach to the fainter telescopic magnitudes that the pool of giants begins to dwindle and the average stars become dominant. This was exploited, for example, in the early studies of globular clusters, those distant associations of thousands of stars, first observed by Herschel, two or three of which are visible to the naked eye simply as luminous stars. These almost perfectly round objects, like colossal bee-hives, are swarming with stars, sometimes numbering several hundred thousand. The giant stars amongst these are few although, clearly, they were the first to be identified individually at such great distances. Under such circumstances, it was not an impossible task to search for Cepheids within those crowded clusters. The young American astronomer Harlow Shapley, whom I mentioned already, decided to pursue this goal when he moved to the Mt. Wilson Observatory at the start of his career. He had turned to the study of Cepheids because of his expertise in computing the orbits of double stars. He was able to show that these stars, being so luminous and with such short periods, could not be eclipsing binaries, for it would have meant that the orbiting stars were revolving one inside the other. Therefore, these were pulsating stars, although the physical processes responsible for this phenomenon were not understood until a few years later. The lengthy project started by Shapley took him several years to conclude, but his published results in 1918 were quite sensational. Using a larger sample of stars, he improved on the calibration of Cepheid variables obtained by Hertzsprung. In some of the nearby clusters he managed to find some of these stars, monitor their periods and, in this manner, deduce how distant they were. For other clusters, where the distance and the crowding of stars made this difficult, he exploited the remarkable similarity amongst globular clusters and, assuming that they were all alike, he compared their sizes on the sky, so that he could estimate how much farther the smaller ones were. He also compared the thirty brightest stars he could find in each cluster and, from their dimming in the farthest clusters, he obtained a separate estimate of their distances. It had been known for a long time that the system of a few dozen globular clusters that are suspended in the neighbourhood of the Milky Way were concentrated in one hemisphere of the sky, particularly in the area surrounding the constellation of Sagittarius. Shapley correctly conjectured that these clusters traced the true structure of the star system, that in the flattened disc of stars presumably forming the Milky Way, our solar system might not occupy a central position, as had been assumed until then, but could well be close to the edge instead, with the distant center lying in the direction of Sagittarius. Furthermore, Shapley found that the dimensions of the star system had been grossly underestimated. According to his own estimates, its diameter was probably close to 100,000 parsec. As it turned out, Shapley had been misled somewhat by his conclusion that there was no interstellar material on the line of sight to the clusters capable of absorbing part of the light, exacerbating their dimming and making them appear more distant than they really were. Nevertheless, these enormous distances were more than an order of magnitude in excess of what had been considered previously. It may now be more convenient if we use a thousand parsec, a kiloparsec, as a unit of distance; the diameter of the star system for Shapley extended then to 100 kiloparsec. We saw a moment ago that Hertzsprung had estimated the distance to the Small Magellanic Cloud to be 10 kiloparsec. According to Shapley's own calibration, the cloud was at 17 kiloparsec; at all events, it belonged well within our own stellar system. Without doubt, Shapley's model for the stellar system stood in sharp contrast with the ideas put forward by Kapteyn. Kapteyn had not only assumed that the Sun occupied a central position, but the entire system of stars barely extended beyond a few kiloparsec; in fact, according to his own publications shortly before his death in 1922, the density of stars was supposed to decrease by 50 percent at a distance, we should now say, as small as 600 parsec and by 4 kiloparsec the density had thinned probably to merely 5 percent of its value in the solar neighbourhood. Clearly, it was difficult to reconcile the two models. By Shapley's account, Kapteyn had only surveyed the region in our immediate neighbourhood. His radical solution to the sidereal problem, as it was then called, immediately found its supporters as well as its detractors. The Cambridge astronomer and physicist Arthur Eddington, whose opinion seemed to count more than most, was soon amongst the supporters, but the distinguished American observer Heber Curtis, at the Lick Observatory, seriously objected to Shapley's calibration of distances. In retrospect, the notion that the center of the Galaxy was located in the constellation of Sagittarius at great distance from the Sun did not seem so far fetched; ever since the skies of the southern hemisphere had been observed in any detail, it had become apparent that the density of stars was not uniform along the Milky Way, but it increased considerably toward the region of Sagittarius. This was apparent even to the naked eye; yet, somehow, it had not been considered to be particularly revealing. Shapley also observed that the globular clusters for the most part avoided the plane of the Milky Way. He attributed this effect, not to dark material obstructing the field of view, but to some dynamical effect instead, suspecting that the gravitational forces in this extended region with high density of stars was unfavourable for the formation of these extremely compact agglomerations of many thousands of stars. In his own grand scheme for the Galaxy, Shapley envisioned an evolutionary process by means of which the globular clusters were disrupted and absorbed as they progressively fell into the center of gravity of the star system. The landscape in 1920 looked very different from what had been anticipated only a few decades earlier. At one time, the sidereal problem may have been seen in the same light as the riddle to solve the vagaries of the planetary motions. However, the prospect of finding a central star, around which all other stars in the Milky Way were rotating, had turned into a problem of a far greater magnitude and complexity. There were not just a handful of stars but many millions of them. The time scales involved were so extended that one could barely infer any motions within it and, what little was perceived, appeared somewhat chaotic. It was not thought at the time that the task was to find some new law of nature, for the stars seemed to be merely obeying the law of gravitation in their motions. Instead, the difficulty resided in applying this simple law to such a complicated and monstrous artifice as the entire Milky Way. Thus, returning again to the task of providing a general outline of the star system, I should point out that we are pursuing two slightly different lines of argument. Firstly, there was the puzzle concerning the scale of distances, the true dimensions of the stellar system. The second problem demanded an explanation of the inner workings of the Galaxy and the general motions of the stars. The problem of the distance scale was solved through a long sequence of steps. As the years went by, new ways of estimating distances were introduced and every new method was compared to the older ones, improving the accuracy of all. Some of the early determinations of distances had not taken into account that there are in fact different classes of Cepheid stars, with similar periods but with different luminosities; this inevitably led to some confusion. Nevertheless, as we shall see, the Cepheid variables always remained an important rung in this cosmic ladder of ever increasing distances. Regarding now the problem of the structure of the Galaxy, it must be said that, for a long time, many astronomers had entertained the notion that some of the nebulae seen in the sky were vast arrays of stars similar to our own Milky Way. It was also suspected that the disc of the Galaxy, when seen from afar, could have the same spiral structure that some of those nebulae had. From their mere appearance, it seemed that these systems sustained themselves by rotating around the central whirlpool, although it was not understood at the time in which sense the spirals were turning, whether the central regions were rotating faster with the arms further out trailing behind or, conversely, that the arms were leading and opening up at the vanguard, as some had proposed. In any event, the investigations of the following years did not proceed completely in the dark; astronomers already had an inkling of the type of answers they wanted to extract from the data they gathered. Two of the major objectives were the detection of the rotation of the galactic system and, secondly, verifying the possible existence of spiral arms. A third problem, still unsolved, was the degree of light absorption in interstellar space. On this matter there had been much disagreement and controversy. The first person to attack the problem of the large scale motion of the stars was the Swedish astronomer Bertil Lindblad. After spending a couple of years at the Mt. Wilson Observatory early in his career, collaborating in the work of Walter Adams to detect giant stars by the intensities of their spectral lines, Lindblad returned to Sweden, first to Uppssala and a few years later as director of the Stockolm Observatory. His interest in the problem of the structure of the Galaxy stimulated later a number of astronomers in northern Europe to devote themselves to the same subject and continue his work. In 1925, at age thirty, Lindblad published an article that contained a discussion of the rotation of the Galaxy, the conception of which was without doubt aided by the new models of an enlarged galactic system with the solar system far removed from its center. In it he found a simple explanation for some observations that had puzzled astronomers in previous years. The great majority of the stars in the neighbourhood of the Sun are seen to move with velocities of a few tens of kilometers per second; however, there is a class of high velocity stars with typical speeds in the hundreds of kilometers per second. These stars, which had been expected to make the phenomenon of rotation manifest, were the least confined to the disk of the Galaxy, contrary to the expectations; they were in fact crisscrossing it at various angles, although, herein lay part of the puzzle, their velocity vectors tended to point toward one particular direction in the sky. This contradicted the predictions stating that the stars with the greater velocities should have the greater scatter in their peculiar motions. Similarly, it was expected that the system of globular clusters, which showed great velocities relative to the Sun, would be constrained to remain close to the rotating disk, but the observations proved the opposite, the globular clusters avoided the Milky Way. Lindblad proposed that the Sun and the majority of the neighbouring slow moving stars were partaking of the rotation of the Galaxy along the plane of the Milky Way, with a velocity perhaps as high as 300 kilometers per second, around a center that lay in the direction of Sagittarius, coincident with the center of the system of globular clusters, which he assumed was at a distance of 12 kiloparsec. According to this view, each of the high velocity stars could very well be rotating around the same center, in an orbit whose plane did not coincide with the disk of the Galaxy, but all these stars were not rotating as a group. The fact that their velocity vectors were pointing toward one direction in the sky was caused by the high velocity of the Sun dashing through this swarm of stars moving in all directions. Likewise, one could now assume that the globular clusters had no coordinated motion of their own, other than their shared attraction toward the galactic center. Lindblad divided the Galaxy in various subsystems, each with its own peculiar rotation. The large disk of stars that forms the Milky Way was itself divided into concentric rings at different distances from the center and each was allowed to rotate with a different speed. Lindblad's thesis proved to be very fruitful and the study of the rotation of the Galaxy was to occupy the rest of his career, as well as that of numerous other astronomers. One of these was the young Dutch astronomer Jan Oort who had started his studies under the guidance of Jacobus Kapteyn. He immediately seized on Lindblad's ideas and set out to confirm the rotation of the Galaxy in its details, in particular regarding the effects of differential rotation on the various rings that form the disc. In a rather simplified manner, the situation is analogous to the one found by a skater on an ice rink where everybody goes in circles; the more experienced skaters, capable of sharp turns, skate fast close to the center, whereas the beginners skate slowly along large circles, close to the hand-rails at the edge of the rink. Think now of our skater of medium skill gliding half-way between the two and wishing to verify this phenomenon of differential rotation. By analogy with the astronomers, we will allow him to do so only by measuring radial velocities from his own position, because he wants to observe the motions to rather great distances, so that proper motions on the sky are out of the question. This is what our skater finds: When he looks straight ahead or behind him, he sees skaters who are on the same circle he is and skating at his own speed, therefore, he detects no relative velocity. If he then looks at right angles directly toward the center or to the edge of the rink, he finds skaters who have different velocities from his own, but these velocities are directly perpendicular to the line of sight, so he does not detect any radial velocity on these directions either. Now, let us say, he looks at 45 degrees between these two perpendicular axes, for instance, ahead of him and toward the center. You can well picture in your mind that he sees skaters who are going faster than he is and who are already turning away from him bending along their sharp circles. He detects now a velocity of recession. Similarly, ahead of him and toward the outside he sees slower skaters apparently bending toward him, he detects a velocity of approach. All this Jan Oort wrote in mathematical form and then searched in the data already available to see whether he could confirm these conclusions. Indeed, all the expected effects were there once one knew what to look for. From the velocities of rotation of the stars he deduced a radius for the disk of the Galaxy of 5 to 6 kiloparsec, about half of what Lindbladt had estimated, and he also verified something that is apparent to the skater on the ice rink as well, namely, that the velocities of recession or approach seen at 45 degree angles increase as we look further away from our own location. In reality, of course, the task at hand was much more complicated than my simplified version may indicate and it was necessary to sort out what was essential from the other peculiarities and random components in the general motion of the stars. I have already mentioned the odd observation that the stars in the vicinity of our solar system seemed to have peculiar motions that depended on the type of stars one was observing. There were also the two streams of stars discovered by Kapteyn, which at one time were thought to be three different streams. Kapteyn had speculated that the effect of the two streams was caused by the disparate velocities of rotation of the stars around the center of the Galaxy, some moving faster, others more slowly or even in the opposite direction, creating these effect of two streams going one against the other. However, if this had been the case, the direction of motion of the two streams would have been perpendicular to the direction pointing toward the center of the Galaxy. Instead, it was now found that the motions of the two streams were aligned toward the center itself, so that a new interpretation was in order. The conclusion reached at the time was that the stars do not orbit the Galaxy in perfect circles, but do so in ellipses, which are randomly distributed. It is as if our skaters on the rink were to skate each on an ellipse of his own choosing; we would soon find numerous skaters constantly crossing our path once toward the outside, then toward the inside of the rink, producing these effect of two streams, one coming from the center of the rink, the other falling into it. In any event, coming back to our main argument, there is one problem that we have not yet mentioned, namely, the determination of a standard of rest for the entire star system. How do we decide that the velocity of the Sun around the Galaxy is of 220 kilometers per second? We could argue, for example, that removing larger and larger values from the Sun's velocity, the asymmetry that was found in the high velocity stars, all flying toward the same direction in the sky, would eventually disappear and that would indicate the true velocity of the Sun. Nevertheless, we are not certain that the system of high velocity stars does not have a rotation of its own, so that we are only measuring relative velocities, and the same can be said with respect to the system of globular clusters and all other subsystems of the Galaxy. Ultimately one would like to find a fixed frame of reference outside the Galaxy. For the time being, from the work of Lindblad and Oort, and those who set out to verify their results, we can draw the tentative conclusion that the Sun is circling the Galaxy with a speed of 220 kilometers per second at a distance between 5 and 12 kiloparsec from its center. Under these assumptions, it would take the solar system roughly 250 million years to complete a revolution. Now, taking into account the velocities of rotation at various distances from the center of the Galaxy, Lindblad and Oort were able to derive a value for the mass contained within each circle, assuming, of course, that it is the force of gravitation that bends the orbits of the stars, preventing them at any time from escaping off the Galaxy along a straight path. Their estimates were very close to what we believe today to be the mass of the Galaxy, in the order of 200,000 million suns. At the same time, they were able to compare these estimates with the counts of stars that had been performed by Kapteyn, Seeliger and their successors. Here they found indeed a great discrepancy. As we saw earlier,

Kapteyn had reasoned that, at a distance as close as one kiloparsec from the Sun, the density of stars was one half of what it is in its immediate vicinity. This conclusions renewed the debate regarding the possibility of considerable absorption of starlight throughout the Galaxy. The problem of absorption had long been an irritant in most astronomical investigations, for it was never straightforward to detect the presence of absorbing gas or dust on the path to distant stars. The effect of a cloud of gas along the line of sight to a star is not only to dim the intensity of the light, making us estimate erroneously the distance to the source, but it also disturbs the balance between the different colours composing the light. This is simply due to the fact that shorter wavelength photons in the blue region of the spectrum are absorbed more easily. However, it is also true that giant stars tend to be intrinsically redder than less luminous stars and, clearly, they can be seen to larger distances. Therefore, if we were to probe deeper into space, without knowing whether there is absorption or not, we would detect a reddening effect and might be misled to believe that there is indeed absorbing gas along the line of sight, when the cause of reddening was altogether different. Kapteyn and von Seeliger had been very much mindful of the uncertainties regarding absorption when estimating the rate of decline in the density of stars and the size of the Milky Way. Kapteyn at first assumed considerable absorption, more than we now know there is, but later dismissed it entirely, which in part led him to underestimate the extension of the disc of our Galaxy. Much earlier, Friedrich Struve had struggled with the problem of absorption and he came to the conclusion that it had to be considerable, but he did not have the means to demonstrate it. Later, when Harlow Shapley was assessing the distances to the globular clusters surrounding the Milky Way, he decided that there was practically no absorption and his exhaustive study had some influence for a number of years. In retrospect, one substantial contribution had been made by the American astronomer Edward Barnard, starting in 1889. Barnard is an interesting character in the history of astronomy. He had received very little formal education, having worked since the age of nine at a photographic studio in his native Nashville, in Tennessee. Nevertheless, through willpower and dedication to science, in the end he had the privilege of being an observer at the great Yerkes and Lick Observatories. He was an enthusiastic comet hunter, but one of his prescient contributions was to attempt to photograph the dark regions of the Milky Way. In this he may have started a tradition, by now proven fruitful on repeated occasions, namely, to observe carefully where there seems to be nothing. From his observations, he argued that the shapes of the dark regions resembled too closely those of the luminous gaseous nebulae within the Milky Way and he suspected that they were clouds of gas, not hot enough to emit radiation. He also pointed to the fact that the presence of stars in front of the clouds seemed to end abruptly and it was unlikely that a hole behind these stars could be so well aligned with the view from the Earth that no stars would be seen at the other end. This work was continued only decades later, after 1920, by the German astronomer Maximilian Wolf, who carefully counted stars in front of the dark clouds, as well as in directions of unobstructed view, in order to deduce the distances to these presumed dark clouds, a method that allowed him to obtain very reasonable estimates. The definitive proof of interstellar absorption came only a few years later through the work of Robert Trumpler, of Lick Observatory. He carried out an extensive study of 100 open clusters found within the Milky Way. Unlike the globular clusters, these consist typically of only several hundred stars and usually have very irregular shapes; they are recognized by the concentration of the stars with a density higher than the surrounding medium. Trumpler used careful spectroscopic and magnitude measurements, as well as sizes and other characteristics of a few nearby clusters, whose distances were known by independent means, as a standard against which the fainter and more distant clusters should be compared. Since he had a large sample of clusters, at the end he was able to reach the conclusion that, on average, the more distant clusters appeared to be larger and more luminous than those nearby. Considering that the Earth does not occupy a privileged position, the more reasonable conclusion was that the distances to those fainter and more distant clusters had been overestimated due to absorption, so that bringing them closer to the observer and accounting for absorption, diminished the requirement to make them excessively large or bright. Indeed, one could tune the degree of absorption, so to speak, in a manner that made all open clusters at the various distances of approximately the same size and absolute luminosity. In subsequent years these determinations were largely corroborated; for instance, spectroscopic observations of stars began to show absorption lines which often did not share the oscillations seen in the lines of the star's spectrum caused by the motions of the star. Clearly, the absorption was produced by gas independent of the star and quite often this absorption was proportional to the distance to the star. Likewise, when in later years astronomers began to conduct extensive counts of spiral nebulae on the surface of the sky, they found that the diminution in the number of nebulae closer to the plane of the Milky Way could be explained quite accurately by the increasing level of absorption as the line of sight pierced a larger section of the Milky Way. By this time, it was almost beyond doubt that the spiral nebulae were indeed extragalactic. In the same manner, the earlier observations of Shapley who had dismissed the existence of absorption could now be explained by the fact that the globular clusters he had studied lie for the most part outside the plane of the Galaxy and are therefore free of excessive absorption.

Plebeius: You mentioned that these astronomers who sought to demonstrate the rotation of the Galaxy and its true dimensions already had, from the observation of the spiral nebulae, a hint, a presentiment, of the answers they were looking for. But it is not clear, from your account, when the Galaxy began to be conceived as an isolated unit within the Universe, as opposed to the Universe itself. Was it not disputed for a long time that these nebulae might all be clouds of gas distributed amongst the stars? And could not the stars have been distributed out to infinite space beyond the reach of telescopes? At least since the time of Immanuel Kant people had discussed this notion of island universes, the vast condensations of stars like the Milky Way, at great distances from one another. Did the astronomers come to confirm this conjecture of external universes before they had a clear understanding of the one we inhabit?

Albertus: Progress on these matters was attained gradually on all these fronts simultaneously and not without hesitations and reversals. I have been emphasizing the elucidation of the distance scale throughout the Cosmos but, as you rightly point out, when I speak of the dimensions of the Galaxy as opposed to those of the Universe at large, I am assuming a distinction that, in historical terms, has been made only recently. However, when assessing the dimensions of the Galaxy and the Universe outside it, we face a situation that is different from the one faced by Bessel and the astronomers of his generation. To break out of the confines of the solar system and measure the scale of distances to the nearest stars, they needed to know the distance from the Earth to the Sun. Indeed, by using the method of parallaxes, the distance to a star was expressed in terms of the diameter of the Earth's orbit. The task of determining the distances within the Milky Way and beyond became a good deal more involved, with a number of intermediate steps, often tentative and contested. Because a variety of measuring rods were used, so to speak, it was seldom easy to disentangle their contradictory conclusions. At the turn of the century, and even after the work of Kapteyn and von Seeliger, as we have seen, knowledge about the distribution of the stars and its reach was still very vague. The dilemma posed by the nebulae, whether they were distant congregations of stars or gas clouds nearby, had not yet been resolved and, in the event they were composed of stars, it was uncertain whether they played a secondary role, as offsprings or attendants, at the outskirts of the Milky Way. The stars were the only reliable signposts that carried our view into the depth of space; it was perfectly agreeable to conceive the Universe as a gigantic disc of stars, becoming more sparse at the edges, until it dissolved into a vast, silent, devastating emptiness. In any case, one has to remember that bridging the gap between the range of distances commensurate with the solar system and the distances amongst stars had required more than a thousand-fold change of scale, without any markings or signs to guide the eye in this immensity that lay between them. Could it not be that on a scale a thousand times bigger than the Milky Way another great structure would be present, well beyond the reach of our eyes, too distant for any source of light to travel across? Kant had speculated that the confused mass of the nebulae were gatherings of stars at enormous distances and that these nebulae, in turn, formed associations on a greater scale, on an ever ascending hierarchy of structures, the end of which we could not fathom. This is appealing as a flight of fancy, but it is only fancy; it is also a flight from the more laborious task of following the lead of our senses and corroborating the evidence they bring to us. The astronomer had to pursue a more cautious route. Today we can celebrate the foresight of those speculations, although they offered few hints on how to go about verifying their truthfulness. In the end, the Milky Way was indeed found to be surrounded by a huge emptiness. Despite its colossal size, only on a scale almost a thousand times its own dimensions was the next structure first apparent, a structure that was not hidden beyond our reach; however dim and evanescent, it was also shining in the night sky. The distribution of the spiral nebulae, well beyond the reaches of our Milky Way, eventually provided the essential clues that led to the resolution of the riddle concerning the large scale distances in the Universe. For a very long time, however, these spiral nebulae had failed to attract the attention of most astronomers. William Herschel had studied all types of nebulae extensively, he had classified and catalogued them and speculated on their role in the birth of stars. During this process, he vacillated many times, considering each type of nebula, whether they were gaseous or made of stars. After his time, nobody paid nearly as much attention to the subject as he had. As surprising as it may seem, for most of the XIXth century, astronomers considered that the nebulae were only of marginal interest or, at the very least, that the problems posed by the stars, and by the Sun and planets as well, were a scientific trough rich enough that made the neglect of the more elusive nebulae completely justified. The less cautious amongst them were not reluctant to state that the nebulae lay outside the mainstream of astronomical research and with their peculiar shapes and structures it was best to leave them to the care of amateurs and curiosity seekers. Indeed, we could single out three amateur astronomers who made notable contributions to stimulate the interest in the nebulae and we should call them amateur only to the extent that they did not depend on their astronomical researches to provide for their livelihood. They were William Parsons, an Irishman, William Huggins, an Englishman, and Henry Draper, an American. All three had in common that they belonged to well-to-do families and each had been privileged with an excellent education. Astronomy was not their nominal profession but they devoted much of their energies to it. William Parsons belonged to the British aristocracy; as such, he held various administrative posts, eventually as a member of the House of Lords, but his main activity remained the construction of telescopes. Emulating William Herschel, he wished to increase the light gathering power of the telescopes available at the time and thought of developing fluid lenses for a refractor. Eventually he settled on a reflector, developing new alloys for large mirrors, of which he constructed a good number, always improving on his grinding and polishing techniques. His crowning achievement was a giant mirror, 1.83 meter (6 feet) in diameter and weighing

4 tons, that he installed in a great telescope in his own state. It would remain the largest telescope in the world for several decades, although perhaps not the most effective one. It consisted of a 16.46 meter (54 foot) tube, slightly over 2 meters in diameter, that could not be steered arbitrarily, although it was possible to swing it upward or downward between two massive piers of masonry. A few weeks after its installation in 1842, Parsons made his most notable observation, the discovery of spiral nebulae, of which he made several drawings that were widely circulated and that brought him well-deserved fame. A few years later he was named President of the Royal Society. William Huggins, a Londoner who did not belonged to the aristocracy nor did he have excessive wealth, managed nonetheless to devote his life to astronomy, in addition to a broad range of interests in the liberal arts. Some fifty years after Parsons, in 1900, when he was 76 years of age, he was also honoured with the presidency of the Royal Society. At the time he was receiving a pension from the government in recognition for his contributions to society. He had married a woman more than twenty years younger than he was although, living a long life, they had many years together and much of his astronomical work was a joint collaboration. Huggins had found his calling in astronomy when he learned of Fraunhofer's spectroscopic observations of the Sun. Although spectroscopy was not at the time a particular field of interest for the majority of astronomers, Huggins immediately saw it as exciting and promising, the path to learn more about the stars beyond the precise determination of their positions in the sky. As a spectroscopist, one of his more important contributions came in 1894 when he turned his telescope to a planetary nebula in the constellation of Draco and discovered that its spectrum resembled the typical spectra of glowing gases that could be obtained in the laboratory. This was the first convincing proof that at least some nebulae were indeed gaseous and not distant unresolved clusters of stars. I also mentioned the name of the American Henry Draper because he was one of the pioneers in photography. Although he lived only to be forty-five, being the son of a prominent chemist and historian of New York, who had taken an interest in photography immediately after its discovery and had himself been the first to photograph the Moon, the young Henry had an early start in his career. He completed his medical studies too soon to be allowed to graduate and visited Britain in the meantime, stopping at Birr Castle where William Parsons had installed his telescope. He was also later to correspond with Huggins concerning the development of astronomical photography. The Drapers belonged to the class of the more prosperous American families who, in those times, not only took pleasure in contributing to the advancement of science but in part derived their social standing because of it. Henry Draper was for a number of years professor of chemistry and dean of the Medical School at the City University of New York; if he was not particularly wealthy in his own right, he married into a wealthy family and he and his wife enjoyed a prominent social role. His many attempts to perfect

the art of astronomical photography eventually paid off; in 1880, for instance, he could photograph the nebula of Orion with a degree of detail never seen before, neither in photographs nor sitting directly at the telescope. To attain this goal it required a fifty minute exposure and he was as proud of his plate as he was of his clock mechanism responsible for the tracking in such a long exposure. It was only after long exposures had become available that photography began to have a greater impact in astronomy, not only for the wealth of details it could reveal in the observation of nebulae, but also in the detection of faint stars that could not be seen by simple visual inspection. Henry Draper's name was later associated to the Draper catalogue of photographs of stellar spectra that I mentioned earlier and that was carried out at the Harvard Observatory many years afterwards through a memorial fund instituted by his widow. A few years after Draper's work, another English amateur astronomer, Isaac Roberts, experimented successfully with astronomical photography, obtaining the first image of the diffuse nebula that surrounds the stars in the Pleiades and, a little later, with a long exposure of the great Andromeda nebula, he observed for the first time that it too displayed a spiral structure. At about the same time, around 1890, William Huggins had also obtained a spectrum of the Andromeda nebula which persuaded him that, unlike the planetary nebula he had observed years before, Andromeda's spectrum was a continuous one, typical of starlight. Thus, the question first discussed by William Herschel as to the true nature of the nebulae still remained open. Meanwhile, in 1885, a new star appeared near the center of Andromeda, reaching 7th magnitude, an appreciable fraction of the luminosity of the entire nebula. At the time, the observation provided ammunition to those who contended that the nebula could not be like the Milky Way, an unresolved system of stars at an enormous distance, if one single star was able to compete in luminosity with a myriad others, perhaps hundreds of millions of them, that gave rise to the extended and seemingly smooth appearance of the nebula. In fact, spirals first attracted the attention of some astronomers because they were conceived as the possible birthplace of stars and planetary systems. A hundred years earlier, Pierre Laplace had sketched an ambitious theory explaining the origin of the solar system from the condensation of a disc nebula. In any event, toward the end of the century, for the first time, a young American astronomer by the name of James Keeler, one of the more respected and authoritative of his generation, began to carry out an extensive project of photographing spiral nebulae. He had studied briefly in von Helmholtz's laboratories in Berlin and later divided his time between the Allegheny Observatory in Pennsylvania and the Lick Observatory in California, where he was the first resident astronomer after its foundation in 1886. It was in California that he did his most important work. From the early results of his survey, he estimated that more than 100,000 nebulae might be found across the surface of the sky. Unfortunately, this project was cut short only two years after its inception, when Keeler died suddenly at age forty-two, and was not undertaken again until a decade later. The rise of astronomy in the United States was quite spectacular. During the early decades of the XIXth century there was not a single observatory in all its territory. However, by the beginning of the next century, not only were there many excellent installations, but the country possessed the most powerful telescopes in the world. In the course of a few years, especially in a field such as the study of the nebulae, the most important advances and discoveries originated in the great observatories of California. The Lick Observatory, on Mount Hamilton, south of San Francisco, was the first to be built at higher altitude. The funds toward its construction were contributed by a wealthy entrepreneur who had no heirs and who had once contemplated building a large pyramid, of Egyptian proportions, as a monument to his memory. Although he was later persuaded to fund a scientific organization, the observatory served all the same as a memorial monument, given that he was buried on its grounds. The Lick Observatory was equipped with a 91 cm (36 in) refractor and later acquired a reflector of the same size, which was the instrument used by Keeler in his survey. The next great American observatory was established almost a decade later, in 1897, in Williams Bay, Wisconsin, outside the city of Chicago. The main force behind this new enterprise was the young astronomer George Ellery Hale, the son of a successful Chicago industrialist and an enthusiastic student and observer of the Sun since his adolescence. At age twenty-five he was appointed professor of Astronomy at the newly founded University of Chicago and the small observatory he had been operating in the backyard of his home became temporarily the University's observatory. Meanwhile, Hale was able to persuade another wealthy Chicago businessman to contribute the funds for a large telescope which became the crown jewel of the Yerkes Observatory and, with a lens over a meter (40 in) in diameter, the largest refractor in the world. Hale was also able to persuade his father to donate the funds for a 1.52 m (60 in) mirror, but the University of Chicago found itself unable in subsequent years to secure the resources for the construction of a new telescope. Meanwhile, Hale was already searching for an appropriate location in California to set up a solar observatory. With his great skills for organization, what started as a modest station of the Yerkes Observatory at Mount Wilson, outside Los Angeles, would develop in a few years time into one of the great observatories of the world. Hale was also able to obtain the funds to build a telescope for the 60 in. mirror, which went into operation in 1908. By that time, Hale had already moved permanently to California and had brought along a few of his associates, such as George Ritchey and Walter Adams. The important contributions of Walter Adams and Harlow Shapley I mentioned earlier were carried out with this telescope. Even before the new telescope was finished, Hale initiated a campaign to build another reflector at Mount Wilson with a 2.54 m (100 in) mirror, which would be financed, once again, with private contributions. The construction and installation of this 100 ton telescope took a decade to complete and was quite a feat of engineering. The telescope saw first light in 1917. Despite all these occupations, Hale maintained an interest in solar astronomy throughout his career. He also devoted a great deal of time and energy to activities that were not perhaps strictly scientific but that make the scientific enterprise more effective. While still in Chicago and not yet thirty years old, he founded with James Keeler the Astrophysical Journal, the first such publication in the United States. At a time when the bulk of the work at most observatories was concerned with positional astronomy, Hale insisted on the importance of astrophysics for the future of the discipline. He was also a founding member and one of the more ardent supporters of the American Astronomical Society. In later years he also became a vocal member of the American Academy of Sciences and, after his retirement from the position of Director of the Mount Wilson Observatory, he took a leading role in the foundation of the California Institute of Technology. The last great campaign he initiated was directed toward the construction of a great 5 meter (200 in) reflector, which he did not see to completion. Appropriately named after Hale, the giant new telescope went into operation some twenty years later, in 1948, at Palomar Mountain, northeast of San Diego. But let us now return to the work that Keeler had left unfinished at the turn of the century. Ten years later, the Mt. Wilson Observatory had already come into operation. George Ritchey, one of Hale's collaborators, was a superb optician who had polished the 1.52 meter (60 inch) mirror and was responsible for its mounting at Mount Wilson, as he would do later with the 2.54 meter (100 inch) mirror. Somewhat of a peripatetic man who was to work for a good many observatories around the world, he was also a skillful observer. Using the 1.52 meter reflector, he managed to obtain some excellent plates of a few spirals that seemed to be barely resolved into stars at their very outskirts. He estimated then that the number of stars appeared to be too small to make these systems comparable to a structure such as the Milky Way. Nevertheless, the fact that the spectra of several spirals had been found to have the characteristics of starlight was an argument that carried its own weight in favour of the island universe hypothesis. Likewise, the German astronomer Maximilian Wolf, taking for granted that the spirals were condensations of stars, set out to estimate the distance to Andromeda by comparing the dark lanes seen along its spiral arms with the similar obscuration that is seen along the plane of the Milky Way. The method was quite sensible although his results were definitely erroneous. According to his estimate, the distance to Andromeda was of 10,000 parsec. This would have made the nebula much smaller than the Milky Way, placed just beyond its edge, if one accepted Kapteyn's own assessment of the distribution of stars in our Galaxy. Moreover, as we discussed a moment ago, only a few years later Harlow Shapley would propose that the Milky Way extended to 100,000 parsec. In that case one would have had to admit that Andromeda was subordinate to the Galaxy and merely a denser cluster of stars contained within it. Using a different method, there was also an attempt to compare the surface brightness of the spiral nebulae to that of the Milky Way, that is, to compare how much luminosity they provide per surface area. If the nebulae were distant galaxies, one would have expected similar values for both. The distance obviously dims the luminosity of the source, but this effect is exactly compensated by the greater number of stars that is now contained within the same element of area on the sky. The results showed, quite correctly, that the nebulae were almost ten times brighter than the Milky Way per unit surface area. As it turned out, the discrepancy was not due to any intrinsic difference but to the fact we see the Milky Way from within and our view is heavily affected by absorption. In any event, these were some of the early attempts to make sense of the spiral structure. Let us now turn to the observations of Vesto Slipher and of Adrian van Maanen that figured prominently in the debate over the nebulae and added to the confusion and perplexity that prevailed for quite a few years. Vesto Slipher, an American trained spectroscopist, was recruited by Percival Lowell to come to work at his newly founded observatory in Flagstaff, Arizona, even before he had finished his studies. He was to remain at the observatory for the rest of his career, succeeding its founder as director, a post he held for decades, and becoming prominent as well in the organization of several astronomical societies. Following the inclinations of its founder, an enterprising and charismatic astronomer, most of the work at the Lowell Observatory was directed toward planetary astronomy with the intent of investigating the origins of the solar system. It was in this context that Slipher was put in charge of obtaining spectra of planetary and spiral nebulae. He was the first to obtain a spectrum of some of the clouds in the constellation of the Pleiades and he showed that it coincided with the spectrum of some of the stars in the cluster, proving that some nebulae, called then reflecting nebulae, were not emitting their own light but reflected and scattered the light of the stars hidden underneath them. Vesto Slipher is remembered today for a more momentous discovery, namely, the recession of the spiral nebulae. Although he was not working perhaps with the best instruments available at the time, he was amongst the first to determine with enough precision the presence of lines in the spectrum of spiral nebulae. This enabled him then to deduce their relative velocity with respect to the Earth. In 1912 he found that the spiral in Andromeda was approaching us at the extraordinary speed of 300 kilometers per second. In the following years, he investigated the velocities of a few more spirals and found that they were all very considerable, in most cases receding away from the solar system. Initially, some astronomers remained quite skeptical of the accuracy of these observations, although in subsequent years they were confirmed by other observers. Slipher was also able to determine the relative velocities of the opposite ends of some of the spirals that were seen along their edges and could in this way infer what was presumably their velocity of rotation. At the time, the very high speeds of rotation seemed to lead to an inescapable conclusion. If the force of gravitation was controlling the motions of the stars, the spiral nebulae were indeed very massive objects and had to consist of very many stars. Meanwhile, Adrian van Maanen, a Dutch astronomer and a meticulous observer, had been appointed to the staff at the Mt. Wilson Observatory, under the recommendation made by Jacobus Kapteyn in one of his periodic visits. van Maanen set out to study the plates of some of the spiral nebulae that are so placed in the sky that we see their discs face on, from one of the poles, so to speak. Then, by comparing the plates taken at different epochs, he attempted to detect any changes in their positions or variations in the configuration of their spiral arms. Clearly, any confirmation of such variations would have been proof that the nebulae were indeed nearby objects; otherwise, adopting the hypothesis that they were distant galaxies, any changes in the positions of its stars in such a short time would have implied that they were moving faster than the speed of light. van Maanen found indeed such displacements, not just in one nebula, but in several amongst the ones he observed. Moreover, the displacements were not random but the arms of the spirals appeared to be extending themselves outward. This coincided, incidentally, with the speculations of the British theorist James Jeans, whose studies of gravitational systems had led him to believe that the spiral arms were emanations from the centers of the nebulae. van Maanen's plates were laboriously analyzed and compared by other observers. Although there was no unanimous agreement in the conclusions, some found variations consistent with those he had described. So it was that, around the year 1920, those who favoured the view that the spiral nebulae were clouds associated with the Milky Way could see some water flowing into their mill, and so too could those who were inclined to believe in the conception of the nebulae as distant island universes. Ejnar Hertzsprung, for example, contended that the high velocities found by Slipher were conclusive proof that the nebulae ought to be extragalactic. Assuming that the globular clusters were not strictly part of the Milky Way, he conjectured that the coalescence of two of them might lead to the formation of such a nebula as the two clusters spiraled toward each other under the force of gravitation. By contrast, James Jeans, who had seen his theories apparently confirmed by the observations of van Maanen, expressed the opinion that the globular clusters hovering around the Milky Way might in fact be the remnants of spiral nebulae after their gaseous arms had dissipated away. One of the more persuasive arguments in support of the view that made the nebulae distant galaxies involved the so called novae, the new stars that suddenly brighten in the sky, to fade and sometimes disappear from view in a matter of a few days. Until the turn of the century, there were historical records of only a handful of such stars; then, in the course of a few years, several were observed. This was due not to any special occurrences in the Heavens, but simply to the fact that a more careful watch was being maintained by astronomers and amateurs alike. Subsequently, it was noticed that by comparing plates taken at different epochs many more novae could be discovered. Stars that surely belonged to this same class were also found in the nebula of Andromeda and in a few other spirals. None of the novae in the Milky Way had appeared close enough to make it possible to determine its distance and absolute magnitude precisely, but it was soon realized that probably most novae had comparable brightness. This provided then a way to estimate the distance to Andromeda, by comparing its own novae to those seen in the Milky Way. The method was stressed by Heber Curtis at the Lick Observatory and used by others as well. It clearly reinforced the view that made the spirals distant and extragalactic. From their apparent brightness, the novae in Andromeda appeared to be at more than a hundred times the distance of their Milky Way counterparts. Curtis also used the argument that the dark lanes seen on the Andromeda nebula were very much similar to the dark clouds of the Milky Way and that the two large structures might be comparable in extent. With the suspicion that these dark clouds obscured the line of sight along the plane of the Milky Way it was now possible to explain why the spiral nebulae, even though they had no association with our Galaxy, appeared to avoid a broad band on the sky along the contours of our Galaxy. Curtis thought it was possible, as it turned out to be correct, that Shapley had overestimated the distances to the system of globular clusters by as much as a factor of ten. If that was the case, the globular clusters belonged well within the neighbourhood of the Milky Way at a distance of a few kiloparsec, while Andromeda was at a distance of hundreds of kiloparsec. The work of Heber Curtis is somewhat peculiar in that he turned to astronomy when almost thirty years old, after spending several years teaching Greek and Latin. After joining the staff of the Lick Observatory he continued the work of William Campbell determining the radial velocities of stars, but later on, over a period of two decades, concentrated on the observation of nebulae. Even if his contributions do not stand out particularly as innovative, one cannot fail to notice his good judgment and the fact that he always found himself on the right side of the various controversies that raged in those times. For some, the high velocities of the spirals and their fast rotations made it all too evident that they were extragalactic island universes. As we have already seen, within a decade the rotation of our own Galaxy would also be confirmed. For others, however, van Maanen's observations also held considerable sway. This controversy had its denouement only a few years later, brought about, as in other cases, by the capabilities of a new and more powerful telescope, this time the 2.54 meter (100 in) reflector that George Hale had brought to Mt. Wilson. In 1923, after months of observing the Andromeda nebula in a tedious and time consuming search for Cepheid variable stars, the American astronomer Edwin Hubble finally managed to find about a dozen of them. After correlating their periods with their luminosities and comparing them to the Milky Way Cepheids, he could estimate the distance to Andromeda. He found a value close to 300 kiloparsec, which put the nebula well outside the galactic system. Edwin Hubble's initiation as an astronomer was somewhat circuitous. A native of Missouri, after some studies in England he returned to the South, still in his mid-twenties, in order to start a law practice, when he decided to give way to his natural inclination and pursue a career in astronomy. It was at the University of Chicago that he met George Hale, who then persuaded him to join the staff of the Mt. Wilson Observatory. Working there for three decades and later at Mt. Palomar, Hubble became one of the preeminent observers, intensely dedicated and never reluctant to undertake big projects. In later years his fame grew above that of his peers, partly due to his accomplishments but also because of his personality that was rather self-imposing. In the beginning, given that van Maanen was also a member of the Observatory, Hubble's conclusions, after his discovery of the Cepheids in Andromeda, created a dilemma and some friction regarding how the same institution could publish results that contradicted themselves. Hubble was very confident of his own results and it took all the skill of the director of the Observatory, Walter Adams, to reach some form of common understanding. In retrospect, van Maanen's observations were spurious and erroneous; however, at the time, even conscientious and careful observers like himself could not arrive at a consensus on the weight of his evidence. Posterity has not been kind to van Maanen and we can always blame his errors for the fate of his reputation, but it so happens that, oftentimes, we applaud and praise the stubbornness and shrewd insistence of an observer who can distinguish in his data patterns or effects when others see nothing but ghosts. Despite the triumph of the island Universe hypothesis, some confusion persisted for a few years. Harlow Shapley, who had claimed that the system of globular clusters extended to 100 kiloparsec, could see the nebula in Andromeda as lying merely at the outskirts of our Galaxy. Even as late as 1930, he resisted the notion that the nebulae were distant and comparable in size to the Milky Way, contending that if they were island universes, then our own system was, as he put it, a continent. Already at that time it had been observed that the external galaxies had a tendency to group themselves in clusters, or supergalaxies, as they were called; such was the case of the very conspicuous one in the constellation of Virgo, which contained hundreds of nebulae. Shapley speculated that the Milky Way, along with Andromeda and the globular clusters, might form a local group of more humble proportions. He contended that the concentration in the direction of Sagittarius might be a galaxy in itself and not the center of our own system, as it was held by the majority of astronomers by then. In any event, it did not take very long before these speculations could be entirely dismissed. In 1925, for example, the Swedish astronomer Knut Lundmark estimated the distance to Andromeda using novae, instead of Cepheid stars, obtaining a value of 500 kiloparsec, almost twice what Hubble had concluded, thereby making the nebula also twice as large. As we saw earlier, the evidence brought forth by Oort regarding the rotation of the Galaxy and Trumpler's demonstration of the degree of absorption within the Milky Way, increasingly made it look very much like a spiral galaxy and of dimensions not too dissimilar to the more distant ones. Milton Humason, working from Mt. Wilson too, obtained the first spectrum of a nova star in Andromeda confirming that it was of the same type as those seen in our own galactic system. In 1932, Hubble presented his observations of a system of 140 globular clusters surrounding the disc of Andromeda emphasizing, despite some discrepancies, the similarities with the environment surrounding the Milky Way. A general overview of the distance scale problem at this time would show that, even when individual distance indicators might give results that were contradictory amongst themselves, it was an entire array of arguments that seemed to converge toward the view that our own Milky Way with its attendant globular clusters occupied a sphere with a diameter between 5 and 15 kiloparsec, while the galaxy in Adromeda was probably of similar dimensions, at least 300 kiloparsec away from us. The final resolution of this dilemma, some eighty years after William Parsons had first observed with his giant telescope the spiral nature of some nebulae, gave a new fascination and urgency to their study. Well beyond the boundaries of our Galaxy, a new frontier had opened inviting the curiosity of a new generation of astronomers. Edwin Hubble eventually found some forty Cepheid stars in Andromeda, which allowed him to calibrate the relationship between their periods and their luminosities with greater precision. The Cepheids are very bright stars but, at the distance of Andromeda, they are already several hundred thousand times fainter than any star visible with the naked eye, almost at the limit of resolution of the powerful telescope at Mt. Wilson. However, despite their luminosity, the Cepheids are surpassed in power by the so-called supergiant stars, which are even more rare, given that they have a rather short lifetime in cosmic terms. Their luminosity exceeds that of the Cepheids by approximately a factor of ten. Hubble still searched for Cepheid variables in a handful of other spirals which, because of their apparent proximity, seemed to be within the reach of his telescope. In this way he was able to determine the distances to half a dozen nearby galaxies. In each of them he was also able to find a few supergiants and, having now access to their absolute luminosities, was able to confirm that they oscillated within a narrow range, comparable to that of similar stars in the Milky Way. In this way, he gained a new tool to assess the distances to nebulae that were perhaps a few times more distant than Andromeda and where Cepheids were already out of reach. He could now begin to chart the environs of the Milky Way, identifying the positions of some forty galaxies up to a distance that he estimated to be 2000 kiloparsec. However, it was found out many years later that the calibration of Cepheid stars available to him had been erroneous and it will be more convenient if we start making the appropriate corrections now to the estimates of distances, before we fall into the same confusion that ensued when these errors were discovered. Briefly stated, Hubble's use of Cepheids underestimated distances by a factor of two, so that the true distance to

Andromeda turned out to be approximately 900 kiloparsec and the range of distances he had access to using supergiant stars reached up to 5000 kiloparsec. Both Andromeda and the Milky Way, without being amongst the most voluminous of galaxies, can still be considered galaxies of first rank. In each case, their extended spiral arms and the surrounding globular clusters could be enclosed in a sphere with a radius of 20 kiloparsec approximately. Given that the distance between the two galaxies is of 900 kiloparsec, we can truly speak of island universes. However, it is also true that the two great galaxies are accompanied by a retinue of lesser ones and even the entire path between them is occupied by clouds of gas, stray stars, a few globular clusters and an assortment of more insignificant material. From the knowledge of the distances to a few dozen galaxies in the neighbourhood of the Milky Way, it was possible to have a first view of their general spatial distribution and, consequently, of the density of matter in the Universe, as well as a glimpse of the various types of galaxies and their range of luminosities. At first sight, it appeared that most galaxies do not differ in brightness from a typical one by more than a factor of two and very seldom by a factor of ten. However, over the years, as astronomers have examined the skies more closely, it became evident that objects of less luminosity are also present, until the point where a dwarf galaxy is indistinguishable in candle power from a globular cluster. The original view of galaxies as island universes, relatively isolated in a vast emptiness, has also required some revision, as more and more material has been found adrift; sometimes even nearby galaxies of decent proportions have gone unnoticed, only because they are extended and their surface brightness on the sky is therefore very low. The classification of galaxies by type was started independently by Edwin Hubble and by Knut Lundmark; their conclusions did not differ a great deal. It was soon observed in long photographic exposures that the spiral arms of many galaxies extended to a distance away from their centers almost double what had been estimated at first. Similarly, it was found that many galaxies did not show spiral arms around their nuclei, not because they were too faint to be detected but because they belonged to a different category, the so called elliptical galaxies, which could range in their ellipticity from being perfectly globular to almost lenticular or disc-like. In the same manner, a continuous range of profiles could be observed amongst the spirals, from those that were dominated by a luminous core with faint arms, to those that essentially lacked a nucleus and consisted merely of their arms. Furthermore, there were the barred spirals with their arms emerging from the endpoints of an extended central bar as well as the irregular galaxies, those defying classification, of which the more conspicuous example is perhaps the Small Magellanic Cloud, one of our immediate neighbours in the southern skies. Long before these peculiar configurations of stars were understood, even in broad outline, there were attempts at placing them in some kind of evolutionary sequence, as progressive steps in the life of galaxies. For some odd reason, the theories never thought of the ellipticals as old, settled, well mixed spirals, but put instead the ellipticals as the originators of spiral galaxies, initially because it was suspected that they grew arms and later because it was thought that only the collision of two ellipticals could lead to the formation of spirals, as they wrapped around each other in falling orbits. Leaving these matters aside and returning to the distribution of galaxies in space, I mentioned that our immediate vicinity is dominated by Andromeda and the Milky Way, accompanied by a couple dozen lesser objects, the most prominent of which are the two Magellanic Clouds next to the Milky Way and two more galaxies, close to Andromeda, one a spiral and the other an elliptical, known respectively as M33 and M32, because they already appeared in the catalogue of nebulous objects published by Messier in the XVIIIth century. Extending the view further out tenfold, one encounters several dozen groups of galaxies, quite a few of which are similar to the local group; they are dominated by one or two galaxies of luminosities comparable to the Milky Way. Because the typical distances between these groups are of the order of one thousand kiloparsec, it is more convenient, when discussing extra-galactic astronomy, to use this distance of one million parsec as a unit and refer to it as Megaparsec.

Plebeius: I am afraid I have long lost any sense of proportion or scale; in any event, it may be futile to attempt to reduce these distances to a human scale.

Albertus: The purpose of drawing maps has always been to avoid dealing directly with the real object of our interest, which is too cumbersome, but still be able to find our way with a reduced and schematic version of it. Earlier, when we began to use the parsec as the typical distance between stars, I attempted to give a description of the vastness of these voids in interstellar space. Now, let us push that image to the back of our minds and draw a map on a different scale. Let us represent a parsec, the typical distance between stars, merely by a millimeter. A millimeter will stand now for hundreds of thousands of times our distance to the Sun. But if we do so, ten or twenty kiloparsec, the typical radius of a sizeable galaxy, will be equivalent to ten or twenty meters, the radius, you might say, of an average house. Of course, if we allow to crowd stars only one millimeter apart within such area and give some depth to the distribution to build a disc like the Milky Way, very soon we will have a hundred thousand million stars, which is comparable to the number of stars in our Galaxy. Finally, a Megaparsec will be equivalent to a kilometer in this reduced scale. If I continue my analogy of making the galaxies comparable to houses and imagine them scattered at distances of the order of a kilometer, then I could very well conjure up the image of traveling through the countryside and encountering a farm here and there. Given that each of these major galaxies is generally surrounded by a few other smaller ones, I can embellish my image providing these farmhouses with a few extra sheds or outside quarters built around the main premises. Next, in this imaginary tour, as you may well predict, we come upon some village, where the houses, perhaps stripped of those additions, tend to congregate much more closely. Indeed, galaxies also tend to congregate in clusters. The old idea of Kant of conceiving nebulae as clusters of stars, which then congregated in clusters of nebulae, was not unfounded at all. Shortly thereafter, when William Herschel began his survey of nebulae, it was already apparent that many of them, of similar size and magnitude and suspected of being at the same distance, tended to congregate in small regions of the sky. One of the more conspicuous clusters is the one in the constellation of Virgo; with all its adjoining groups, it includes hundreds of galaxies. Edwin Hubble made the first attempt to determine its distance using the method of supergiant stars, but he was deceived in their identification and wrongly concluded that the cluster was at a distance of 8 Megaparsec, when the actual value is closer to 20 Megaparsec. Our local group of galaxies is at the edge of this large association, as if we were living at the outskirts of this town. The Virgo cluster is somewhat irregular in shape; others, such as the Coma or Perseus clusters, although much more distant, are quite apparent in the sky because of their colossal size and their compact structure. In their central regions, almost the same volume that contains our Milky Way and Andromeda can contain more than a thousand galaxies. These are the true analogues of the towns or villages I was conjuring up a moment ago. Edwin Hubble, in his first reconnaissance of selected areas of the sky after 1930, concluded that clustering of galaxies was quite common and estimated that perhaps as many as a thousand clusters could be found across the sky. Nevertheless, he also emphasized the general uniformity in the distribution of galaxies throughout space, thinking that only a small percentage of the total mass of the Universe could be associated to clusters. More or less simultaneously, Harlow Shapley, who was then already at the Harvard Observatory conducting a survey of clusters of galaxies, began to emphasize the opposite point of view, pointing to the unevenness in the distribution of galaxies on the surface of the sky. These observations were not easy to quantify, but the subsequent history has probably tended to favour Shapley's view. The general distribution of galaxies has been found to be far from random and clustering is more ubiquitous than Hubble had suspected. Of course, a more accurate account had to await the development of new methods to assess the distances to remote galaxies. On this long standing problem, let us come back to Vesto Slipher's great discovery of the recession of the nebulae. By 1917, he had laboriously recorded the velocities of twenty-five spirals. After the initial and somewhat incredulous reception of his findings, some astronomers began to suspect that the velocities of the nebulae were somehow related to their masses or their distances, given that the smaller ones on the sky appeared to have the higher velocities. The German astronomer Carl Wilhelm Wirtz, who worked in relative isolation and is perhaps for this reason not very well remembered, was one of the pioneers in the study of the nebulae. He sought to establish this correspondence between distance and velocities in 1921, two years before it was finally confirmed that the spiral nebulae were indeed extra-galactic. Clearly, he was not dealing with the absolute distances to these galaxies, but using only relative ones, derived from their apparent magnitudes or sizes, and matching them with their corresponding velocities. To address this question properly we have to return to an older problem that still remained unsolved, namely, the motion of the Sun, not just amongst the neighbouring stars, but relative to an absolute frame of reference in the Universe. The discovery of the rotation of the Galaxy and the determination of the motion of the Sun within it still lay a few years in the future. The spiral nebulae, distributed across the sky, even if they were not unrelated to the Galaxy, provided a frame relative to which one could study the Sun's motion. Perhaps each individual nebula was moving in its own peculiar way but, taken in aggregate, their random motions would cancel and one would be able to infer the velocity of the Sun relative to the ensemble. However, it had now been verified that there was a systematic component to the motion of the nebulae relative to the Sun, in the sense that they seemed to be flying away from it with larger velocities at greater distances. We must keep in mind that the speed of the Sun within the Galaxy, as it would be determined soon, was slightly over 200 kilometers per second, while the velocities of the nebulae under consideration were also of the order of hundreds of kilometers per second, so that the task was to disentangle these velocities of comparable magnitude. One of the first investigations of this nature was carried out by Knut Lundmark who, as we have seen, was another pioneer in the study of galaxies. If we imagine then that all the nebulae are at rest, we would need only to observe three of them at different locations to deduce the three constants that determine the motion of the Sun, two for its orientation on the surface of the sky and one for its speed. Given that the nebulae are in motion too, we observe many of them and adopt for the velocity of the Sun the one that, when subtracted from the velocities of the nebulae, cancels any motion they have as a group. In other words, we interpret any motion of the ensemble of nebulae as the motion of the Sun amongst them. However, because the nebulae are seen to be expanding away from the Sun, we could attempt to subtract from their motions also a radial velocity so that their residual motions appear now smaller and even more random than previously. This is similar to what William Campbell had done in studying the motion of the Sun in the midst of the neighbouring stars that also seemed to be expanding away from it. In the case of the nebulae, we can assume that the velocity of expansion is in each case proportional to the distance. At the same time, we can determine this constant of proportionality, essentially by observing many nebulae and choosing the value that provides the best fit. Lundmark even suspected that the velocities might not be strictly proportional to the distances, that instead they declined after a certain range or grew more slowly; in such a case a parabola, instead of a straight line, might describe the correspondence better. As it was, the velocities of very few nebulae were known at the time; furthermore, their distances did not reach very far and it is obviously rather fruitless to distinguish between a line and a parabola when comparing them over a very narrow stretch.

Plebeius: Was it not sensible to expect that some curve like a parabola would provide the more appropriate answer? How could the velocities remain proportional to the distances? By reaching into the depth of space one would come upon nebulae with velocities greater than the speed of light, which would have been unacceptable.

Albertus: It is true, although the distances and velocities that could be observed at the time were still very far from approaching that limit and such considerations had little impact in practice. Nevertheless, we ought to examine this matter in more detail. Allow me first to add that, after Knut Lundmark, the second study along the same lines was published by Edwin Hubble in 1929. One decade after the work of Slipher, the number of galaxies for which the velocities of recession had been determined had not increased substantially. Moreover, the uncertainties associated with the determination of extra-galactic distances were still quite considerable. Nonetheless, Hubble was able to deduce the motion of the Sun relative to the nebulae and then find a satisfactory relationship of proportionality between distances and velocities, despite the considerable scatter in the data. The largest velocities that Slipher had observed reached 1000 kilometers per second; they corresponded to galaxies in the Virgo cluster, which Hubble had initially placed at a distance of 2 Megaparsec, so that, in his first assessment, he concluded that the velocities increased at the rate of 500 kilometers per second for each Megaparsec. Later it was realized that the Virgo cluster was almost ten times more distant; consequently, the rise in velocities is more gentle than he assumed, closer to 50 kilometers per second per Megaparsec. Returning now to your comment on the velocities that might increase to surpass the speed of light, there will be no need in fact to make use of a parabola and a line is all that is necessary. As it turns out, the whole state of affairs will be characterized by this universal constant of proportionality, which some astronomers began to denote by the letter H, associating it to Hubble's name, although we might as well call it instead Lundmark's constant or perhaps Wirtz's constant. We should point out that the astronomer does not observe the velocity of recession of a galaxy directly but only the displacement of a characteristic wavelength or frequency in its light spectrum, caused by the Doppler effect, and it is from these displacements that we can readily infer the velocity. To be more precise, the velocities of recession make the colours of the spectrum shift toward the red and the astronomer singles out one natural frequency and measures the extent of its shift. Its fractional change, the magnitude of the shift relative to the magnitude of the frequency itself, is what he calls the redshift z; it is the same quantity as the measure of the change in the wave's period relative the period itself. You will recall now that we discussed these matters earlier. If two successive pulses leave the source at an interval of time T and they are received by the observer within an interval of time T', according to his reference frame, we know how to relate these quantities using the transformations of Lorentz, which depend only on their relative velocity v. Thus, we had seen that

$$T' = \sqrt{\frac{1 + (v/c)}{1 - (v/c)}}\, T.$$

Here we are interested in the redshift z, or the fractional change in the intervals of time, namely, $z = (T' - T)/T$, which we shall write for convenience as $1 + z = T'/T$, or

$$1 + z = \sqrt{\frac{1 + (v/c)}{1 - (v/c)}},$$

which is equivalent to

$$1 + z = \frac{1 + (v/c)}{\sqrt{1 - (v/c)^2}},$$

the latter form only to make explicit that when the velocity is a small fraction of the speed of light, the denominator is very close to unity and we can pretend that the numerators are equal, so that the redshift z is the same quantity as v/c. If a galaxy is receding with a velocity that is one percent of the speed of light such as those that Vesto Slipher was observing, then, roughly speaking, the redshift will also amount to one percent and we can write $z = .01$. Furthermore, since we had concluded that the distances are also proportional to the their velocities with a constant of proportionality that we denote H, then we write the following approximate equalities for the distance D of a given galaxy, its velocity and its redshift,

$$v \approx H.D \approx c.z.$$

Plebeius: In truth, however, from the formula that you wrote before we know that, as the velocities increase, they will approach their limiting value c, whereas the redshift z will keep growing to infinity. What law is the distance D now to obey?

Albertus: Indeed, as the velocities begin to increase, the fractional displacement of the frequencies become comparable to the frequencies themselves and we can obtain redshifts $z = 1$, $z = 2$ and so on. The question that you posed did not find an easy answer. It was perhaps for that reason that astronomers gradually began to substitute the notion of redshift for that of distance and spoke, rather loosely, of a galaxy that is located, for example, at redshift $z = 1$. If we were to insist on associating a distance to a given velocity, given that these increase at the approximate rate of 50 kilometers per second per each Megaparsec, we would reach the limit of the speed of light at a distance of 6000 Megaparsec. This could represent the edge of the Universe although, in such a case, we would have placed ourselves at its center. May I mention, incidentally, that if I were to carry further my earlier comparison of the association of galaxies to the crowding of houses in villages and towns and given that I reduced the scale of a Megaparsec to one kilometer, I could contend that, in this representation, the diameter of the Universe would have been reduced to the diameter of the Earth. Thus, having set the distance to the nearest star at one millimeter when I started this analogy, making the size of the Milky Way comparable to the size of a house, now I realize that my map would have to be as large as the Earth to represent the full extent of the visible Universe. Be that as it may, it was in an attempt to answer some of these questions regarding the size of the Universe that Edwin Hubble undertook a vast project to measure the velocities and the distances of the farthest galaxies that could be reached with the 2.54 meter reflector at Mt. Wilson. In this endeavour he had the help of Milton Humason, a skillful and dedicated observer. The story of Humason's lure to astronomy is a happy one and pleasant to relate. When still an adolescent, he had abandoned school to become a mule driver at Mt. Wilson during the construction of the Observatory and later managed to be hired as a porter and maintenance worker, rising slowly through an assortment of jobs until he became part of the scientific staff and an administrator of the Observatory. By 1931, Humason's measurements had reached redshifts amounting to $z = .08$ and velocities of approximately 25,000 kilometers per second, ten times greater than those reached by Slipher. These galaxies were one hundred times fainter than Andromeda and, although photographs could still record galaxies a few times more distant, it was already a very laborious task to obtain a spectrum, with tens of hours of exposure without appropriate automatic guiding devices for the telescope. These galaxies were selected amongst the brightest in the most populous clusters, mainly because it was in these environments that one could predict with some reliability their absolute brightness and, consequently, their true distance. At one tenth the speed of light, the proportionality between redshifts, velocities and distances still held satisfactorily. Before proceeding any further, we must address the origin of these mysterious velocities of recession. Some people, even to this day, express their suspicion regarding this phenomenon of expansion that puts us, the observers, at the center of it all. But this aspect of the mystery was answered in various ways a long time ago. When you inflate a balloon that has dots painted on its surface, each dot sees the others recede and can consider itself the center of the expansion, as long as his entire universe consists of this two dimensional space, the surface of the balloon. I see my neighbour move away with a certain speed, my neighbour sees his neighbour move away with the same speed, so I see his neighbour, twice removed from me, with double the speed. This simple fact accounts for the proportionality between velocity and distance. Now, if we were to adopt the naive view that the velocities remained always proportional to the distances, then we would be approaching the edge to the Universe at approximately 6000 Mega-parsec, when the velocities finally reach the speed of light. Similarly, if we were to travel backward in time to a sufficiently distant past, considering that the galaxies are seen moving away from each other, we would come to a time when they all had their origin at the same place. Regardless of the interpretation that we adopt, you see that we have come to something that appears to be fundamental. Merely from the evidence of the observations, we have deduce a universal constant of proportionality that leads us into conceiving a natural length to the Universe and a natural time scale to its history. Because Edwin Hubble had underestimated the distances to nearby galaxies and, therefore, had assigned to the universal constant a value ten times too large, in his estimation the edge of the Universe was near and its time scale was very short, of only 2000 million years, whereas it was already known at the time from the geological record that the Earth was at least twice as old. Clearly, something was amiss. It is part of the mysterious phenomenon of the recession of the galaxies that these motions were first being observed at the same time that a new and unrelated theory was being formulated, which soon would provide the means to their interpretation. In 1916 Albert Einstein published his revolutionary theory of gravitation. Far from being concerned with the velocities of the nebulae, he had sought for years to recast Newton's law of gravitation into a more fundamental theory. We have taken a rather long tour today, Plebeius, visiting on a good number of theories but this gives us a glimpse, at least, of the historical context which gave rise to each new idea. According to Newton's law, masses attract each other with a force that is inversely proportional to the square of their distance. It is the same law that Simeon-Denis Poisson had established for the attraction of electric charges of opposite sign. Einstein was very much influenced by the way in which James Maxwell had encoded the information regarding electric and magnetic forces in the so called electromagnetic field, whose oscillations generated the waves, carriers of light and energy. Given any distribution of masses throughout space, we should certainly be able to feel, at each location, their combined force of attraction, so that one could say that these forces were expressed by a pervasive gravitational field. This implied that the information regarding the forces did not need to be fetched from the distant masses doing their pulling, but could be read locally from the configuration of the field itself, as if there were an all pervading gravitational aether, similar to the fictitious and already abandoned electromagnetic aether. Despite his daring and imaginative way of speculating, Einstein had the ability to concentrate on some of the simplest facts of experience and extract from them far reaching and unanticipated consequences. He started by equating uniform gravitation with uniform acceleration. This corresponds very much to our intuition. When we are shot upward in an enclosed capsule, we feel our knees bending although, without access to looking outside the capsule, we cannot confirm that we are under constant acceleration or, alternatively, at rest but subject to a heavier gravitational force pulling us from underneath. Einstein transformed this insight into a law of physics. He stated that our observer inside the capsule would not be able to conduct any physical experiment capable of telling him under what circumstances he is living. For instance, if he throws a projectile on a horizontal trajectory from one wall to the other, the projectile will tend to fall if a gravitational force is pulling from under the capsule; it is irrelevant whether the capsule is at the time moving with uniform speed, for the projectile shares in the same inertial motion of the capsule when our observer sends it on its trajectory. But imagine now that there is no gravitational pull; instead, the capsule is accelerated upward constantly and uniformly. Then, as the projectile is flying, without it ever knowing, the capsule is speeded upward. The observer, who is riding with the capsule, sees the trajectory of the projectile bending downward and falling. The two effects are indistinguishable. What would happen now to a light ray going through the capsule from one wall to another in the same fashion? Let us remind ourselves that the law of gravitation states that the attraction between two bodies is proportional to their masses. We think of the light rays as a bundle of photons flying like our projectile across the capsule. However, these photons carry no mass and they should be undeflected in a gravitational field. On the other hand, if the capsule happens to accelerate when the photons are in transit, the observer inside it would certainly perceive them to be falling. Clearly, this simple experiment would be the end of our principle of equivalence. Albert Einstein took the opposite point of view. He insisted that the principle should remain valid and that photons would be equally deflected by the action of gravitation. If we ignore now the massive object that is exerting the gravitational pull and we look simply at the empty region of space through which the photons are flying, we would have asserted that light travels along straight lines. In fact, when we speak of straight lines, beyond those that we define abstractly in our geometry books, but straight lines in physical space, we can only conceive them as the shortest path between two points which is the path followed by a light ray. Herein lies the remarkable conclusion that Einstein was able to reach. If the gravitational attraction bends the path of a light ray, then it will be this curve that is the shortest path between any two of its points. There will be no straight lines other than these straight lines, even though the principles of Euclid may no longer apply to them. We might encounter one of the curved geometries that we discussed earlier. Hence, by this reasoning we come to the conclusion that the field equivalent to Maxwell's, the field of gravitation that Einstein was seeking, is indeed something more fundamental than the simple law of Newton, it is the underlying geometry of the world we live in. Less than a century earlier Carl Friedrich Gauss had attempted to test the true nature of the geometry of space, motivated only by the conviction that what exists in a mathematical sense can very well have a counterpart in the material world. Einstein had now provided a physical argument for the existence of curvature, based on the notions of mass and of gravitational attraction.

Plebeius: Let me pause at this moment. When we discussed the electromagnetic field and the waves associated to it, I could still imagine the underlying framework of space supporting them, in the same manner that I can see the waves on the surface of the sea rolling along relative to the fixed framework provided by the shoreline. Now, however, you have canceled that underlying framework and there are no fixed directions of space to orient me. The gravitational field will provide the framework and the waves at the same time. It is as if I am on a desert of shifting dunes; without some reference, any changed configuration would be defined only in relation to the one it is replacing. When I say that the shortest path between two points is now curved, am I not making reference to some notion of what was straight before, some piece of Euclidean geometry that is lingering in my mind and has still some degree of reality? You explained these things earlier and I understand that each geometry can be defined consistently on its own terms. However, I wonder whether one could not still hold on to the familiar and intuitive notion of space and impose on it the structure that allows me then to speak of bending the light rays and everything else.

Albertus: We could start a lengthy discussion on these matters but I am afraid that they would be more engaging from the point of view of philosophy than fruitful from the point of view of physics. In fact, Henri Poincaré, the brilliant French mathematician, argued in favour of the opinions you have just expressed. To some extent, the argument is about the perspective we adopt and about conventions. Are we to speak of curved lines in straight space, or of straight lines in curved space? Furthermore, we also use planispheres to study geography. We can represent the surface of the Earth on a plane as long as we follow the appropriate procedures and learn how to interpret our results. There is, however, something that goes deeper than mere convention. Poincaré insisted that one could always retain the notion of an underlying geometry satisfying the postulates of Euclid, while allowing that our measurements, being affected by gravitational forces, would lead to discrepant results. In any case, these effects are real and we have ways of determining what type of geometry explains them in a most succinct manner. So real are these effects that we cannot avoid them; even if there was an underlying reality which retains its Euclidean character, we do not have access to it. So, instead of saying that the shortest path between two points is a straight line, but that the effects of gravitation cause us to measure something different, we say outright that the shortest path is now something different and this is economical in thought. Let us now make our way back to the still unresolved astronomical problem of the recession of the nebulae. The theory of relativity has taught us very conclusively that space and time form one single construct and that we cannot separate one from the other. If, in fact, a gravitational field distorts the nature of the underlying space, making the distance between two points different from what we expected it to be, then it is only one step removed to accept that time measurements are also distorted. If space is bent, so is time. To put it differently, if gravitation is equivalent to acceleration, we know that the latter produces changes of velocity, which in turn changes the measurements of lengths and time intervals. Let us think now of a light ray, a train of electromagnetic waves undulating with its characteristic wavelength and frequency, as it travels through space. The wave carries with it, so to speak, its own standard of length and time, or the standard of the source where it was produced, which may not coincide with the local standard of a different region the wave is now crossing, because of the distortions caused by a gravitational field. Another way of describing this situation is to say that the wave only carries information about oscillations of the fictitious electromagnetic aether. Thus, as the wave arrives, every point receives orders to oscillate with a given frequency, but the period of oscillation is interpreted according to the local clock. Of course, if we think of the electromagnetic wave as a bundle of photons, whose energies are determined by their corresponding frequency, we are saying that the energy we can extract from these photons depends on where and when their energy is caught and transformed. This effect had its first and dramatic confirmation in 1925 when Walter Adams at the Mt. Wilson Observatory analyzed the spectrum of the faint companion to Sirius. You recall that its existence had been inferred at first from the oscillations in the proper motion of Sirius. This motion had led to the determination of the distance to this pair of stars. Knowing now the absolute luminosity of the faint companion and its temperature, one could deduce the radiating area and, hence, the size of the star. On the other hand, from the size and period of their orbits around each other one could also learn the masses of the two stars. Combining these pieces of information one learnt also the density of the stars which, in the case of Sirius' companion, proved to be enormous, giving rise to a powerful gravitational field on its surface. Under such circumstances, clocks run more slowly and frequencies are reduced or, if you wish, a light signal climbing out of this gravitational well to the outside world will be perceived as if it had been redshifted; it is precisely this effect that Walter Adams was able to confirm observing a redshift of 20 kilometers per second in the spectrum of Sirius' companion. Of course, you understand now, we speak indifferently of redshifts and velocities because we can associate one to the other, although in this case, for the first time, the redshift was not caused by motion but by the distorted geometry of space in the neighbourhood of the compact star. Now, what happens to a light ray as it travels from a distant galaxy to us? The wave clearly propagates past the combined gravitational field of all the surrounding matter and it is not guaranteed that our measurement of its wavelength or frequency will coincide with the measurement made by the source that dispatched it, either because our measuring rods follow different standards or because something happened to the wave train on its long trip. Thus, a shift in the frequency, a redshift that increases progressively with the distance to the source is not something that appeared utterly inconceivable in the light of the new ideas that were being entertained. The theory of Einstein had the advantage that it was presented in mathematical language and there were precise ways to make things quantitative. It is most remarkable that one year after Einstein's first publications appeared, in 1917, as the measurements of the redshifts by Vesto Slipher were being announced, the Dutch astronomer Willem de Sitter found a particular solution to Einstein's equations that could come to explain Slipher's observations. Willem de Sitter, a former student of Kapteyn, had started his career studying the Moons of Jupiter, but he had a good grounding in physics and mathematics and followed with interest the more adventurous speculations of the theoretical physicists of his day, speculations that were somewhat removed from the concerns of the practical astronomer. It must be said that the equations that Einstein had brought forth were quite complicated and, in addition, they could be interpreted in more than one way. Much confusion prevailed for at least a decade on how they were to be applied on a cosmic scale, beyond simple examples such as the gravitational field of a single star, as in the case that I just pointed out. Nevertheless, the suspicion persisted that some related phenomenon, occasionally referred to as gravitational drag, was the source of the redshifts observed in distant nebulae. If one adopted the most naive view, according to which the redshifts were caused by the Doppler effect and the motion of the galaxies, one had to look for the source of all the kinetic energy impressed on them and was faced all the same with the riddle of the galaxies having originated at the same location in the distant past. One particular interpretation of Einstein's ideas led to a slightly different view. Instead of alluding to the motion of the galaxies, one spoke of the expansion of space, which carried the galaxies along with it, producing the same effect as if they were moving through space. The galaxies were now at rest but space grew between them. To be sure, one was trading one mystery for another. The expansion was supposed to be all pervasive and it affected also the rulers one would have used to verify its existence. The devices used to measure expanded at the same rate as the objects to be measure, so that the expansion itself was not subject to experimental verification. It is curious that these attempts always sought to explain the observed phenomena through transformations in the underlying space and changes in the length scale, but never appealed to modifications in the standard of time. We could as well have contended that space remains invariant but it is instead the pace of clocks that has been quickening throughout the history of the Universe. If that were the case, the light signals we receive from the distant past would appear to have lower frequencies compared to our present standard, which is tantamount to say that we would observe redshifts that are proportional to the distance of the source. In short, regardless of our interpretation of choice, the expansion remained more an article of faith than an effect we could observe directly. Edwin Hubble, for example, was often careful to speak of redshifts, which is what the astronomer measures, and set aside the use of velocities, suspecting that perhaps the redshifts were caused by a yet undiscovered physical effect. Nonetheless, he proceeded with his project to obtain distances, redshifts and, if you wish, velocities of the most distant galaxies he could reach. By 1935, Milton Humason, working at Mt. Wilson but using a new spectrograph designed for the future 5 meter telescope of the Palomar Observatory, already in the planning stages, could measure more than a hundred redshifts reaching galaxies with velocities in excess of one tenth the speed of light, at 40,000 kilometers per second. The proportionality of redshift or velocity and distance still appeared to hold unchallenged. By this time, Hubble had undertaken a new large project which would take many years to carry to completion and would not, in the end, lead to any conclusive results. It was designed to test the geometry of space directly, according to Einstein's theories. The task required to count the number of galaxies within a given aperture and up to a chosen apparent magnitude. If we assume for simplicity that all galaxies are equally luminous and that each occupies the same volume of space, then we can take the count of galaxies as a substitute for measuring the volume of a cone, piercing into space from the position of the observer at the vertex. This is a variation on the experiment suggested by Gauss. If the number of galaxies turns out to be low, then we are probably living in a spherical universe; if the number is larger than expected under the assumptions of Euclidean geometry, the curvature of space would be negative. For this project Hubble had also the help of Nicholas Mayall, an astronomer at the Lick Observatory, who obtained some of the photographic plates. These deep exposures reached galaxies of 19th and 20th magnitude, of which there are already several million across the entire sky. The interpretation of the results was not entirely straightforward. For instance, the redshifts seen in distant galaxies displace the main bulk of the light emitted toward the red, but the sensitivity of the photographic plate remains fixed, concentrated toward bluish colours. As a consequence, distant galaxies would appear less luminous than they should, which might lead us to deduce that there are fewer of them and that they are more sparsely distributed. Hence, we might arrive at the conclusion that space is spherical but for the wrong reasons. There are several similar effects that must be carefully taken into account. For the comparison of his observations to the theoretical predictions, Hubble enlisted the help of Richard Tolman, a physicist and professor at the California Institute of Technology, who probably had the motivation and the skill that Hubble lacked to pursue these matters more thoroughly. At one time, Hubble found that his data implied a spherical universe with a radius shorter than the distance he was able to reach with the telescope at Mt. Wilson. It was perhaps experiences such as this one that caused him to doubt the interpretation of the redshifts in terms of the Doppler effect and led him to abandon the count of galaxies for more than twenty years as a means to elucidate the geometry of our Universe. The commission of the colossal 5 meter reflector at Palomar Mountain in 1949, with its capacity to reach fainter magnitudes and more distant galaxies did not have the impact that one might have expected on the problem at hand, although the deficiencies resided not in the instrument itself, but in the fact that the long exposures required often spoiled the photographic plates with diffuse emission from the atmosphere. Instead, the Hale telescope proved to be invaluable in solving the inaccuracies in the distance scale that had been accumulating over the years. The earliest fruits were harvested by Walter Baade, an astronomer from Westphalia who had started his career at the Hamburg Observatory but later emigrated, like many other German scientists of his generation. Settling at the Palomar Observatory and over a period of three decades, through the impact of his personality and the results of his studies, he influenced the work of a large number of his colleagues. He was the first to resolve some of the bright stars at the core of the Andromeda nebula and to notice the differences between these stars and those in the outlying arms of the spiral. This led to a similar distinction in the diverse star populations composing the Milky Way, a distinction that is based primarily on the abundance of various chemical elements in the composition of the star. Furthermore, Baade was unable to find in Adromeda some of the shorter period variable stars that are present in the globular clusters of our own Milky Way and that were supposed to be within reach of the new and powerful telescope. As a consequence, he was led to postulate the existence of two classes of Cepheid variables, loosely associated to each one of the two populations of stars, one of them a few times more luminous than the other, with the shorter period variables associated to the less luminous ones and, for that reason, beyond reach at the distance of Andromeda. These stars, now confirmed to be less luminous, put the globular clusters of our own Milky Way at a distance much nearer than what Shapley had estimated around 1920. Indeed, the old calibration of Cepheids had mixed stars of the two types and, since those stars that Hubble had identified in Adromeda to estimate its distance were all in the spiral arms, where only the bright class of Cepheids is present, it was now realized that the distance to this neighbouring galaxy was more than twice the distance previously estimated. This, in turn, made Andromeda larger in real terms and not only comparable to our Galaxy but perhaps even slightly larger. The second important adjustment to the distance scale was carried out a few years later by Allan Sandage, a young American astronomer who was then completing the big survey of galaxies initiated by Edwin Hubble. He discovered that the brightest stars that

Hubble had used to estimate the distances to galaxies in the Virgo cluster were not stars in fact, but much more luminous clouds of hot ionized hydrogen. The combination of these errors explains the gross miscalculation in the initial estimate of the distance to the center of the Virgo cluster, which Hubble had put at slightly over 2 Megaparsec when it really is at almost 20 Megaparsec. The next push to reach greater distances in the cosmos came from an unlikely source. It was a feat of the emerging astronomy in the radio band. Ever since radio waves had been produced in the laboratory almost a century earlier, it was suspected that the Sun might be a source of electromagnetic energy in radio frequencies, in the same way that it produced visible light or infrared radiation, as William Herschel had discovered. But the technique of detecting radio waves was slow in developing; the earliest attempts at doing this type of astronomy were in the hands of amateurs, who did not receive much encouragement even after they demonstrated, rather surprisingly, that diffuse radio emission from the direction of the Milky Way could easily be detected. In a few years time, however, the development of radar technology produced a revolution in instrumentation and electronics that made the prospect of radio astronomy much more appealing. These achievements were the result of an intense and concentrated effort, of the type that men seem to be able to deliver, as history has often proved, only when trying to find ways of destroying each other, instead of pursuing knowledge for its own sake. In any event, the new techniques soon made radio astronomy a viable discipline and the design and construction of appropriate antennae advanced quickly with the support of government funds. It was suspected at the outset that most of the brighter sources of radio emission would be found amongst nearby stars. It was indeed a surprise to find out, once some of the very brightest were linked to their optical counterparts, that they corresponded to exploded stars, far in the midst of our Galaxy as well as to distant galaxies, several Megaparsec away. The earliest radio telescopes were not very precise in identifying the positions of discrete sources. By 1960, a few catalogues had been compiled containing hundreds of the brighter radio sources in the sky. At that time, one of the first stars to be optically identified with one of these radio sources was set aside as unusual because of its uncharacteristic spectrum. However, three years later, when another identification was done to a similar star, taking advantage of an occultation by the Moon, a closer examination of the star's spectrum revealed an extraordinary redshift, $z = 0.158$, comparable to the redshift of the most distant galaxies then known. A check on the star that had earlier been dismissed revealed an even larger redshift, $z = 0.367$. These were stars dashing through the Galaxy at fantastic speeds or, alternatively, they could be compact sources at cosmic distances with luminosities of the order of billions of stars. For some years, a lively debate ensued, not unlike the earlier debate concerning the spiral nebulae. Were these sources galactic or extra-galactic? Meanwhile, more of these quasi-stellar objects were discovered; in barely a couple of years, one was found with a redshift $z = 2.01$. Their search was now pursued in a more systematic way. Optical telescopes became very useful again. Because of the bright emission at ultraviolet wavelengths, these star-like objects had bluer colours than most stars and could be found using multi-colour photography. The optical discoveries demonstrated that not all of them were radio sources. Subsequently it was confirmed that these Quasars, as they came to be known, were gigantic cauldrons confined to the innermost center of galaxies, usually surpassing them in luminosity to such an extent that the galaxy could barely be recognized in its glare. The Quasars became now the most valuable beacons to probe the deepest regions of the Cosmos and a tacit race started to find as many objects of these type as possible, at redshifts $z = 1$, $z = 2$, $z = 3$ and beyond. Allow me to describe succinctly at this moment the setting in which this search is conducted. What does the astronomer see when he peers into a typical square degree cut out on the sky? This is approximately the area covered by your thumbnail when you extend your arm. The brightest galaxy in this area is most likely a galaxy comparable to the Milky Way, at a distance of fifty Megaparsec, whose diameter covers only a small percentage of the total aperture you are peering through; its luminosity is 2,000 times fainter than the great Andromeda. Within the same area, even when looking into a direction away from the Milky Way, there are probably a couple hundred stars of similar luminosity, whereas one would be lucky to find a Quasar as bright as any of those stars. Setting your threshold now ten times fainter, you discover there are now, let us say, seventy galaxies scattered throughout this square degree, but the number of stars comparable in luminosity has also increased to seven hundred and you are probably sure to find one Quasar amidst these many stars. Let us now take a more powerful telescope and set the threshold ten times fainter still. The number of galaxies keeps multiplying; there are now perhaps a thousand of them, at distances typically of the order of 500 Megaparsec. Nevertheless, the galaxies continue to be outnumbered by the stars of equal brightness in our own Galaxy. At these fainter magnitudes we are likely to find a few dozen Quasars amongst the one or two thousand stars. Hence, you see that a good deal of sifting has to be done to find these Quasars and even amongst these few, perhaps only one of them has a truly exceptional redshift beyond $z = 3$. We could proceed and search for objects ten times fainter within the same area of the sky. Our targets would be now a thousand times fainter than the ones we noticed at first. Now, for the first time the number of stars begin to decline and we begin to find a multitude of galaxies, ten thousand of them perhaps, whereas the number of Quasars also increases moderately to one hundred, distributed at all distances throughout the Universe. And there is still room, Plebeius, within that single square degree to go ten times fainter and find now a hundred thousand galaxies reaching into thousands of Megaparsec, each of them barely a second of arc across the sky. The number of Quasars does not rise substantially now. This is in essence a consequence of two factors. We have already reached the edge of their distribution and, secondly, the number of Quasars that are intrinsically faint is not so large. If we were to go to magnitudes ten times fainter we would now encounter a million galaxies, apparently crowding together in our small piece of sky, although, of course, these galaxies would be, for the most part, relatively small and scattered at all distances, over several thousand Megaparsec, across the length of the Universe. It may appear as a most remarkable coincidence that, on this square degree we have been examining, we begin to run out of sky, so to speak, at the same time that our probe reaches farther into the confines of the Universe, exhausting the sky in depth as well. Something is afoot, only a person with little imagination could think of coincidences. In any event, when we speak of the confines of the Universe we should be careful to emphasize that we are alluding strictly to the encounter of large redshifts, to the fact that light, figuratively speaking, tires as it journeys through space and, therefore, it has only a finite reach. We are definitely not speaking of the confines of the Universe as a material edge beyond which there is nothing, nor of the possibility that our view from Earth is obstructed beyond a certain distance by the accumulation of galaxies on the surface of the sky. I should add on this subject, incidentally, that our view of the Universe as a vast empty ocean dotted with galaxies here and there has changed as dramatically as the old view that made the Galaxy merely a collection of isolated stars. In the same manner that interstellar space has been found to be a rich reservoir of gas, dust, cosmic rays, magnetic fields with a wealth of information on all manner of cosmic processes, so too intergalactic space can seldom be said to be really empty. To mention but one example, it was discovered around 1970, examining the light from distant Quasars, that the ultraviolet region of their spectra appeared corroded, gnawed at, you might say. In principle, one expected to encounter a continuous distribution of frequencies or energies amongst the photons composing the spectrum; instead, it was found that there were many troughs, sometimes dozens of them, at very precise energies, where very many photons were missing. Only one explanation for this phenomenon seemed plausible. The photons had been absorbed by intervening matter during their trip to the observer. Indeed, each of these ultraviolet photons was progressively redshifted during its long journey, so that its energy would come to coincide with some excitation energies of the hydrogen atom for a brief moment, but if this occurred when the light ray was crossing a cloud of intergalactic hydrogen, the photon had a great probability of being absorbed. A more energetic photon would survive untouched, until perhaps further down the road, when it too was degraded and the light ray might be encountering a different cloud of hydrogen. Hence, by examining the Quasar's light, upward in energy in the ultraviolet portion of the spectrum, one could reconstruct all the vicissitudes of the light ray's journey from the source to the observer. It was soon verified that there is hardly a line of sight to a distant Quasar that does not encounter several or many of these diffuse clouds, sailing almost unnoticed through the great voids in between the galaxies. Similarly, it has often been found, when looking into some dark patch of the sky, that a galaxy is discovered, not even very far away, but so distended and of such low surface brightness that it had previously gone unnoticed. One of the more fascinating aspects revealed by this close scrutiny of the skies has been the enormous diversity of environments encountered throughout the Heavens, from those tenuous clouds of diffuse hydrogen that are barely noticeable to the huge concentrations of galaxies in clusters and the linking of these clusters in even greater structures. Amidst this diversity and taking into account also, within each category of objects, the peculiarities found in each individual example, which makes for even greater variety, it is just as astonishing to discover certain uniformities, scrupulously observed, sometimes expressive of general physical laws, sometimes the expected outcome of a myriad random events. It is these uniformities which have been put to great advantage in the task of improving the determination of distances to far away galaxies. It has been found, for example, that the rate of rotation of the arms in a spiral galaxy are tightly correlated to its absolute luminosity, so that by observing this rotation, and knowing the apparent luminosity of the galaxy, one can infer its distance. Another ingenious method, used for galaxies which are not too far, although perhaps far enough to make the search of Cepheid stars very difficult, exploits the fluctuations in the surface brightness of the galaxy. As the galaxy recedes in distance, its light dims at the same rate that its image shrinks, so that the luminosity per unit area on the sky remains the same. But since the light is composed of individual stars, the graininess of the image is clearly perceptible in nearby galaxies, whereas it becomes more diffuse and uniform in distant ones. The fluctuations in luminosity within a sample area are so precise that they are also sufficient to determine the distance to the galaxy. The number of methods used to ascertain distances grew with every new astronomical discovery. Their comparison and mutual calibration has become so involved that the occupation of cosmic surveyor can be said to be a profession in itself. The Greeks used the word geometer for those who measured the land and we may be in need of a new word for those who measure the Heavens. For the most distant galaxies, the rare but stupendous exploding stars, or supernovae, have proven to be invaluable in estimating their distances, although it has been a cyclopean task to reproduce these explosions on paper and with analytic formulae, in such a way that their output power and subsequent decline could be estimated and their distances determined. For the rich clusters, one can observe, perhaps not the first, but the third or fourth ranked galaxy within the cluster; here, once again, the uniformity in the Universe is such that the absolute magnitude of these galaxies oscillate within very narrow ranges, so that from their apparent magnitudes we can also deduce their distances with remarkable confidence. But before we lose our way considering these matters in greater detail, we must return to our central concern regarding the overall dimensions of the Cosmos. What is the true nature of the redshifts? How does the proportionality between distance and velocity resolve itself? Can we really speak of having reached the edge of the Universe? I do not know whether we could say that every age in the past deceived itself into believing that it could provide a reasonable account of the nature of the Cosmos and make a sensible estimate of its true dimensions. The encounter of these cosmic redshifts has brought us now against a physical limitation. It is the exhaustion of light that impairs our view, not the narrowness of our minds. Our time has finally come to see the end of this ancient quest to reach the farthest extremities of the world. But have we really come to the last horizon? It is most interesting that whenever our search seems to have thoroughly investigated a given frontier, a new one springs up, totally unsuspected, and we marvel at our naiveté or at our inability to have noticed it earlier. Even in relatively recent times, within the field of astronomy, it was realized that, for several millenia, the observation of the Heavens had been conducted exclusively by visual means, namely, by paying attention only to those photons with wavelengths in a narrow range around half a micron to which the human eye is sensitive. This may very well be like extracting a few ounces of gold from a vein in the mountain and leaving the rest of the mountain behind; a conscientious and dedicated geologist cannot afford to be so prodigal. Likewise, electromagnetic radiation exists in a colossal range of frequencies or energies and most astrophysical processes are very generous in producing most of them. Just as in the case of radio astronomy, which I mentioned earlier, new astronomies have since emerged in various energy ranges, providing new frontiers of knowledge, completely unsuspected a century earlier. One of the more remarkable discoveries took place in the microwave range of the electromagnetic spectrum, perhaps because it came about in a rather accidental way. A discarded antenna used for satellite communications was being refurbished to be used in radio astronomy when it was noticed that it was beset by a persistent background noise that could not be accounted for, either by sources in the sky, by ground emissions or by noise in the receiver or the antenna itself. As it turned out, this observation, performed at a wavelength of seven centimeters, was the first detection of the diffuse cosmic radiation that permeates the Universe. The observation was consistent with the interpretation that it was thermal radiation at 3.7 degrees Kelvin, although the demonstration that the radiation was indeed thermal came much later. At such temperatures the radiation peaks at wavelengths in the millimeter range, with minor wings in the longer wavelength radio band and in the infrared. Of course, if we think of the Universe as a big receptacle, where the stars and galaxies, like scattered cauldrons here and there, have been burning of all eternity, it is not inconceivable that this great immensity, after such a long time, could have been warmed up to a certain temperature, however modest. But the original interpretation of this cosmic background radiation pointed in a different direction. Indeed, such a radiation had been anticipated and even its temperature had been roughly estimated. Ever since the discovery of the redshifts of distant galaxies and their interpretation in terms of an expansion of the Universe, many believed that its implied origin at one concentrated point in the distant past had produced a colossal explosion whose remnant ought to be still reverberating throughout the Universe after billions of years, relaxing as thermal radiation in an ever expanding space. In a sense, this tenuous glow, this ever present sea of radiation was seen as the final curtain beyond which our eyes could not penetrate. The attempt to reach greater and greater distances was also the attempt to reach further into the past; the greater the distance to a light source, the longer its light signal has been in transit and the older the image it conveyed. This is also what motivated the search for the most distant Quasars. Astronomers saw them as the first manifestation of the emerging world that we would come to inhabit and, like archeologists digging into the Earth, they rummaged amidst thousands of stars and galaxies in each patch of the sky to find those that had the highest possible redshifts and could reveal the oldest secrets. There was a certain irony in the fact that our efforts to observe the farthest regions of the Universe took us also to the most remote past, since the idea that the Universe has been expanding continuously brings us all to a common origin and the farther out we looked, the closer we came, in a way, to where we are. Could we say then that this situation, in which we now find ourselves, precludes any knowledge of the Universe as it is at present? What could its origin in a point-like explosion, however fanciful a proposition, teach us about its configuration today? We still wish to ask whether the Universe is finite or infinite at this very moment. I am sure you recall the way in which the great Greek mathematician Archytas had phrased this dilemma. If the Universe were finite, he asked, what would happen when I come to the edge and throw a spear beyond? If there is an impediment that stops its flight, that too, whatever it is, must be part of the Universe. This seemed to make an infinite Universe easier to comprehend. Nevertheless, Aristotle, thinking of the Earth at rest and making the Heavens rotate daily, defended the view of a finite world, arguing that the greater its size, the more violent would be the revolutions of its outer spheres. The acceptance of the Earth's rotation made the notion of an infinite world perhaps more tolerable but hardly appealing. Kepler saw it with a secret, hidden horror, as if we were left to wander in an immensity that had neither limits nor center. In fact, Kepler had concluded that the stars were much closer amongst themselves than they are to the Sun, which gave our solar system a distinguished, central position in the Heavens. Perhaps only Giordano Bruno, with his bountiful imagination, found the proposition of an infinite plurality of worlds seducing and to his liking. Descartes, by contrast, was considerably more ambiguous. For him infinity had no specific attributes, one could not say whether an infinite number was even or odd; thus, only God could be infinite, but the world was either finite or indefinite. At a later epoch, when it was discovered that the force of gravitation ruled the motions of the Heavens as much as those on Earth, the notion of a finite Universe suffered a new reverse, for it was thought that any finite assembly of stars would be unstable and would collapse under its own weight. But no sooner had one thought of an infinite world populated by an infinite number of stars that another difficulty seemed to spring up. One expected in such a case that any line of sight would eventually encounter a star, so that the sky, far from being dark, should be alight in all directions. Because the light of a star is radiated evenly in all directions, the power received at any one location decreases like the square of the distance to the source; on the other hand, if we count the sources shining on us, assuming they are uniformly distributed, their number increases at the same rate as the surface area over which they are distributed, that is, like the square of the distance, so that the decrease in power of each one of them is compensated exactly by the increase in their number. As a result, if we divide all the sources in successive shells, at increasing distances, the light we receive from any one of these shells is the same as the light we receive from the next one; hence, an infinite distribution of stars would imply that we ought to be awash in infinite light. Of course, the stars have finite sizes and we could not claim to have a direct line of sight to an infinite number of them. The light from the more distant stars would certainly stumble upon the nearer ones before it reached us; nevertheless, the sky should be lit up at night all the same with as much light as with many thousands of suns. One could certainly argue that many dark clouds distributed amongst the stars might absorb much of this light, but this subterfuge is not effective in the end, because the clouds would heat up and radiate in turn what they had absorbed. These arguments were already discussed in the XVIIIth century; they could have been raised again many years later, after the finite extent of the Galaxy was demonstrated, but this time in reference to the nebulae. Why were we not awash in the light of an infinity of galaxies? In due course, it was shown indeed that the galaxies do crowd in every patch of the sky as far as the eye can see, but on this occasion the discovery of the cosmic redshifts provided a more convincing escape from our dilemma. The light from distant galaxies tires, as it were, on its long journey across the Universe and, for all practical purposes, it has only a finite reach. But is this really a resolution of our dilemma? Perhaps it only deepens it, for it prevents us from reaching to the edge of the world or from proving that there is none. Have we come any closer to deciding whether we live in a finite or an infinite Universe? Or, should we believe, as Galileo Galilei believed, that this is one of those questions, happily inexplicable, that will forever remain hidden from human knowledge? As it turns out, the cosmic redshifts have provided us with more of an opportunity than a hindrance to extricate ourselves from our quandary. By expressing these things in the language of geometry and gently yielding to the power of mathematics, we have finally come to answer our age old questions, although, as I wish to demonstrate to you presently, we have had first to realize that the questions, in the manner that we had once asked them, made little sense and had little relation to the world we inhabit. Let us therefore examine the shape and design of our Universe in greater detail. Given that I am going to concentrate exclusively on the abstract architecture of space and time, if we were to restrict ourselves to the most mundane conclusions drawn from common sense, we might decide that there is not much to be examined. The immensity of the world extends from one end to the other and one season follows another. All the events that have taken place or will ever take place anywhere are thus ordered in this sequence of days that we call universal history. But we have granted that the language of Nature is mathematical and we should like to inquire to what extent the laws of geometry preside over the architecture of the Universe. I have the impression that we have now assembled all the necessary ingredients to make sense of these matters and I wish to offer you a glimpse of the way in which we have come to understand them in recent times. By providing you with a broad outline and pointing to a few of the details that make our contemporary view of the Universe so startling, perhaps you will feel enticed to read my book and study these questions to greater depth. Geometry furnishes us with the rules to compute distances, volumes and time intervals, but we are also interested in the more general contribution of geometry to the task of composing a map of the Universe, a map that is not necessarily a replica, but a representation of it that we can call accurate and faithful. To this end, we will resort of course to the natural tools of geometry, to charts and to systems of coordinates, that will allow us to label the events we wish to represent. We are also interested in ascertaining to what extent the vantage point of the observer, the location and time from which he contemplates the Universe, may affect his conclusions. In this regard we are very much constrained, given that we can only observe the Universe from our own location and time. Even if we were to include the entirety of human history, it would not amount to more than an instant in the evolution of the Cosmos. Therefore, we must remain uncertain about our conclusions, not knowing whether they depict the true nature of the Universe or they are conditioned by our circumstances and reflect our peculiar point of view. However, we shall transform our weakness into our strength and assume that the position we occupy is not remarkable in any way. Then, as it has already occurred in the past, we shall gain access to a more privileged view of the world at the very moment we relinquish any claim to have been granted a privilege. The hypothesis that all observers have a similar view of the Universe, regardless of their position in space or time, we expressed earlier in mathematical language using the concept of symmetry. Thus, we shall search for a geometric representation of the Universe that is fully symmetric. According to the meaning we gave to this expression, it requires that any observer in possession of a system of coordinates, who sees the world with certain characteristics to his left or to his right, to his past or future, could be transported to any other location, could cast again his system of coordinates and he will see exactly the same characteristics to his left and to his right, to his past or future. To put it differently, there are transformations of our geometric construct of space and time, which bring it to coincide with itself despite the fact that in the interim the location of its points has been reordered.

Plebeius: I could also say that two copies of the same space are brought to match with each other, even though each point is not matched to itself on the other copy but to some other point.

Albertus: Yes, the net effect is that two observers at different locations could exchange information in such a way that, as one of them relates his observations regarding the geometry of the space expressed in his own system of coordinates, the other observer could interpret these observations as if they referred to his own environment and not to that of his interlocutor, and still verify that they are satisfied exactly. This is precisely what we mean by a symmetric Universe. Of course, we know that our Universe is not symmetric in its details. The stars tend to congregate in galaxies and there are vast regions of empty space. However, from our vantage point we see that, throughout the Universe, every few Megaparsec there is a galaxy and we contend that the same observation could be made from every other point of view; that is, our Universe is symmetric in the aggregate. To proceed further we need to incorporate two ideas that we discussed earlier. One of them we called curvature. We learned that the presence of matter causes the underlying space to curve and we expect that the overall distribution of matter throughout the Universe will have some effect on its geometry. However, you will recall that in our discussion of curved surfaces there was no mention of time and we treated them as abstract objects, as if they existed outside time. We know now that in physical space the coordinates of space and time cannot be disentangled and the two form a single geometric entity. Therefore, we must also incorporate the principle of relativity into our account. We are quite fortunate that Nature is very economical in the number of tools and principles it employs to govern its processes. We call it economy and elegance although, perhaps, it is only our lack of imagination, or little ability to discriminate, that makes us reduce everything to a few principles. Be that as it may, let me give you an example of what I have in mind. When we discussed the symmetries of the sphere, we spoke of the group of rotations. If I consider the case of the unit circle, for simplicity, I can draw a pair of perpendicular axes of coordinates x and y, through the center of the circle, so that the points on the circumference have coordinates always satisfying the condition

$$x^2 + y^2 = 1,$$

again, a restatement of Pythagoras' theorem. I could express the rotation in the following terms. I hold the axes fixed, so that a point having coordinates (x_1, y_1) in relation to those axes, and satisfying the previous condition, will move after the rotation of the circle to another point that satisfies the same condition and whose coordinates in relation to the same axes I can easily compute. Here, however, we shall express exactly the same thing in a slightly more complicated manner. We let the axes rotate with the circle and introduce new axes x' and y' where the old axes were. Therefore, those new coordinates we computed, say (x_1', y_1') refer to these new axes, which have taken up the old position, while our point on the circle even after rotation retains its old coordinates (x, y) in relation to the x and y axes that have rotated with it. We do this so that the situation we describe is perfectly symmetric, as can be seen graphically,

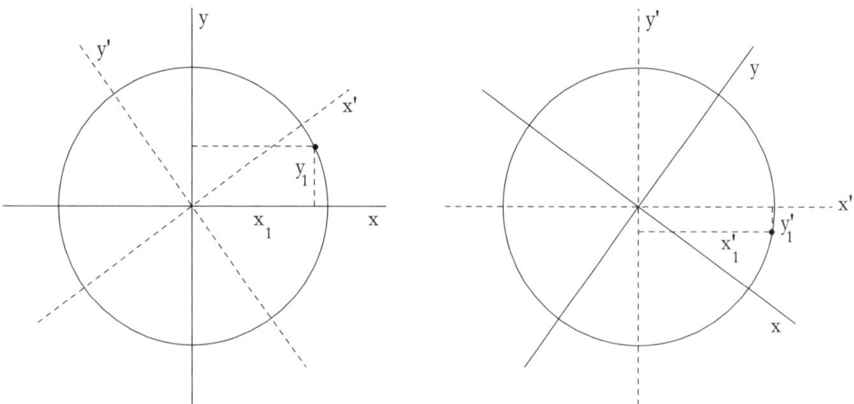

relating events before the rotation in coordinates (x, y) and after the rotation in coordinates (x', y'), where each set refers to different coordinate systems. This is very much reminiscent of the transformations of Fitzgerald and Lorentz that we introduced when we discussed the principles of relativity. In that instance, we had introduced the coordinates x and t on a plane representing space and time and we concluded that an observer in motion with respect to our own frame of reference would choose coordinates x' and t', subject to the constraint that the validity of the equation $x^2 - c^2t^2 = 0$ would also imply $x'^2 - c^2t'^2 = 0$. Although we are in relative motion, during the brief instant when we happen to be at the same place we can compare our systems of coordinates superimposing one on the other or, if you wish, I can sketch what his system of coordinates are in my chart and he can do the same in his. It is not that one observer is at rest and the other is in motion. We say that these are two observers in relative motion and their standings are perfectly symmetric.

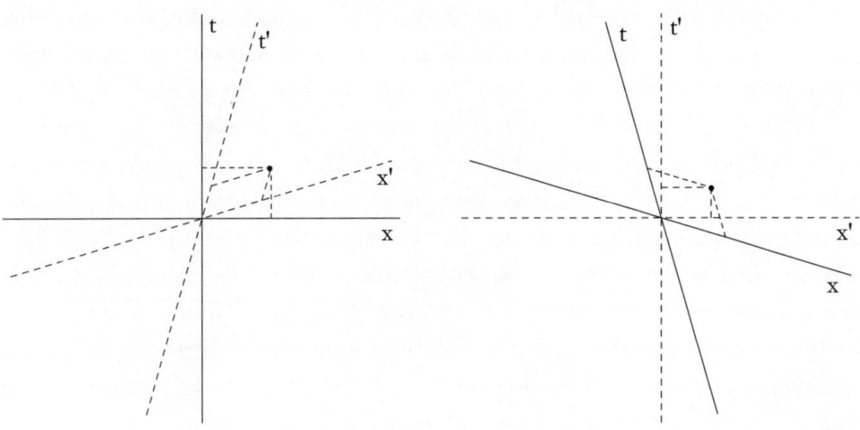

You notice that in one case we sought transformations of the plane preserving the sum of the square of the coordinates, this is a simple rotation; in the other case, when transforming from one system of coordinates to the other, we seek to preserve the value of the difference of the squares. Despite the similarity, we must also point out the difference. In the first case there is no real distinction between the x and y axes and the rotations we seek can take us all the way around the plane. In the case of space and time, we certainly cannot transform a time-like axis into a space-like axis or viceversa. The two diagonal lines representing the trajectories of light rays, $x = ct$ and $x = -ct$, also represent the limits beyond which an axis cannot be tilted or rotated. Nonetheless, the formal procedures to treat these symmetries mathematically are quite similar. Let us now come to the description of the Universe we inhabit. At the time we discussed curved surfaces, I mentioned that the condition of total symmetry reduced our choices to only three surfaces, either of positive or negative curvature, or to the case of curvature zero, namely, Euclidean space. The same can be said of spaces of three dimensions and the same, once again, of space and time, having four dimensions, which is the case we are interested in. The space and time of curvature zero is the one we have used in our descriptions, albeit in only two dimensions, one to account for space and one time dimension; the symmetries in this case are the obvious ones, realized by successive translations along the axes; they express the fact that, on a plane, we can take any point as the origin of coordinates. Regarding time, we speak of resorting to time translations, which is to say that we can choose any moment as the initial time to set our clocks running. We should also remind ourselves that the transformations of Lorentz also form part of the group of symmetries. This is a consequence of the fact that two observers in uniform motion with respect to each other have equal right to use their coordinate systems to account for all physical processes in a similar fashion. The necessity of curvature turns our attention to the other two spaces although, of these other two, only the space of positive curvature deserves serious consideration. In the space and time of negative curvature, time folds onto itself and becomes cyclical. This might make for a rather fanciful Universe, but we do not have any evidence that we have already been what we are or that our actions today might influence our past. Let us consider then the space and time of positive curvature, the only remaining instance of a fully symmetric Universe available to us. Here we face once again the difficulties of providing a satisfactory graphical representation of the object we wish to describe. Firstly, we realize that we have to contend with a space of four dimensions and, in the same manner that we can only represent a two-dimensional sphere by embedding it in three-dimensional space, the fact that our space is curved forces us into demanding even more than four dimensions in order to make its curvature explicit. Hence, we shall restrict ourselves to depicting a curved two-dimensional universe, embedded in a space of three dimensions. Although we shall have at our disposal only one spatial dimension, in addition to the dimension corresponding to time, this will suffice to describe the process of exchanging light signals between distant observers. I shall label the coordinates of our three-dimensional space x, w and t and the two-dimensional surface representing space and time will be composed of those points (x, w, t) satisfying the condition

$$x^2 + w^2 - c^2t^2 = R^2.$$

Once we have understood this simpler case, we should return to the full four-dimensional space and time represented by the similar equation

$$x^2 + y^2 + z^2 + w^2 - c^2t^2 = R^2.$$

What type of surfaces are represented by these constraints? If we consider first the points, or events, as we prefer to call them, taking place at the time $t = 0$, in our two-dimensional universe, we clearly find the equation of a circle $x^2 + w^2 = R^2$. In this context, R represents the radius of the world; space is indeed curved, we are dealing with a finite, spherical world, in our case a circle, although in reality it is a three-dimensional sphere. At a later time $t = T$, we still have the equation of a circle that we can write in the form $x^2 + w^2 = R^2 + c^2T^2$. You realize that this process of slicing our surface by considering all events taking place at the same time T produces a stack of circles of ever increasing radius, as time evolves to infinity. The same result obtains if we proceed toward negative values of time into the past.

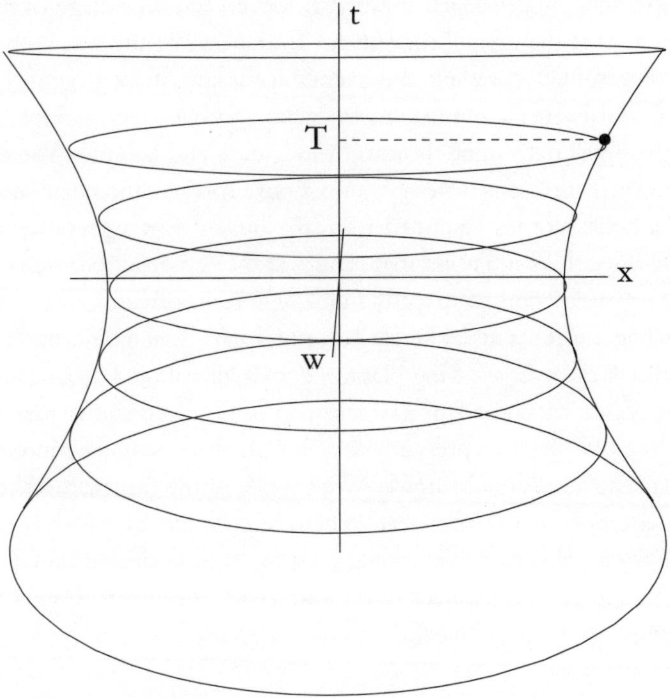

It is not too great a leap of the imagination to conceive that the representation of space and time in four dimensions would result in a stack of three-dimensional spheres of increasing radii, if we could ever draw a picture of it. Even with our representation of space and time in two dimensions we have to contend with certain difficulties of a different nature. Our first reaction in seeing this representation is to say that each circle represents a stage in the history of the Universe, that the world appears to have contracted until the time designated as $t = 0$, when it reached a radius R, and thereafter, it began to expand again. Hence, it goes without saying that this Universe does not seem to have the type of symmetry that I had advertised, it varies from one epoch to another. An observer could identify the time in which he is living by measuring the radius of the world. This misunderstanding derives from the fact that the symmetries that concern us here are not entirely the same symmetries that we perceive with our eyes. We have already encountered a similar situation when we discussed the symmetries of surfaces. The symmetries of a sphere are readily apparent to the eyes and the group of rotations makes them completely explicit. In the case of a surface of negative curvature the symmetries were less evident, for they were to be seen with the mind but not with the eyes. You recall that we represented this surface by the interior of a circle and we stated that an appropriate transformation could place any of its points at the center, so that the geometry was intrinsically symmetric and no location was particularly privileged. Notice, however, one advantage of this type of representation over the case of the sphere. In the case of the sphere, because the geometric representation retains the overall symmetry, if we desire to introduce a system of coordinates, we must choose what we shall label the north pole, for instance, which will determine the arcs of longitude and latitude. The case of the negatively curved surface is different; any of our representations, by means of the interior of a circle, comes equipped from the outset with a preferred system of coordinates, since there is a point that resides at the center of the circle. Returning now to our representation of space and time, you realize that it must resemble this case somewhat, for it seems to single out, not a particular point, but a particular time, namely, all the events taking place at $t = 0$, distributed around the neck of our surface, points which occupy a privileged position, separating past and future in equal halves. We shall see presently that this privilege is only fictitious, as in the case of the negatively curved surface. What is true is the fact that our representation comes equipped with a preferred system of coordinates, which is particularly appropriate for an observer performing his observations at that time $t = 0$. The events distributed around the neck of the surface are the events he would call contemporaneous.

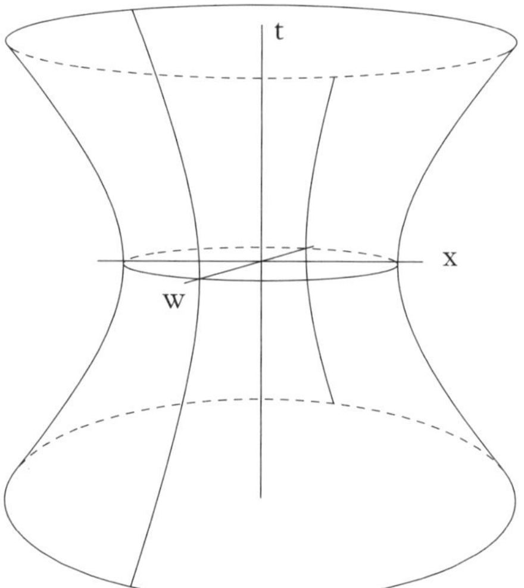

Notice that if we cut this surface along a vertical plane that passes through the origin of coordinates by setting, let us say, $x = 0$, then their intersection singles out the two branches of a hyperbola, the same type of curve that we obtain when we cut a standing cone along a vertical plane. For this reason, our representation of the Universe is called a hyperboloid. One branch of this hyperbola we have just singled out represents the life history of an observer, from the remote past to the most distant future, an observer who happens to be at rest relative to the frame of reference we have chosen. The second branch of the hyperbola obviously represents the life history of another observer, at the antipodes of our world. If we were to tilt the intersecting plane slightly, we would still obtain a pair of hyperbolae, each branch of which would represent an observer moving with uniform velocity toward the right or the left, as the case may be. To increase the speed of our observer we need only incline the plane further, until the velocity of our observer approaches the speed of light. Of course, an almost horizontal plane would intersect our surface along an ellipse and not an hyperbola. At the transition between the two cases, the intersection will coincide with two lines. The same situation arises with a cone when the intersecting plane runs parallel to its edge. We can also generate those two lines easily by intersecting our surface with a vertical plane that passes through a point at the neck of the hyperboloid. If, for instance, we single out the vertical plane determined by the condition $w = R$, it is clear from our equation $x^2 + w^2 - c^2t^2 = R^2$ that we ought to have $x^2 - c^2t^2 = 0$, that is, the coordinates must satisfy $x = ct$ or $x = -ct$, while w remains constant $w = R$. These are indeed the equations of two lines representing the trajectories of two light rays dispatched by the observer at time $t = 0$, or else they represent light rays arriving at our observer, at time $t = 0$, from the distant past.

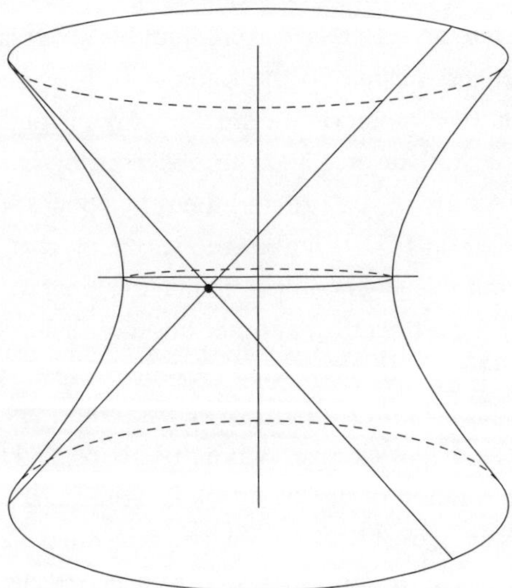

Interestingly, the curved surface of the hyperboloid can be generated by one of two families of straight lines, which represent the paths of light rays traveling toward the right or toward the left, clearly all the paths a light ray could follow in a one-dimensional world.

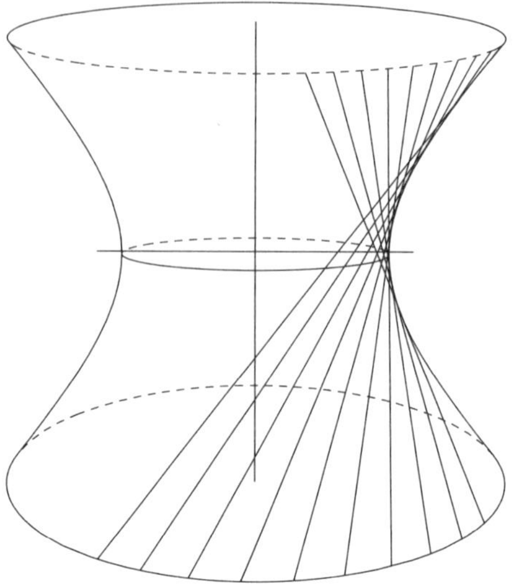

Let us now turn to the symmetries of this Universe, which provided the original motivation for us to turn our attention to it. We shall discover that these symmetries reveal to us a remarkable and unexpected landscape. I wish to emphasize something I have already alluded to. We adopt the point of view that each representation of this Universe by means of a hyperboloid, with its own system of coordinates, is associated to an observer carrying out his observations at time $t = 0$, an observer who is at rest relative to the framework provided by that particular system of coordinates. Each point on the surface of the hyperboloid represents an event in the history of the Universe and our observer, by drawing the horizontal circles composing this surface, has ordered them by epochs; the events along the neck of the hyperboloid, at $t = 0$ represent everything that is happening at the time that he is laying out this all-encompassing chart. Now we ask ourselves, what system of coordinates is pertinent for another observer living somewhere else and at some other time? How does he order the events of history? Firstly, it is quite clear that, if the second observer happens to be contemporaneous to the first one, we need only make use of the circular symmetry of space. Here the eyes do not deceive us and a mere rotation of the hyperboloid would carry one observer to the other. As long as they are at rest with respect to each other, we can almost say that they share the same system of coordinates. If the second observer finds himself making his observations at some other time, according to the principle we just stated, we must find a system of coordinates in which he finds himself once again at a point on the neck of the hyperboloid. The total symmetry of this space and time assures us that no point occupies a privileged position and guarantees that a rearrangement of the points of the hyperboloid can be accomplished without altering the underlying geometry, in such a way that, in the new coordinate system, our observer is placed indeed on the neck of the hyperboloid. How do we accomplish this? Let us consider, for convenience, an observer A, whose life trajectory lies on the plane that runs parallel to my figure, so to speak, where the coordinate w vanishes, $w = 0$. Thus, our observer initially has coordinates $x = R$, $w = 0$ and $t = 0$. For reference I shall also single out the observer B residing half way around the Universe with coordinates $x = 0$, $w = R$ and $t = 0$. Let us now meet observer A at a later time. We climb along the corresponding hyperbola and encounter him at position A'.

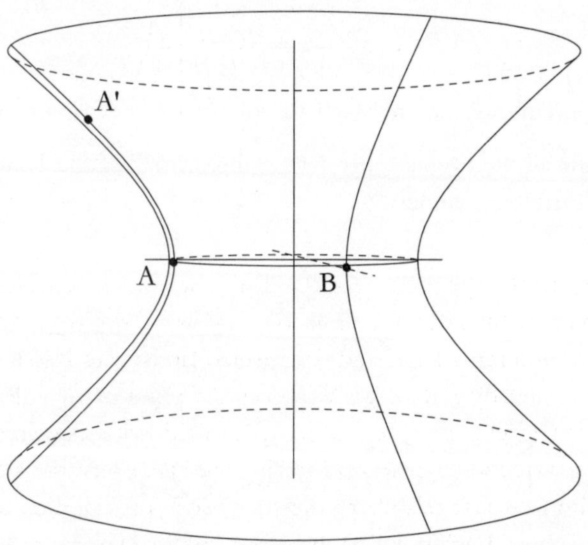

Fortunately, we have already introduced the tools necessary to accomplish our task. The hyperboloid is embedded in a three-dimensional space and time and we can have recourse to the transformations of Lorentz that we discussed in a different context, namely, for the exchange of information between two observers in relative motion in a flat space of curvature zero. If you wish, this is another instance where we witness the economy of Nature, using the same tools for two different objectives. In this case, leaving the coordinate w undisturbed, we can find a transformation of the coordinates x and t into the coordinates x' and t', while $w' = w$, so that the observer at A' will find himself along the x' axis. After all, the hyperboloid is singly identified by the condition $x^2 + w^2 - c^2t^2 = R^2$ and if a point (x, w, t) satisfies this condition, it will also satisfies this condition after we perform a Lorentz transformation. To put it differently, earlier we characterized the transformations of Lorentz as being exactly those that preserve our quadratic contraint. Now we can say that the Lorentz transformations are exactly those that transform the hyperboloid into itself. As you surely recall from our previous discussion, the necessary transformations depend on a parameter that we labeled v

because it was then interpreted as a velocity. Here the same parameter enters into our equations and it is obviously related to the time we have waited to meet the observer A again; the higher up I go along the hyperboloid to catch the observer at A', the larger the value of v. I shall write once more the explicit form of these transformations, although we shall not need to carry out any explicit computation with them. We have

$$x' = \frac{1}{\sqrt{1 - v^2/c^2}}(x - vt) \qquad ; \qquad w' = w$$

$$t' = \frac{1}{\sqrt{1 - v^2/c^2}}\left(t - \frac{v}{c^2}x\right).$$

Notice now the curious way in which the points on the hyperboloid rearrange themselves, how all the events representing the history of the Universe are now reordered in a different manner.

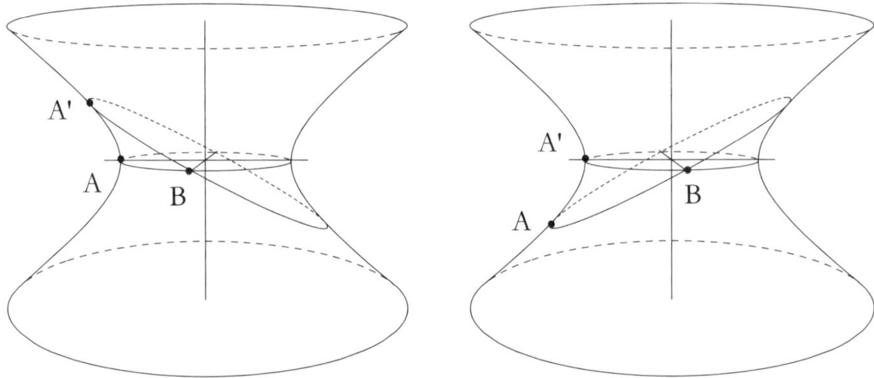

The events that our observer at A' will consider contemporaneous to him formed an inclined ellipse following the inclination of the axis x'. These events move now to form the neck of the hyperboloid and the original position of our observer at A naturally slips into the past. This is most peculiar; when we described the ordering of events according to A, we stated that each horizontal circle represented an epoch in the history of the Universe, as if the representation was a record of how events had taken place in the past and would take place in the future, something on which everybody could agree. Now we realize that the ordering of events is done differently by observers occupying different points of view. For example, we must face the following disconcerting fact, that events contemporaneous to A' taking place on the other hemisphere of the world are events that the same observer, earlier in time, when at A, had already deemed as past. Indeed, from his point of view, it is as if on the other side of the world time was flowing backwards.

Likewise, events that A held to be taking place in his own time are now thought to remain still in the future. Notice also that, despite the passage of time, both A and A' consider the event B to be contemporaneous to them. Suppose now, in order to make sense of these extraordinary occurrences, that our observer at A decides to neglect past and future events, but concentrates in compiling a record of events taking place now, at the time he is living, and he repeats the same procedure at each stage throughout his life, when he arrives at A', A'' and so forth, doing the same thing at all times in his past history. This procedures seems to produce an entirely different account of the history of the world. Instead of piling up one circle upon the next, as was presumably indicated by the time coordinate t in his original system of coordinates, he now produces a sequence of ellipses that have progressively greater inclination and such that they all intersect at two events, B and its antipode.

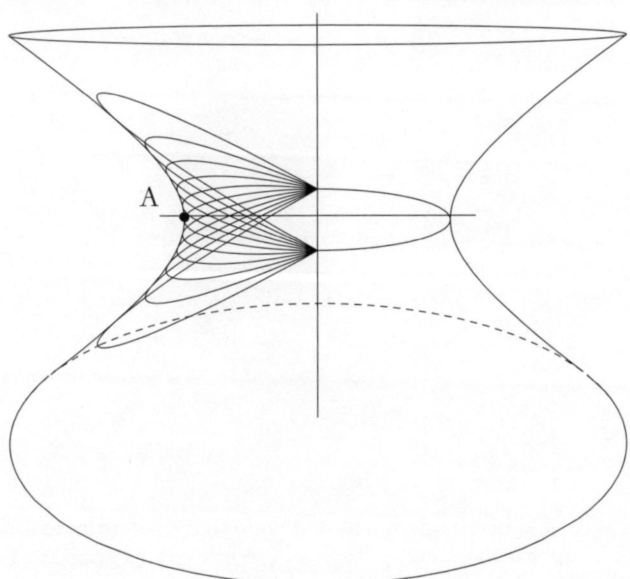

As long as our observer does not stray too far, reaching to these points, or trying to account for events occurring on the other hemisphere of the world, this new way of charting all events in history does not seem to contradict common sense in any way. Your realize that even if our observer at A were to travel several million light years to his right or to his left or if he were to return into the past or travel millions of years into the future, he would be exploring a very minute fraction of our hyperboloid and within that region all events would appear ordered more or less in our conventional way. It is only when we attempt to comprehend the totality of the Universe at once that we perceive the trickery and limitations in our senses and we must let the good order of mathematical reasoning speak directly to our minds. Now, notice that when we let our observer sweep all of history in the manner I just described, recording all present events at each stage in his infinite life, from the remote past to the distant future, we come to realize that the Universe is much larger, that there are events he has failed to take into account. Oddly enough, by insisting on recording only events that were at some time contemporaneous to a moment of his own life, and we assume his life to have been infinite, our observer has left out of his chart events that seem to exist in other regions of time inaccessible to him. This is most easily seen by observing our hyperboloid from above; the ellipses will sweep all events on a band stretching along the x axis, leaving more events beyond their reach than have fallen within it.

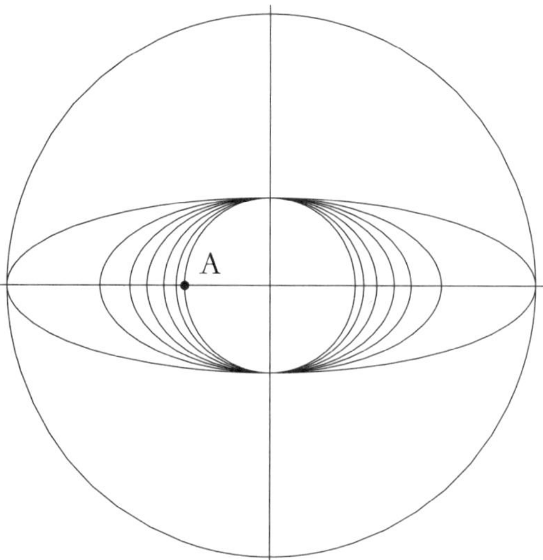

Nevertheless, it would be erroneous to conclude that the Universe is composed of disjoint regions, as if there were several separate Universes in one. We can easily see that the light signals that our observer can dispatch at time $t = 0$, if allowed to travel unimpeded, will reach some of those points or events in the Universe that he will never have considered as having been contemporaneous to him. Furthermore, this relation is not reciprocal, for some of those events, according to the frame of reference appropriate to them, will consider some moments in the life of our observer to have been contemporaneous to them.

Plebeius: Thinking of the hyperboloid as a stack of circles, one for each time in the history of the Universe, a point of view that, according to what you said, is perfectly valid for this instant, if I were to communicate instantaneously with any other observer at this very time, he would agree with me on the times and places that I assign to each event in the Universe, given that he conceives of the hyperboloid in exactly the same manner, composed of the same stack of circles. Furthermore, we would have accounted for the totality of events in the Universe. However, despite this agreement, we must acknowledge that this is not the way things will unfold. Therefore, we adopt next a contrary point of view. Instead of seeking agreement with the rest of the world at the present time, our observer seeks agreement with the viewpoints he will adopt at various epochs, that is, he seeks a coordinate system that he can claim will remain valid for all times throughout his life. This is the system of ellipses you just described. The price he must pay is that now his chart of the Universe is somewhat restricted and, indeed, as he tries to reach half way around the world, his system of coordinates collapses into a point. Thus, one system of coordinates did not remain valid as time flowed and the other does not remain valid when I attempt to extend it to all of space. Before we draw any further inferences regarding the peculiar nature of our Universe by using these charts, would it not be advisable to conclude that the mathematical tools we are employing are deficient and not quite up to their task?

Albertus: I would be inclined to agree with you, Plebeius, but the evidence has been mounting in their favour and we have had to concede that, despite their strangeness, these ideas carry a good measure of the truth. We are able to explain with their help quite a few things that would seem very puzzling otherwise. The computations of distances, volumes and such matters, when applied to the astronomer's observations of the Universe on the largest scale, have come to match their predictions very well. This agreement is more satisfactory than the one we had been able to reach following the prescriptions of other geometries that, at first sight, had seemed more reasonable. Therefore, we have been forced to conclude that the shortcomings you alluded to are not limitations of our coordinate systems; quite the contrary, they reveal the true structure of space and time. We have had to accept that such concepts as the history of the Universe make little sense, that is, we cannot speak of the evolution of the world as a whole, of its past origin, of its fate in the future; to do so only betrays our ignorance. Let me now describe a few of the consequences we have been able to draw by following the logic of this new geometry we have adopted. Consider two observers at rest, A and B, at time $t = 0$, a certain distance apart, but sharing otherwise a common view of the Universe. We shall ask them to carry identical clocks so that, after a pre-arranged time, they can both pause and take stock of the Universe once again. Notice, before we allow the clock to start ticking, that, at time $t = 0$, both observers agree on the precise location of those events A' and B' where they will pause to draw a new chart of the world.

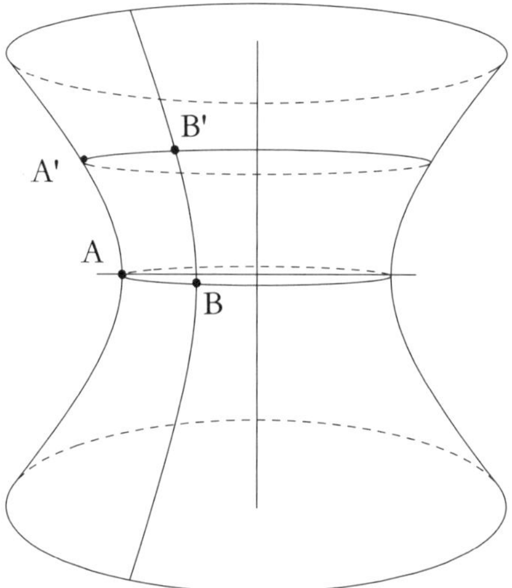

Then, as they arrive at A' and B', respectively, our observers draw new charts so that they see themselves once again at the neck of the hyperboloid. But now they could not agree that they are doing it at the same time.

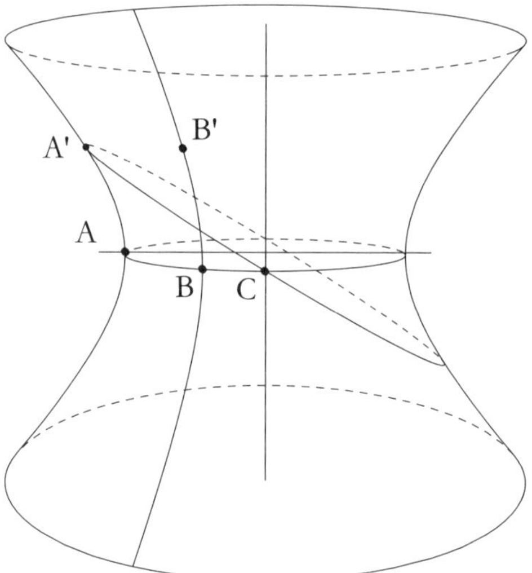

Indeed, as one of our observers pauses and prescribes what event is contemporaneous to him in the life of the other observer, the other observer, waiting to reach the prescribed time, goes past that event without marking it, so that our initial observer concludes that his colleague's clock runs more slowly than his own. We cannot escape the conclusion that the notion of contemporaneity is ephemeral and cannot be preserved, by cosmic standards, over any sizeable length of time. Any two observers depart into incommensurable futures, so to speak, and the same is true of their past. The situation I have just described regarding the perceived delay in each other's clock is perfectly symmetrical and, once again, does not imply any contradiction. We are not setting the two clocks side by side to see which one runs faster. We have only used the clock of observer A to mark the events in the life of observer B and the clock of observer B to mark the events in the life of observer A. In each case the local standard of time seems to be delayed in relation to what the distant observer would have prescribed. We had encountered a similar situation when we discussed the case of two observers in motion with respect to each other according to the principles of relativity. In fact, the two phenomena are very much related. We started with the assumption that the two observers saw themselves at rest at time $t = 0$; at a later time, this is no longer the case. When the observer A considers his new reference framework, after arriving at A', he sees himself still at rest in the new system of coordinates, but the original life trajectory of observer B going past B' reveals now that that he is moving away from the first observer at A'.

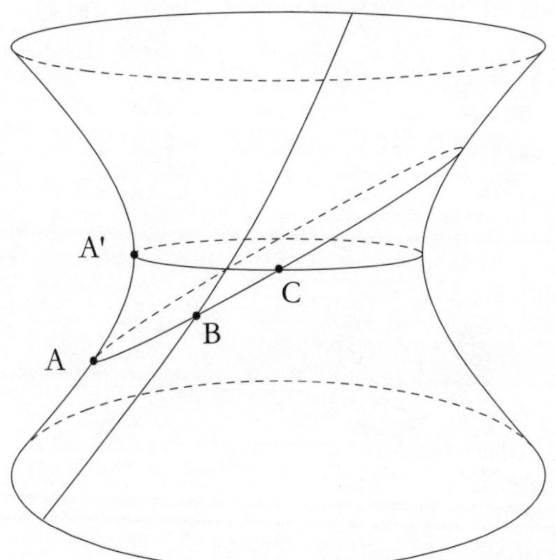

Rather surprisingly, these motions are generated by the mere passage of time and do not involve any applied forces. Similarly, if we were to investigate how the two observers had perceived each other before the time $t = 0$, we would find that they had been moving toward each other. In each case, the velocities we find are so much greater the more distant time we consider away from $t = 0$. Hence, we conclude that any distant galaxy that is at rest with respect to the Milky Way at the present moment, must have been approaching us until the present time and will move away along the line of sight from now on, with ever increasing speed. The present time, as defined by our frame of reference, is not peculiar in any way and there must be at the various distances throughout the Universe, at this very moment, galaxies in various state of motion in relation to us. Nevertheless, I wish to emphasize, before I throw you into too much confusion, that I am describing things to which, strictly speaking, we do not have direct access, since all the information we receive is transmitted by light signals and, at any given time, the regions of the world that we are examining at various distances are also seen at various epochs. In this regard, it is a most happy circumstance that we have been able to elucidate completely the nature of the redshifts of galaxies that had puzzled astronomers since early in the XXth century, when Vesto Slipher made the first measurements. I mentioned earlier the Dutch astronomer Willem de Sitter who, shortly after the publication of Einstein's equations, provided a solution that might explain the observations made by Slipher. In fact, his solution contained the same ingredients of the geometry I have been describing to you. His intuition led him in the right direction indeed, but he presented his results from a specific point of view and failed to exploit the full consequences of the theory in front of him. In the following years several other distinguished mathematicians and physicists attempted to develop de Sitter's model without greater success.

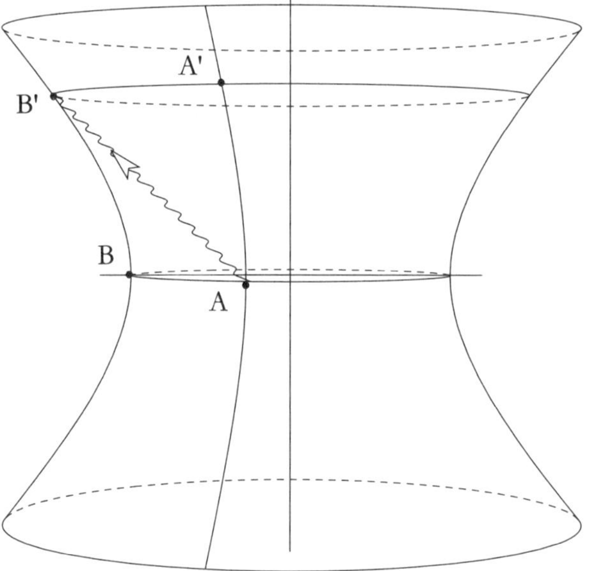

Amongst them, the American scientist Howard Percy Robertson pointed out a decade later that the geometric framework introduced by de Sitter possessed a full group of symmetries, precisely the notion I have been stressing in this discussion. But somehow we have had to wait all these decades before these ideas were put in the appropriate perspective and we were able to explain the redshifts seen in distant galaxies in a fully convincing manner. Thus, let us consider, once again, two observers A and B, at rest at time $t = 0$, when A dispatches a pulse of light in the direction toward B. I emphasize that A and B agree on the time the signal is dispatched. They also agree that they are both at rest. These are two observers caught at the instant $t = 0$ and they essentially share the same frame of reference. However, by the time the target observer receives the signal at B', he must adopt a new system of coordinates and it is now apparent that from his new vantage point, the source A is perceived as having been in motion away from observer B at the time the light signal was dispatched.

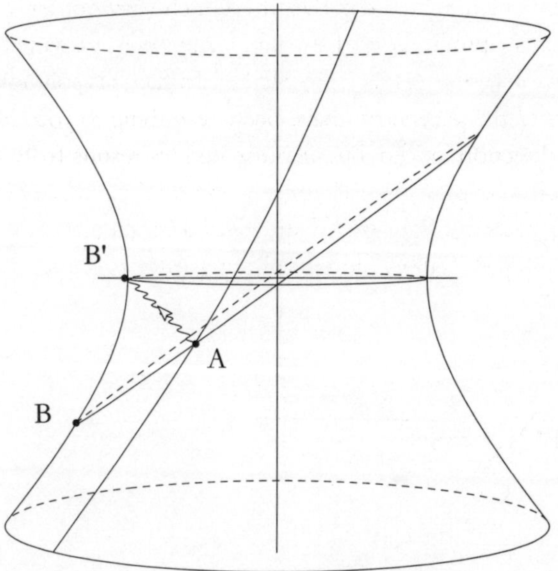

As long as A is not too far away in the most distant regions of time or space, a simple computation shows that its apparent velocity of recession, as measured by B', is proportional to the distance that separated them at time $t = 0$, a relation many times verified in observations but never convincingly explained. We have already talked about the hypothesis of the expansion of space which seemed to carry the galaxies along, without telling us what was expanding in relation to what. Now we finally have the clearest explanation of how these redshifts come about. Earlier we discussed the proportionality between distances and velocities, or redshifts, without coming to a conclusion about how these relations were to be continued when the velocities approached the velocity of light. It is not necessary for us to write down in any detail how these relations have now been made precise in the context of the geometric ideas I have just presented. But I think that we should examine, however briefly, the manner in which any observer perceives the farthest reaches of the Cosmos to which he has access. At any given time, in his own frame of reference, the information he receives from distant galaxies is collected, we might say, by two light rays arriving at his location at time $t = 0$. Of course, we are using a one-dimensional world in our description and it is obvious that one source of light would obstruct the view of any sources behind it. In truth, we should think of this light ray as summarizing all the possible directions of space and collecting information from all possible distances.

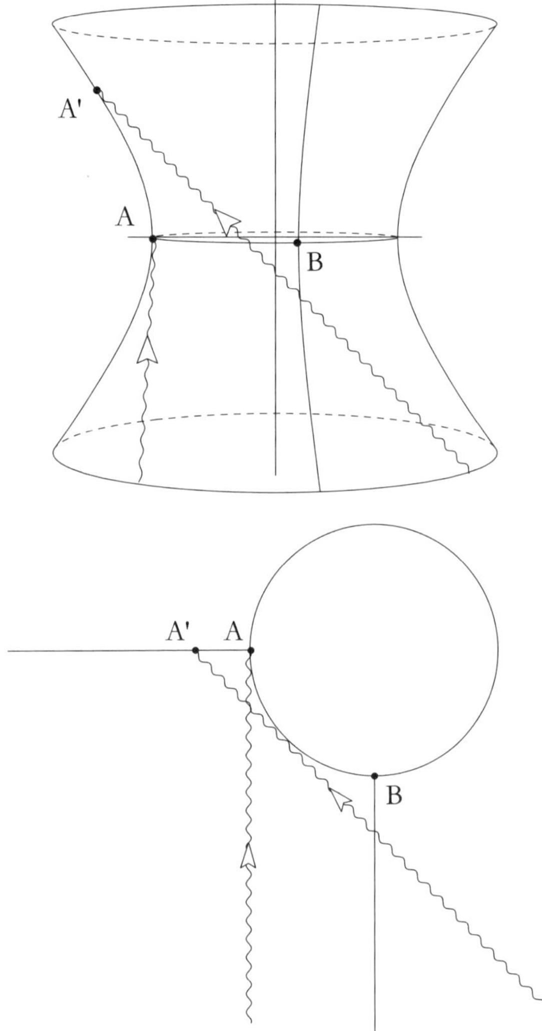

In any case, looking now at how these things are depicted on the surface of our hyperboloid, we verify that our light rays could never retrieve information from sources that are, at present, on the other hemisphere of the world, as if the two points exactly half way to the antipode, precisely where our systems of coordinates break down, represented somehow the horizon beyond which we could see nothing. Yet, at the very next moment, when our observer inexorably climbs up a few steps along the hyperbola of his life trajectory and he adopts a new system of coordinates, we realize that the light signals he is receiving intersect in the distant past the life trajectories of observers that were previously thought to be on the other hemisphere of the world and beyond reach. This is most clearly seen when we contemplate the hyperboloid directly from atop the time axis, although now the hyperbolae representing the life trajectories of our observers at rest appear as straight lines emanating from the neck of the hyperboloid. Thus, we come to the surprising result that our past can grow, in the sense that new galaxies come into view, which had remained unseen until the present time. If we examine these matters from the opposite point of view, we realize that they are related to our previous discussion of the necessity of motions at all distances at any given time. Let me explain. The horizon I spoke of, half way around the world, is not unlike any other region of the Universe; it is only our system of coordinates that collapses at that point. Now, due to the motions of the galaxies, I am compelled to accept that, at the present time $t = 0$, there ought to be galaxies moving across that horizon, in and out of the hemisphere I am occupying. The light emitted from these galaxies are the ones that I shall consider, in my most distant future, to be the most remote, and it is precisely that traffic of galaxies in and out of a hemisphere at all times that allows for new sources to come into view along successive light rays without ever exhausting themselves.

Plebeius: I have been following your description with some bewilderment and a touch of exhilaration, for I am not sure I have understood every detail and I do not feel I would be able now to reconstruct on my own what you have just said. However, I feel as if I have glimpsed, through a door slightly ajar, a landscape which I did not know existed, although now, having briefly encountered it, I think I could recognize amongst others. I speak of exhilaration because it resembles having studied the map of a region we have not yet visited. We experience only the anticipation and a vague sense of orientation about the territory we are about to explore, but we lack the familiarity and ease that come with knowing it first hand. There is a very peculiar appeal in the way that mathematics presents its arguments, a cold appeal perhaps, but in its very indifference to our emotions or our normal ways of thinking, it takes us outside ourselves, beyond our expectations and misconceptions. In the end, however, considering that it never ceases to be a human creation, it seems almost miraculous that it can force the external world into its own molds and stratagems and say something meaningful about it.

Albertus: Perhaps it is not so much that we force the world into our own ploys or contrivances, but that our minds, after many trials and errors, yield and are bent into Nature's ways. Allow me to make one final observation on this matter of retracing the path of a light ray into the past. To put things in a different perspective, suppose we travel backward in time along the trajectory of the light ray, from its destination at time $t = 0$, on the neck of the hyperboloid, to its source in the distant past. By prolonging the light ray sufficiently far, we could assume that the source that produced this signal occupies today, after all this time, a position very close to what we have labeled our horizon. In fact, however far we prolong the light ray, it never intersects the life trajectory of the horizon and we might conclude that the geometry of our hyperboloid is stretching out to infinity the very small region between the source point we just considered and the position occupied by the ancestor of the observer at the horizon today. Despite these considerations, nothing prevents us from bringing any source point along the light ray to the neck of the hyperboloid and adopt a system of coordinates appropriate to him. Then we would repeat the same process as before, realizing that the light ray can be continued without end, encountering new regions of space that never exhaust themselves. This brings us to a new paradox. We started from the assumption that the Universe at the present time is a three-dimensional sphere of radius R and, therefore, of finite volume. This was supposed to account for the entirety of our Universe. Now, instead, we realize that by delving into the past, if it were not for increasing redshifts that cripple our view, we could peer into an infinity of galaxies along every line of sight. The light rays we are chasing into our past along the surface of the hyperboloid are indeed straight lines; obviously, we cannot be saying that the infinity of worlds we encounter arises from circling the hyperboloid again and again. The paradox is resolved, once again, by the fact that it is unreasonable to separate space and time, as if they were different entities. The hyperboloid is finite in the spatial direction but is infinite along the time axis and when we compute, according to the prescriptions of our geometry, the volumes that our sight might be able to reach, all the way to infinite redshifts, we find indeed that they grow without bound to infinity. Thus, to the age old question, whether we live in a finite or infinite Universe, you might answer that it is finite and infinite at the same time.

Plebeius: Sometimes, by wishful thinking, reason tricks our senses into seeing things that are not there to be seen. Now, quite to the contrary, I feel as if reason nudges my senses, persuading them to stop believing in something so obviously real that they are rather reluctant to give in. Perhaps they just refuse to surrender what is false but seemed familiar and comprehensible until now, in exchange for something true but new and disconcerting. By your account, each and every event in our Universe is represented by a point on that so-called hyperboloid. If I understood things correctly, the life span of any being or any object is described by a line wandering upward along its surface. Therefore, from my present vantage point on the equator of the hyperboloid, I know that everything that has ever existed or will ever exist is, at this very moment, crossing the equator and shares with me this instant in the history of the Universe. The world is spherical and bounded, consisting of this finite set of occurrences I am holding in my imagination right now. However, you ask me then to follow a light ray that has just touched my eye, backward as it traveled from the depth of space and time and I must now accept that there could be no end to this trip. This light ray might, in principle, have seen an infinity of worlds that have never been, nor will ever be, contemporaneous with the world in which I am living. I cannot overcome my perplexity. I know that often, once we have understood things properly, it is that very perplexity which we cannot understand, for everything has fallen into place and we think that we have always believed what we now believe, as if it could never have been otherwise.

Albertus: It is very true, Plebeius, once we see things in their new simplicity, instead of delighting in the discovery we have just made, we lament how foolish we had been in our old ways and we marvel at the pains we took and the tortuous path that we have followed. But your perplexity is not unwarranted. All knowledge is fragmentary, tentative and temporary. We illuminate something at the cost of putting something else in the shade. Notice that the contradictions that you just exposed do not reveal any disproportion or incongruity in the world or even in our own perception, but only in the way we used to describe them. Far from telling us that we are on the wrong road, they caution us to ponder carefully which statements are licit for us to make and which are not. We imagine the history of the Universe from the remote past to its distant future as a stately procession in which we all march at the same pace, but we fail to realize that this notion is a fanciful shibboleth, a contrived creation of our imagination that does not correspond to anything real. We used to think that all the spheres of the heavens rotate around us but we later learned that this was not a good way of describing the order of things. However, we have not ceased being parochial and we have kept thinking that our clock is the clock of the Universe. It is time once again to take stock of the more cumbersome furnishings in our representation of the world, discard them and make way for a simpler and more faithful description of it. Our coordinate systems may appear contrived in their own way, they warp and are awkward to use. Nevertheless, they teach us the rules to treat problems of space and time in proper order, according to the true laws of Nature. Perhaps when we speak of the laws of Nature we are abusing the term and applying to the natural world names that were designed and are appropriate only for our social conventions. Clearly, when describing a law of Nature we are merely observing that things go the way they go and not otherwise and we call that a law. We use this name because we are accustomed to the fact that respecting a law is rewarded whereas any attempt to go against it is forbidden and punished. Thus, Heraclitus says, the Sun will not transgress its measures. This is all very well, it stands as a law of Nature. However, we must recognize that science aims not so much at prescriptive laws but descriptive laws instead. We observe the regularity of certain occurrences and we try to predict when and how the same patterns will manifest themselves again. Aristotle came closer to state a proper law of Nature in our modern sense when he stated: The fire burns here as it does in Persia. We observe a regularity and we attempt to determine how it comes about, what is fundamental to it and what is accidental. This is a rather delicate question and I am not sure I could make it precise, but when I say that a law in science is not prescriptive, I mean to say that, as the law describes how a sequence of events will unfold, it does not imply that Nature has made an intelligent choice amongst various alternatives. Sometimes we may state the law in that way but the most profound laws are those that state the necessity of a certain order in such a way that there could be no other alternative. In this sense, the law is descriptive and complete. To a first approximation, a normative law may be the best we can hope for. Then, as we begin to insert the phenomenon we are describing into a larger context, we often find that such a prescriptive rule is superfluous, that each element is ultimately necessitated by its relations to others. Yet, as we strive to find an organizing principle to accommodate a wider range of phenomena, our laws become more vague and less effective. It is sometimes argued that centuries ago the Chinese could have been more successful in advancing their scientific knowledge if they had not been so fond of searching for the most basic and grandiose principles ruling the world. From ancient times they were interested in this elusive notion of the harmony of Nature and they were aware that one could not possibly arrive at describing its processes with merely a code of laws and regulations, that in some sense the order of the world was uncreated and spontaneous or, as they would say, things follow their course out of their own nature, like a servant follows a good master, out of cooperation and not coercion. Yet, however deep and appealing this concept may appear to us, it is also true that it is impossible to describe the fine details in the workings of the natural world starting from such a wide-ranging principle.

Plebeius: In their traditional thought, the Chinese never found much use for a law-giver in the Universe. Even if not entirely reflected in their social organization, from a philosophical point of view at least, they have shunned the notion of hierarchies in describing the natural order, preferring to think that the whole rules the parts as much as the parts determine the behaviour of the whole, each acting by consent, not by compulsion. I am reminded of an interesting remark of Aristotle, which is similar in character; he stated that Nature is like a man who is his own physician. Nature, he said, is agent and patient at the same time. It is steered and does the steering all at once.

Albertus: Man has never ceased looking for images that would make the natural world more comprehensible, always drawing analogies to the way we do things and, inevitably, resorting for our comparisons to objects that are man made. When clocks became more commonplace and their mechanisms more elaborate, men began to refer to the Creation as a stupendous, fantastic piece of clockwork, instead of saying plainly that the clock was a delightful, little piece of the Creation. I suppose that this type of thinking is for the most part harmless and may, on occasion, be useful, but it can also lead us into blind alleys when carried too far. We so often marvel with incomprehension at the miracles of Nature, how could the human eye be so well crafted for the tasks it must accomplish, that we soon begin to evoke some intelligence, not human intelligence but something like human intelligence, without knowing what we are referring to beyond the word itself. When we see a light source reflected on a mirror, we discover that the point of reflection is such that the length of the path the light ray has followed, from the source to the mirror and then to our eyes, could not have been shortened by the choice of any other path. How could the light ray have known at the outset in which direction to travel? Of course, this is not the right question to ask, but at one time it was believed that the phenomenon revealed some kind of foreknowledge, of intelligent order. This is an invitation to speak of purpose in the Universe, of final causes, and very soon we are thinking of a master plan, a grand design in the order of things, which we intend to unravel.

Plebeius: From an historian's point of view, the slow rise of a scientific tradition in the West over a period of several centuries took root in fact within societies whose beliefs and modes of thought had been shaped by an even longer tradition of Christian teaching. There was already an order in the Universe, it had a purpose and a meaning, and the new skeptical and inquisitive way of studying the natural world could only come to celebrate that view or to undermine it. Was it not precisely this view of the Creation as a magnificent piece of clockwork that slowly weakened the belief in a superior force, a hand that gave it life and direction? I would venture to say that, in the very early days, those who compared the world to a great machinery did so in wonderment and in reverence, delighting as much in the craftsmanship of the watchmaker as in man's ability to comprehend it. They probably did not understand Tertullian's assertion justifying his faith by its utter incomprehensibility. Far from withdrawing from the world, they found their faith within it. They saw the world as the Lord's vineyard and themselves as happy labourers, helping to bring forth a good harvest and eager to celebrate it with their own discoveries. It is the spirit that must have animated Kepler, considering the way you portrayed him. But that view gradually disappeared; perhaps it was Descartes' mechanistic views that gained ground, the world as machinery was taken in a more literal sense, seen now as a contrivance, following its own inexorable course, and man as a detached, distant but fair observer. Would you not agree that, ever since that time, most students of Nature have felt compelled to adopt this cloak of objectivity and to set aside their personal convictions and beliefs, as if there was something unseemly in revealing how one's desires and preferences steered one's endeavours? There must be many examples to the contrary but it seems to me that, on the whole, the earlier attitude, spontaneous and unaffected, never came back to life again.

Albertus: Isaac Newton is probably a representative example of the transition you are referring to. His scientific works are written in a dispassionate style, never alluding to his labours or to the path that led him to his conclusions. He is reverential and does not hide his faith. The planets, he says, can go on safely in their orbits on their own, but they are where they are because of the counsel and dominion of a powerful and intelligent being. Yet, he seldom ventures a statement more revealing or explicit. What is puzzling and mysterious about Newton is the fact that, despite this public attitude, his private papers, not intended for publication, were found to contain thousands of pages with various investigations related to sacred history and the chronology of ancient kingdoms; they contain many commentaries on Church doctrine and on prophecies, on heresies and on many other subjects. The passion and dedication with which he pursued these studies do not let us doubt of his earnestness. However, his apparent secrecy and reticence convey a sense of awkwardness or embarrassment. Regardless of the tasks to which he applied his enormous energy, Newton seems always to be animated by an exalted and indefatigable spirit; he saw himself as an interpreter and messenger of divine wisdom. This is also apparent in his studies on alchemy, which he never divulged. His attitude, one would be tempted to conclude, was marked by a curious view of the world that made it one part riddle and one part conspiracy, a view that prized all the more a rational explanation of its workings, making of faith and revelation something undesirable or a poor substitute to knowledge. In any event, we may not be justified in prying too much into his private papers, which he did not wished to have published. His writings in mathematics and physics were more influential than those of any of his contemporaries, but few scientists seem to have followed him in his many other wanderings and none did it with his zeal. Newton was once chided for suggesting that if the orbits of the planets were not stable, as the application of the laws of mechanics seemed to indicate, then divine intervention might be required to restore them to their due course. Many years later, the French mathematician Pierre Simon Laplace adopted quite a different attitude when, in a nonchalant manner, upon hearing of the opinion held in antiquity that the Moon's purpose was to illuminate the Earth at night, he quipped that, if such were the case, the Lord had been quite careless, for he could devise the equations of an orbit that produced a full Moon every night. Laplace belonged to a different epoch in which, for many at least, Providence had been replaced by the laws of Mechanics. When Laplace visited Napoleon Bonaparte, the future Emperor asked him where was God's place in his exposition of the system of the world, to which Laplace is reported to have answered that he had had no need for such a hypothesis.

Plebeius: But this new mechanical conception of the Universe has thrown us into a different dilemma, from which I do not think we have extricated ourselves yet. Either as a vast machinery or as a collection of atoms, blindly following the laws of physics, science has taught us to view the world in a deterministic manner. The very success of this viewpoint has made us reluctant to see it in any other light. Casting ourselves as external observers, pursuing our inquiries still in possession of our free will, we seem to have made an exception with ourselves or to have withdrawn from the picture that we are attempting to describe. Otherwise, is the success of our enterprise going to be so complete that we will come to see that we are not the rulers of our actions? I have always sympathized with the Greeks of antiquity who thought that we could never escape the course that Fate has drawn for us, although they believed at the same time that the gods atop Mount Olympus were capricious and had changes of mood, not unlike human beings. In this manner, through the unpredictability of their actions, a degree of randomness was introduced into the world. It was then the role of Fortune to conspire with this element of chance in order to shape our lives. We could not fight against Fate but we could nibble at its edges and, acting judiciously, help bring unto us more good fortune than bad. However arbitrary was this view, it was an optimistic one. By learning about the ways of the gods, we could exercise some control over our lives. Knowledge had the power to set us free. Science, on the contrary, seems to have taken us in the opposite direction. The more knowledge we acquire about the Universe, the more its course seems to be preordained and constrained by necessity. If our investigations lead us to the conclusion that our free will is an illusion, could it not be that the investigations themselves had been less than impartial and that our conclusions had been preordained too? Could we not have been deceived into believing we are not free when in truth we are? Could we not assert our freedom and make it true simply by an act of will? What sort of entanglement do we find ourselves in?

Albertus: I am afraid I could not be of much help in answering those questions, but allow me to state how things stand from a mathematical point of view. I should return to Pierre Simon Laplace, whom I mentioned a moment ago, for he was one of the first to emphasize the precise mathematical language of determinism that had just sprung from the new developments in celestial mechanics. The laws of dynamics make explicit how the forces of Nature impart accelerations to all bodies. If we have a collection of particles and we know their positions and velocity vectors, we can let them proceed for an instant, guided by these vectors, to their new positions. But now a new configuration realigns the forces, which confer new accelerations, altering the velocity vectors. Then, from the new positions and guided by these new vectors we can let all particles proceed one step further and we can repeat this process again and again. This implies that the knowledge of the initial positions and velocities, along with the laws of dynamics, are sufficient to determine the future course of events, provided that no external agents comes to disturb our particles. If we now think of all the atoms in the Universe and all the forces, thereby discounting any external agents, and provided we could record the positions and velocities of each atom at any given instant, at least in principle, it seems that the future of the Universe could be considered irrevocable and sealed. This is the view eloquently expressed in the writings of Laplace. However, if we turn now to the geometric representation of space and time that I described to you today, we find that the version of determinism that I have just discussed is incompatible with the nature of the Universe we live in. We saw that it does not make sense to speak of one instant in history, then the next, and so forth. We have eradicated the notion of a universal clock. This is very far from saying that we can no longer formulate any laws of physics. We can always confine ourselves to what happens within a box where, despite some insignificant distortions, a single clock might allow us to proceed from one instant to the next and, provided we keep track of what goes in and out of the box, we could predict all future events within it, according to the laws of mechanics. Therefore, as a mathematician, I can speak of causation in the sense that, within a limited context, past events influence the outcome of future events. However, when it comes to determinism applied to the Universe as a whole, I must say that it is an ill-defined concept and it is fruitless to discuss whether it is true or false.

Plebeius: Your position may very well be unassailable, but I suspect that it does not cancel a dilemma that we still find to be meaningful from the point of view of our immediate experience. I am reminded of Saint Thomas addressing the difficulties concerning the freedom of the will in the face of God's all-encompassing knowledge. He imagined himself at the top of a mountain, seeing a convoy of men traveling along a road, down in the valley. He could see the road behind them and the path they would follow next, but in no way was he influencing their actions. In the same manner, he contended, God knows all of man's actions. Necessity is not coercion, he contended, foreknowledge is not foreordination. This, of course, seems to be more an escape into wordiness than a solution to our quandary. Nonetheless, considering that your mathematical theories have so much distorted and confounded our naive notion of time, I suspect that Saint Thomas might have offered a sympathetic ear to them. His contention that free will was part of the doings of

God was based on the belief that the mind of God exists in other dimensions than ours and is not bound to know things in succession, but can contemplate things from a vantage point outside time. The necessity of the future following the past is only an illusion of our perception, he would have said, caused by our confinement to lead lives tied to the flow of time.

Albertus: I would respond to him that we ought to play gods ourselves and step outside time, not out any concern to preserve our free will, but merely to understand the true nature of the world we live in.

Plebeius: Are you not playing god already in a different way? You refuse to consider whether we actually have a free will or not. You conclude that the dilemmas posed by a deterministic world simply do not arise when your equations fail to make sense of them. You are putting into the hands of your mathematics to decide what is meaningful and what is not. Is there not a dogmatic element in that attitude as well? We can never profess to be truly skeptic and to be swayed only by objective evidence for, even when you claim to be led by reason in a dispassionate way, there is an underlying allegiance to the type of evidence you will admit. In this sense, there is an element of faith, if you wish, that precedes anything we can claim as knowledge.

Albertus: I would not present myself as a paragon of skepticism nor would I claim to have blind faith in reason. I would have faith in the power of reason if faith itself was reasonable, which I think it is not. What avenues are open to us when confronting our ignorance and our shortcomings? Are we to resort to rebellion, to resignation or to indifferent acceptance? Clinging to beliefs is not the way. If we learn anything investigating the sciences, it is that we need not face ignorance with discomfort or fear. Instead, they teach us to accept ignorance and, by exploring it, turn it into knowledge, for it is a form of knowledge to know the extent of one's ignorance. If, in my pursuit, I allow reason to be my guide, how could I test whether my confidence in it is well placed? Could it not be that to evaluate whether reason is a reliable tool I might need to use the tool that needs to be examined? In my eyes, reason is nothing more than prudent common sense. I am not at all inclined toward philosophical arguments. As a scientist, I will attempt to conduct my arguments with as much rigour as possible but, when it comes to philosophy, I will grant to you that I am somewhat of a dilettante.

Plebeius: You seem to find some delight in preserving some mystery within reason, the unintelligible instinct, certain above all of its own uncertainty, the power of the mind that cannot be justified by anything but itself.

Albertus: Curiously, reason does not make you feel that you are at its whim. Perhaps what is peculiar about it is that itself exposes its rules, its reach and limitations. As it is said of the good sailor, the man of reason always knows the length of his anchor line and knows how far he can venture with confidence and when he is out of his depth. Of course, when mathematics is at the center of your concerns, you want to discard everything but the arguments of logic and reason becomes all powerful and unassailable. But you soon realize that you have entrenched yourself in a world that is somewhat bare and rarefied, the very citadel of reason and, as you venture outward, you perceive its shortcomings and inadequacies. If you listen carefully, it is reason itself that points to you the need for other ingredients to arrive at a sound judgment. Nevertheless, against anyone who would dismiss reason on behalf of emotions and intuitions, or even premonitions, revelations and dreams, as a true vehicle to knowledge, I would certainly jump in defense of reason, even if it was a lost battle, for reason relies on arguments and these might be dismissed by someone disdainful of it. It is precisely against those who presume that reason is the only true path to knowledge that I would direct my criticism with more relish for then, trying to argue reasonably against them, I would be using the very same tool that they trust most. I am thinking in particular of those who have followed Descartes to the letter. They think that once the rules of logic have been discovered and settled for all ages the final obstacles to prejudice and ignorance have been overcome, as if by disassembling each problem into its elementary constituents and thinking clearly about each one of them we will always arrive at a satisfactory and conclusive answer.

Plebeius: Francis Bacon, another of Descartes' contemporaries and a great propagandist declared, in all his exuberance, that he was intent on creating a new logic of discovery built, as he said, on the rubble of the old ages. Perhaps at the time, after man had begun to think that he might not have been created in God's image but was merely a creature of Nature, after the Earth had been dethroned from the center of the Universe, he could now, relying on his own skills and the resources of his ingenuity, become the master of the world of which he was no longer the center.

Albertus: Considering the predicament they found themselves in, Bacon and Descartes set out on the right path; it is enough for us to look at the degree of superstition they were fighting against, the dependence on dogmatic doctrine, on ancient authority. They were justified in rejecting any and all inherited knowledge, starting from the beginning, examining each truth on its merits and following the dictates of reason. However, once we adopt this point of view and travel this path long enough, we begin to have our doubts. Can we really speak of a logic of discovery? Is reason indeed the only source of all the knowledge we acquire? Are we allowed to include within its province our errors and mistakes, our free associations and analogies, all matter of riddles and paradoxes? For they are often what leads us to discovery. Like an explorer in a new territory, we advance somewhat at random, by intuition, suspicion and guesses. Doesn't reason come afterwards, when a full reconnaissance has been made, to clear the terrain, to build the shortest roads and to provide us with accurate maps? Truth has often many sides. A set of logical rules by itself cannot lead us to discovery. The answers we arrive at are always as good as the questions we are capable of asking. The meaning of things is commensurate with our limited understanding and what is meaningless to us today may be due only to our ignorance. By all means we should not fall into the belief that reason is a tool that exists on its own, independently of ourselves, that we happened to suddenly acquire and that has made us proud owners of a master key, a key that will enable us to solve the riddle of the Universe. At the very least we should suspect that this tool may have some uses we have not yet discovered, that it may wear off with age or that we might be able to polish and improve it. Those who insist on reason being our only path to knowledge sometimes believe that we have now arrived at the final secret code, that it is no longer a matter of asking questions. These have already been asked. Now we stand on solid foundations and can begin to build piecemeal the pyramid of knowledge. All we need is at hand; we can proceed to write all the answers one by one in the great book of Truth. It is not difficult to share in the exhilaration. But this exhilaration in the power of reason had its costs and led to a certain arrogance and pedantry. I say arrogance because it led to the illusion of objectivity, to the belief that now, at last, we had a set of rules and principles to which anyone could resort, in all ages and circumstances, to answer the fundamental questions of Nature and express them in a universal language. But it also led to some pedantry for it contained within it the assurance that any problem could be tackled by the same method of reduction, with the deleterious effect that many complex and intricate questions were reduced to such simplified and abstract versions that they have lost their character and led to solutions that have little relevance to the original problems. Descartes certainly fell prey to this tendency. I suppose that philosophy has always courted this risk, of becoming too narrow-minded and engaging in excessive wordiness, of straying from the spring that brought it into existence, more at ease lingering in the knowledge of its own knowing than in the knowledge of the world as such.

Plebeius: I am afraid it has been a long time since philosophy has abandoned the common man. It no longer lures us into reflection and self-examination. At one time, I imagine, philosophy was still an art and in its task of illuminating the human experience made its fruits accessible to us all. Today, unfortunately, we have professional philosophers, who spend a good part of their time scrutinizing the writings of other philosophers. In this regard, I think you would be unfair to Francis Bacon. He insisted forcefully in engaging in a dialogue with Nature, against what he called the first distemper of learning, when men study words and not matter. There is a passage where he speaks against those who, knowing little history, either of nature or time, out of no great quantity of matter and with infinite agitation of wit spin out unto us those laborious webs of learning which are extant in their books. The wit and mind of man, he continued in his florid style, if it works upon matter, which is the contemplation of the creatures of God, works according to the stuff and is limited thereby; but if it works upon itself, as the spider works his web, then it is endless, and brings forth indeed cobwebs of learning, admirable for the fineness of thread and work but of no substance or profit. Bacon and those who shared his enthusiasm were not merely extolling the power of reason because it provided the surest road to knowledge but, above all, because they saw it as the best tool for action. Denis Diderot, more than a hundred years later, showing perhaps less exuberance than Bacon but more practicality in his accomplishments, was not just interested in understanding the world but also in changing it. The idea of filling our cities with indolent contemplators, as he wrote, did not seem very appealing to him. He eloquently expressed the view that reason should not necessarily occupy the first rank in the order of invention. He was particularly fond of artisans and he asks why it is that history forgets the names of the great craftsmen of the past while those of warriors that have only wreaked destruction are immortalized in great monuments. Diderot and his colleagues, who were shaping with great enthusiasm the climate of opinion of their age, were not so much interested in theories but in action, action perhaps guided by reason.

Albertus: I would not wish to count myself amongst those indolent contemplators. I have great respect for the artisan even though, like Diderot, I must recognize that I am not one of them, but I say it with regret and not with pride. The artisan has his own way of apprehending the world, quite different from the manner of he whose learning and inventiveness come from his mind and not his hands. But before we heap too much abuse on our philosophers, if we praise action instead of contemplation, let us concede, when we consider reason as a tool, that taking good care of it and learning how to use it is an activity in itself. Quite apart from this distinction between the desire to understand the world and the attempt to act on it, it seems to me that mankind has lived through different eras, at times submissive to its lot and complacent in its narrow modes of thought and, at other times, a subversive and creative spirit has prevailed, eager to upset the preexisting order. We hardly need to say where our sympathies fall. We do not wish to be content in ignorance and doubt or to surrender our desire of exploration; a lack of curiosity or a docility of will are qualities that we find distasteful. The optimism of Diderot and those who shared his views was healthy and productive. Their confidence that a combination of thought and action could make man master of his own life was based on experience and on the fruits of their labours. But we should think of the benefits of confidence as does the acrobat on his tightrope, it leads to ease and excellence in the performance and sometimes also to our downfall. I am not sure we could characterize our own times in the same terms, as subversive and eager to replace an old order. We have kept the subversive spirit, but in these convulsions any reference to an old order has lost all meaning. We have lived through an age that has brought unprecedented material change, to the extent that our capacity to act has exceeded our capacity to comprehend how to act. At the same time, our powers have given us a false sense of security and we have acquired the vanity of those who always know best. I think this is as much true when we consider the present condition of mankind from the perspective of its entire journey of thousands of years on the surface of the Earth, as well as when we consider our own plight in the immediacy of this historical moment. Firstly, if it is true that man became dazzled by the magnitude of his achievements, he also lost the sense of his own fragility, confusing this seducing varnish of civilization with a true advancement in human nature. We do not pause to think that man has remained the same he has been for all of recorded history and much beyond. Do we realize how extraordinary and delicate is this edifice of learning and well-being we have acquired, how some unforeseen catastrophe, not even of the magnitude of the plague you spoke of earlier, or far from a cataclysm of cosmic proportions such as a new ice age, but perhaps an almost imperceptible misfortune might throw us out of balance, destroy our social order and leave us wandering without a purpose in life? I cannot help this feeling, when I consider our own times, that the transformations we have brought on ourselves and our sense of confidence have almost detached us from history. We find its lessons somewhat distant and foreign. We have the attitude of adolescents, racing to a future that we think we know how to make ours, although we run away from the present as soon as we have conquered it. There is a frivolous element in this chase toward a promised land that we do not know what it consists of. Is this merely a blind faith in the inexorability of progress?

Plebeius: I would agree with you entirely on our apparent disdain for the value of history, but I am less certain we could speak of a frivolous element in our strive for progress. I think our confidence and feeling of superiority are born of an objective assessment of the goods we have procured ourselves, without having to lose in the process a reasonable sense of our reach and limitations. I have always considered myself an optimist, which is not to say that I believe in the inexorability of progress. Concerning our disdain for history, I am reminded of the case of Voltaire, who never held back his sharp wit to flaunt the prodigious superiority he felt his own age had achieved, but devoted a great deal of energy to study and to write abundantly about the many ancient and foreign cultures of the past, at the same time that he was disparaging them. In his days there was a vigorous debate as to whether the ancients or the moderns had achieved a higher degree of civilization. Although the arguments often centered on weighing their respective literary merits, those who favoured the ancients maintained that their opponents were like children, rebelling against the nurses that had raised them, but these countered that the moderns were the true ancients, the old and seasoned by time, who could benefit from the accumulation of past wisdom. I imagine that those who held a more balanced opinion believed that the ancients could not have competed against the modern advances in techniques and in the sciences, but perhaps in the arts, in eloquence and the general civility of life, the ancients might have acquired a higher degree of sophistication and, for some, even if they saw their present time much richer in ideas, they also saw it poorer in taste. It would seem very uncharacteristic of our age to be holding such a debate. We believe that we can stand on our own; from the point of view of history we have become somewhat parochial. Perhaps, as you said, more than thinking that only the present holds anything of interest, we are constantly chasing an unknown future in the name of progress. However, it strikes me that if you were to express any enmity toward progress in the present circumstances, not only would you be viewed with suspicion, but you would find yourself in rather strange company. For there are always those who are suspicious of any new knowledge, who believe that man, seeking what does not belong to him, is guilty of some form of trespassing. This is as old as the stories of Prometheus stealing the fire from the gods, or of man eating from the forbidden fruit. There is also a group of people who see in the intricacies and difficulties of modern life, even merely in the pretense of decorum or the cynicism behind all ceremony, the dark and corrupting influence of what we call civilization and they advocate a return to what they think is man's original nature, perhaps more intuitive and less censored by reason, not so contrived and artificial but simpler and more primitive, something that they do not perceive as discrediting, much less ruinous.

Albertus: You are right, Plebeius, and I would not want to see myself associated with those who speak of the Fall of man, condemn our loss of innocence or bemoan the advancement in knowledge, for the same reason that I do not sympathize with those who believe in the inexorability of progress. I tend to be distrustful of those who see some great design or organizing principle giving meaning and direction to our doings. We have already had a long cast of philosophers who, with elaborate theories, have attempted to explain the course of history toward the fulfillment of man's destiny, a fulfillment that, by coincidence, was taking place at the time they were living. We must be skeptical of these great simplifications.

Plebeius: Indeed, but precisely because we should be suspicious of any grand organizing principle, it seems dubious to me that merely a blind belief in progress could be the guide to our behaviour. When it comes to the advances in the sciences, we have had an unprecedented success in applying every increment in knowledge to improve the conditions of our lives, to ease pain and discomfort, as well as to free ourselves from tedious tasks, liberating our energies for more rewarding and fulfilling enterprises. How should we expect our scientists to behave, for example, when we have good reason to believe that a concerted effort in research might lead to the discovery of an effective medication against some illness still plaguing humanity, or that we might discover a method of predicting when and where the next earthquake will strike? We would be remiss if we did not concentrate our efforts and apply our ingenuity to bring forth these advances. It has nothing to do with a vague, specious expectation of progress, but with concrete and often attainable goals. To the extent that we can identify our own good, the way forward is clear. We give ourselves to progress, not because of its inexorability, but because of its goodness. It is true that sometimes the benefits we acquire transform us and make us seek other goals, but in this sense it would be unfair to say that we are permanently escaping the present, racing toward an unknown future. Most of the fruits we derive from these advances stay with us forever; this is certainly true, in particular, of the knowledge we acquire. We cannot dismiss it and step back; if there is something we cannot learn, it is to be ignorant again. Only from this point of view we might say that progress is indeed part of our fate.

Albertus: What you say is true and history bears proof indeed, although we must also acknowledge that it seems most true when we consult the written history for, in recounting the past, we tend to erase all the dead ends, its uncertainties and contradictions. Each age feels the need to rewrite the past as if it was a path toward the present, but when that past was still current, it was probably seen as the present is always seen, ripe with confusion and open to many possible futures. Judging from my experience, I would have to conclude that the goodness pointing our way forward and guiding the progress in the sciences has little to do with the way our scientific knowledge advances. I am not expressing the cynic view that many of us are spurred by material rewards, fame, or the eagerness to do better than those who are chasing the same objective we are chasing. I am thinking, instead, that the advancement of science is usually quite chaotic. We start aiming at a particular destination, but we are soon diverted pursuing other goals, led sometimes by premonitions, by the sense of adventure, sheer curiosity or the thrust of pure speculation. At other times, we do not even look for answers to specific questions, but labour to understand what are the right questions to ask. It is true that a good many tasks in science can be methodically charted in advance, like the reconstruction of a disassembled puzzle, the loose pieces are at hand and one can foresee the contours the end result will follow. But it is quite different at those decisive moments when something new and unanticipated is found, perhaps by accident, perhaps following an intuition against all common sense or useful purpose. These are the instances of great consequence that mark the real advancement of science and it seems inevitable to conclude that, when it matters most, the desire to do good was irrelevant in the final turn of events. In what concerns the inevitability of progress, I would be inclined to agree with you that we cannot learn to be ignorant again. However, I think that we lose knowledge at the same rate that we acquire new one. We do not really need to await a calamity that would annihilate our civilization in order to render us ignorant in any way. As we learn new things and refashion our theories, we dismiss old ones, theories that may have explained particular details the new ones do not, only that we now deem them uninteresting. Also, much of what we call knowledge is experience, the exercise of knowledge, and in the same way that it is very difficult for us to feel the way we felt when we were children, society as a whole loses its past experience and older ways of apprehending the world. Even within the sciences, it is very difficult to repossess the knowledge Archimedes or Kepler had, in the way they experienced it. The knowledge we write in books might be difficult to unlearn but the knowledge we cannot write down is most evanescent and fragile.

Plebeius: It may very well be, as you say, that the scientist often sees the good he can do as remote and uncertain or indifferent to his task, caught as he is in the logic of his own investigation, being led in whatever direction it may. But should he not pause once in a while to examine the course he is following? It would be an interesting, though harrowing, experiment if we were to proclaim that, being the inclination toward the good of little consequence for the advancement of knowledge, we might henceforth let it drift wherever it pleases. As it is, science has already followed a dangerous path on more than one occasion, bringing us nightmares as much as comfort, and I would not be willing to subscribe to a science that is scornful of morals. Could we afford a science that is unaccountable and irresponsible?

Albertus: The answer should be no. Nevertheless, not simply out of a contrarian spirit, I should venture to say that we are more in need of advocating the irresponsibility of science, perhaps not to advocate it, but to proclaim it without shame. If, in practice, that is the way it behaves, it would only be a matter of honesty. We may set out with responsible goals and good intentions but, once embarked on our investigations, we are often carried by their own logic, perhaps by their own illogic, since there is no logic of discovery. The paths we think we must follow end sometimes in blind alleys, in accidental detours, and things we stumble upon by chance prove occasionally to be decisive. The desire to be responsible may even interfere with our enterprise and send us far from where we intended to go. Mere curiosity or adventure, the pleasure of entertaining daring hypotheses, or an idea pursued only by a sense of taste or elegance, may prove to play a bigger role in pointing our way. The making of science is somewhat anarchic and most of the time, in its very details, does not know where it is headed. It is only a posteriori that we invent a plausible path to chart its progress.

Plebeius: If you proceeded as if you were persuaded that striving for the good is not in vain, even when you did not believe in it, then I would feel more at ease. One thing is how the scientist perceives himself and the goal he is pursuing, quite another the impact his activities have on society at large. The effects that the activities of science have had on our lives are so considerable that a call for science to assume its responsibilities seems to me not only sensible but even imperative. The drive of curiosity and adventure that you may experience with pleasure may also become dangerous in its consequences. We might consent that our scientists depart on their trips of exploration, fishing for ideas in distant seas but, upon their return to the harbour, we should inspect their catch and evaluate its worth. We have seen sometimes in the past some of our very best scientists put their skills and ingenuity at the service of those who wished to find new and more grotesque ways of killing and causing devastation. The very least we must demand is that our scientists consider the fruits of their undertakings, that they understand that the burdens imposed on science are comparable to its achievements.

Albertus: I also decry the power that science can wield and the harm it can cause. Yet, while I sympathize with your misgivings, I cannot share your desire to make science more trustworthy. Truly, what responsibility are we asking science to assume? If it is the responsibility to do no harm and tend to the good, we must accept that it seldom knows in advance where the good is; that is the purpose of its search, but it often must proceed blindly. We recognize that, once acquired, knowledge is power, but are we willing to add to this power an aura of responsibility it does not possess? Why then invest science with an authority it does not deserve? Moreover, it is not the enterprise of science that would exercise this authority, but the individuals who pursue it, individuals with their own personal interests, preferences and tastes. Should we invest science with authority merely because of its power, rather than its wisdom? I do not wish to draw a sharp distinction between science as abstract knowledge and its applications, between science as an exercise of the mind, confined to observing, to building theories, at most to experimenting and, on the other hand, to the use we make of this knowledge, to techniques and tools capable of transforming our lives. Very often one arises from the other. However, when we contemplate the colossal effects that our scientific knowledge has ultimately had in changing the material conditions of our lives, we realize that these activities lie at the edge of science and, usually, have little to do with the efforts of the scientist as an individual. They require a different kind of effort, organization and industriousness that involves society as a whole. This is one more reason to insist on the fact that science at its core is still irresponsible and should not feel embarrassed to expose its irresponsibility.

Plebeius: You make me think of Leonardo da Vinci and I imagine that you would find him very congenial to your ideal of scientist, not only for his inventiveness and self-assurance, but also because of his daring and his lack of apprehension or prejudices, as he demonstrated by not having any scruples in dissecting the human body and violating the morality of his age. He always seems to have been driven by the passion for detailed observations, to capture the facts of the world, unadulterated and without interpretations. As he says quite explicitly, we can only love or hate what we know. It is clear from examining his writings that he felt he was celebrating the magnificence of the creation by learning more about it and making it his own, without in any way feeling the need to be unduly reverential or sanctimonious. Nevertheless, at the very same time, in his own private life, he had no qualms to court the most powerful of his day, however unsavory, peddling his skills and services to win their favours. In this sense, I think that Leonardo da Vinci would make a good example of the type of irresponsible scientist you advocate.

Albertus: Your portrait makes him all the more likable. But, truthfully, of course that is not the type of attitude I advocate. Firstly, when I proclaim the irresponsibility of science, it is not because I think that it must be that way, but because I think it cannot be any other way. I exaggerate my argument because I do not think it is sensible to demand that science be responsible. Moreover, while we are honestly irresponsible as scientists, we can still be responsible as citizens. What I want is that we avoid deluding ourselves into believing that, only because our scientific results can have a considerable impact on our lives, we could demand from science to proceed responsibly. Quite to the contrary, because of its possible impact, we must expose its irresponsibility. Some argue that knowledge is inherently good, that on those occasions when it may deliver us, in good conscience, into the hands of a calamity, it may be due only to the fact that all knowledge is partial and incomplete; they argue that the only solution to dangerous knowledge is more knowledge. I would not be so sanguine about adopting this as our maxim, but I concede that in practice we often find ourselves constrained to follow that logic. Nevertheless, I would contend that the ultimate responsibility lies not with knowledge, which is what science delivers, but in something more vague that we might call wisdom, or the knowledge about how to use knowledge. I am not really facetious in making the case for the irresponsibility of science. If we were to disregard its effects, its services as well as the ills it brings upon us, we might be led to consider the value of science on its own merits. It seems to me that, at present, the argument for the value of science is so overwhelmingly a utilitarian one, that we do not feel inclined to consider it in any other light. The entire scientific enterprise occupies an ever larger number of people and plays an ever increasing role in public life. We have been reaping its fruits with great zeal and dedication. The scientific outlook of our civilization has taken shape slowly over long centuries, but the vertiginous changes in the material aspects of our lives that it has facilitated are rather recent and abrupt. Like the discoverers of a new continent we have been racing to plunder this new Eldorado with voracity and impatience. Is science not a good in itself, regardless of its utility? If not, we might well conceive that our civilization, once the novelty fades and the fruits of science become harder to come by, could at some time impose the task of making new discoveries on those confined to labour camps or prisons, as is the case with other menial and uninspiring tasks. If we are less extreme and decide that science is a good in itself, then we should ask ourselves what are its virtues, how much of our activities should be pursued as an end in itself and how much for the benefit we derive from them. Just like a gardener must decide how much land to assign to his orchard and how much for ornament and pleasure, so too a scientist must decide how much space to allocate to flowers or to fruits. This comparison is not too far-fetched, for in the garden of science one is often confronted with the same difficulties and assortment of tasks as in a regular garden; there is overgrowth to contend with and there are weeds, there are the crawlers that never rise very high but tend to cover everything, there are the vines rising at the expense of others, there are the one-day blossoms, the perennials and all other varieties of plants. Instead of seeking our justification only in the unconditional goodness of its products, we should also consider, as with any other human activity, how science promotes a good life. I do not mean to imply that these questions ought to have definitive answers but we should entertain them more often. One would like to think that the general scientific outlook, its modes of thought and argumentation, its unprejudiced weighing of evidence, the way it exposes the vanity of wishful thinking and the easy deception of our senses, all these things are important in themselves as much as for the uses we make of them. However, after we have witnessed the rise of this enormous army of professional scientists, it seems to me sometimes that the true spirit of science is more alive amongst those who practice it as amateurs, for distraction or enjoyment, and less amongst those who practice it in order to earn a living. The end products of science seem to be today more important than the means to obtain them. The spirit of competition, even the zealotry with which some pursue their goals, and also the pedantry and self-agrandizement one encounters all too frequently, all this is rather appalling and discouraging. I am not speaking of dishonesty, the bending of rules, or the lack of principles, but sometimes it seems that taking oneself too seriously is an equally unfortunate disease. One could argue that acknowledging the limited responsibility of science might also help in reducing that sense of self-importance. At first sight, it seems odd that one would want to put less emphasis on the utilitarian aspect of science and even boast that our pursuits need not be oriented toward contributing to the welfare of society in any direct way. But it often happens that the defense of our undertakings on account of their utility does not have other purpose than to offer us an alibi. We justify our altruism by working toward a goal that lies outside ourselves and we measure our virtue by asking how we contribute to the advancement of science, but it would be equally sensible to ask how the activity of science contributes to make us more accomplished and civilized human beings. Science is, above all, a way of probing and understanding the world that surrounds us, it is a language with its own grammar and syntax. In this regard, it is very much like music, it allows us to understand and express things that we could not communicate otherwise. Of music we do not demand anything more than the joy it brings us, but science is entangled with many other things in life and affects us in more complicated ways.

Plebeius: If you allow me to interject, I will concede, for the sake of argument, that society as a whole should not believe that it can procure itself a science responsible of its actions. Even if this is not true, I tend to agree that we should exercise our responsibility as citizens, retain some control in our own hands and not grant to science any more power than is necessary. However, you are also arguing now that science is a good in itself and I deduce that it would be most desirable to make it a part of a good education. This is all to the good, not only will an acquaintance with science make us better citizens, but it will also allow us to understand better that which we must control, except perhaps that we might find ourselves in an ironic position, that being wary of the authority that science might exercise on us, we are, in a way, by desiring to control it, allowing it in through the back door, giving it a preeminent position in education. Some people perceive that this change has already taken place and they complain that the arts and letters have lost ground in education in favour of the sciences, whose traditions, stories and goals begin to dominate our mental landscape and provide a frame of reference for us all.

Albertus: Yes, it is a delicate matter to achieve the appropriate balance between opposing needs. Firstly, we do not wish to fall into any kind of dogmatic doctrine that would defend the preeminence of scientific thinking, pushing us into some kind of orthodoxy and obliging us either to conform or to be ostracized. On the other hand, the need to learn the rudiments of what has come to permeate the world we live in, must be defended; even if we never become proficient in science, we must gain some familiarity with its principles and procedures. In this regard, science has fortunately some merits, for it is not self-deluding or gullible; it teaches us to be cautious and critical and never to accept indiscriminately what we are taught, but to scrutinize things and make a reasoned judgment. It is precisely when we want to be critical of science or perhaps even to disengage from the influence it has on us, that an acquaintance with the true nature of science is most welcome, for science is best criticized in its own territory and with its own tools, with clear and precise thinking. It is rather distressing to find that much of the opposition that the sciences occasionally encounter is for the most part misguided, as if it originated in something visceral and atavistic, formulated in specious and muddled arguments that one would rather ignore or cast to the wind. An informed citizenship will not be uncritically acquiescent to the dictates of science nor prejudicial against it. Furthermore, if we wish an enlightened citizenry it is also to stimulate the quality of the science it procures itself. Much of the scientific enterprise these days is of a vast scale and requires a great deal of organization and planning and cannot rely on the entrepreneurial spirit of a few enthusiasts; it requires the support of the society at large, which must also be patient sometimes to await for the very lengthy work to bear fruits. I spoke earlier of the pedantry we often encounter amongst those peddling their own scientific work, but it is also true that our research is often tentative and exploratory and we must be tolerant of what at first sight might seem nonsensical or pedantic, but may turn out to be significant and profound in the end. Not always do we have the ability to distinguish one from the other. For this reason, we should feel compelled to present our views in a manner that is as clear and intelligible as possible. Science, by being sometimes very difficult and, at other times, making itself too difficult, has been accessible to less people than it should. This impenetrability has contributed to its aura of authority and, frequently, it has lent this authority to the established authority of government, whether legitimate or not, producing too cozy a relationship between knowledge and power, between those who know and the ones who can. One would have expected that, through education, science would have had a beneficial effect on civic life, but this has not always been the case. In some measure, it has acted to undermine democracy. By democracy I do not mean just the institutions of representative government but, more generally, the rather loose understanding that we have collectively of the relationship between the individual and society as a whole, of the unstable balance we search between self-determination and consent in the administration of the common good. The fact that the dictates of scientific progress present themselves to us as imperative and inexorable, along with the messianic zeal with which some of its advocates often come to us, helped in no small measure by the undue preeminence we have given to expertise, all these things have contributed to harm democracy. I think that science should occupy a preeminent role, but I also believe that it has no right to usurp it; instead, it should be granted to it daily by the citizens it serves.

Plebeius: To have a true democracy you wish that the scientists would descend to the center of town on the day of the fair, so to speak, and that they present their wares, that they compete with astrologers, charlatans, tricksters, soothsayers, counterfeiters and all their ilk, confident that a well educated citizenry will have no difficulty discerning where to go when searching for genuine goods. I suspect that you are right, that it is in this constant confrontation that lies the beauty and hardship of democracy, that it gives error a chance and is willing to compete with falsehood and deception on equal terms, confident that a mixture of good sense, self-interest and reason will eventually prevail. Nowadays, we praise excessively the virtues of democracy, indulging too much in its vocabulary but not committing it necessarily to practice.

Albertus: That may be because the practice is laborious and sometimes tedious, but we do not have many alternatives if we wish to remain rulers of our own lives and not yield to power any more than we must. Science, for its part, need not use devious or deceptive methods. The fact that we deny to it any special authority or decline to submit to the tyranny of expertise, does not preclude us from presenting forcefully the contributions it has made to society, or from explaining its traditions and the requirements it imposes in order to be successful. For this purpose, it is most desirable that the scientific enterprise be conducted in an open and transparent manner. Herein lies a grievance of mine with Isaac Newton, that throughout his life he seems to have pursued two kinds of science, one of them public, for which he gained our admiration, the other one secret and, as it turned out, unprofitable and ill-conceived. It is quite clear that he held this belief in some kind of arcane, esoteric knowledge, which I think is very detrimental to science, the suspicion that there is a hidden knowledge, to which only the elected few can gain access.

Plebeius: You prefer to agree with Robert Boyle, whose words Newton apparently did not heed, for he stated that the works of God are not like the tricks of jugglers or the pageants that entertain princes, where concealment is requisite to wonderment, but the knowledge of the works of God, he said, proportions our admiration of them, they participating and disclosing so much of the unexhausted perfections of their Author that the further we contemplate them, the more footsteps and impressions we discover of his perfection.

Albertus: It is one thing to say that art and ingenuity are needed to reveal the inner workings of Nature and quite another to hold that cunning is required to solve a riddle that has been purposefully designed to deceive us, or that the truths that we see manifestly are only a clue and a cipher to a hidden mystery, as if there were two tiers of truths, the second and more profound one only accessible to those who have been initiated by special means. Science should stay away from this type of nonsensical ideas; it should remain open and contemptuous of secrets. It should seek the interest and approval of the common citizen, not by the power of authority or through intrigue, but with no other means than honest persuasion and eloquence. We submit to the rule of common sense, perhaps trained, informed and intelligent common sense, but common sense nonetheless. Science may have its own language and its own traditions, but it forms part of the culture of a community and it should never stray too far from the concerns of the common man, even as it tries to educate him and shape his views. This desire to give preeminence to the common man, the belief in the fundamental equality of all men and women, we have come to value very highly and we have placed it even above the rule of science. Indeed, even if it came to light that some race or group of people were superior to the rest, in whatever way this superiority might be measured, we might still choose to live by our ideals and not by the truth. When such proposals have surfaced in the past and someone has suggested that we direct the composition of the human race by favouring the breeding amongst certain groups, discouraging it amongst others, we have not only found it undesirable and unacceptable but we have almost felt ashamed of ourselves. Our reticence is manifest in the reluctance to interfere with the composition of the race as a whole, but also in our desire to respect the integrity of each individual. This is evident in the treatment of mental illnesses, in the general caution when using methods or prescribing medications that might alter the character of a patient, his or her own identity, and, ultimately, in our ambivalence about interfering with the genetic makeup of an embryo. Although it would be very difficult to point to something specific that we could call human nature, we wish to be very protective of it. Ironically, this has made itself more evident as we have become increasingly aware of the constant process of evolution taking place around us and within us. Whether it has been in the depths of the sky or deep in the fossil record, we have been confronted with the impermanence and mutability of everything in this world. Even with our own hands we have done our share by causing an enormous upheaval in the environment where we live. Granted, this has taken place in an insignificantly thin layer on the surface of this remote planet, but the consequences could well be ominous for our own future. We have been reluctant to alter our human condition directly, but we have done so indirectly, since man as an individual, or as a species, is constantly adapting to the environment he is transforming. We adapt in our customs and habits, sometimes merely in our mental representation of the world; we adapt our culture, as we might say, but slowly perhaps we adapt our own physiology, as our organs adjust to changes in our surroundings or in our diet and, eventually, our biological make-up may vary and adapt too. For a very long time we searched for the impulse of life, the elusive ghost separating us from inanimate matter, but this goal gradually receded, transforming us more and more into the prosaic mechanisms of chemistry. We became aware of the inevitability of random mutations that are a consequence solely of the missteps of Nature. However, as our knowledge about these processes has increased, we have felt ambivalent and reluctant to interfere with what Nature has made us to be. By contrast, even after recognizing plainly that we are seldom the masters of our passions and desires, we insist on our malleability and the capacity to purposefully change our behaviour. We consider it is a transgression to interfere with Nature's order, presuming it is permanent, knowing it is not, and at the same time, believing in our own perfectibility, we find it not only admissible to steer through actions what we want to become, but we think it is praiseworthy to do so. We do it by exhortation, by example, uncertain of the agent that should assume this task. Some say it is the role of education, other assign it to Providence, or perhaps it is just the inner motions of our souls that will lead the way. We think of a more perfect condition but we do not know what shape it will take, either in our individual faculties maturing to a more perfect state, to call it virtue or happiness seems so inadequate, or perhaps it will lead us to a more perfect social organization. However noble our efforts and struggles may be, sacrificing the present for a more perfect future, at each stage we realize the goal we might want to reach is not what we had anticipated, putting us always at the first rung of the ladder we wish to climb.

Plebeius: It is striking that however much we speak of our efforts and resolve to become more perfect, we have so little intimation of what our destination will be. Have you noticed that when you ask young children what they want to become when they grow up, they invariably choose professions that seem colourful and full of adventure? They wish to be firefighters or captains of big ships. Of course, as they grow older their preferences change, but it could not be said that they did not direct their actions toward realizing their original intentions. It is just that the very process of attempting to become a firefighter acts on their will and perception of the world, so that their goals change accordingly. It occurs to me that, despite our pretense of having become adults and having understood the inner workings of the world, our notion and aspiration of a more perfect man may simply be, in the light of what we shall become, a childish desire to be firefighters. Our goals always transform themselves as soon as we take a few steps pursuing them.

Albertus: Well said, Plebeius. If we kept that thought in our minds, we would cease to take ourselves so seriously. Of course, we do not delude ourselves entirely believing there is a state of perfection to be attained; it exists only in our fantasy as a chimera but, if the chase is all that matters, we should pay more attention to the means we employ rather than the ends we pursue. We make so many concessions in life, always in the name of a sacrifice, pretending to forsake the present in deference to a promised but dubious future. Proceeding with your analogy to children, we often impose tasks on them with the argument that they might be meaningless now but will be justified when they become adults. We use the same argument in our solliloquies, expecting to justify our actions by the goals we pursue, which remain largely unknown to us until we achieve them. We should recognize that our differences in relation to children are a matter of degrees. We are always in the process of becoming and each stage in life should be measured against its own standards. Therefore, when we think of our perfectibility, it seems clear to me that striving to be perfect should be made part of being perfect.

Plebeius: Perhaps no one has expressed this belief in our perfectibility with the candor and enthusiasm displayed by the Marquis Jean Antoine de Condorcet in the famous Sketch he wrote while in hiding in Paris, during the terror, just after the Revolution, awaiting to be captured and sent to the gallows. As the title he gave to his Sketch indicates, it is a historical chart of the progress of the human spirit and Condorcet tries to show, in light of the historical record, that the drive toward perfection is a law of Nature. Although man is confined to live in a minute corner of the Universe and for a fleeting moment, he thinks that by yielding to this drive to perfect himself, man becomes an active participant in the great drama of Creation. Condorcet certainly fell victim to some of the attitudes that you criticized, believing that mankind was now arriving at a stage in which it could finally free itself from tyranny and oppression, from prejudice, ignorance and idolatry. He thought that a universal language of reason was within reach, applicable not only to the sciences but to all human activities, that would also make it possible to map the road to progress in advance. Nevertheless, he did not place his optimism in the accomplishments of a few enlightened men, but championed the cause of democracy and the preeminence of the common man. For him the drive toward perfection was a collective enterprise that society as a whole expresses in the form of civilized life. But this effort acted in the same way as in an individual who, by education and training, improves his faculties as he grows up. Condorcet thought that our ability to understand the natural order allows us to put it to work to our advantage, sometimes even evading the constraints of Nature, or pretending to do so. In any event, either by doing its bidding or by trying to free ourselves from it, we are only in more subtle ways conspiring with it to advance the cause of progress and arrive at a more perfect condition.

Albertus: Perhaps the idea of perfectibility has come about because we have fallen into this illusion of self-awareness, as if this was a quality peculiarly human. We observe the world at large stumbling along on its way, so we say, either by instinct or following the laws of Nature, whereas we, through our faculties, have the capacity to reason, to remember, to anticipate, and having understood the process of evolution can, for the first time in history, steer this process consciously and purposefully. This brings us back to the questions you raised earlier regarding our apparent free will in a deterministic Universe. However, it is not our ability to choose a course of action that I want to stress but simply our awareness of what is taking place. Why should we assign ourselves such a special faculty? Why could we not say that everything in this Universe has this capacity of consciousness? Why is a butterfly not conscious of being a butterfly? Of course, we are not speaking of human consciousness but butterfly consciousness. And when the atoms in a diamond fall into line to form a perfect crystal why do we not say that they do it, conscious of what they are doing, with diamond consciousness, as it were.

Plebeius: We do feel indeed that part of being human is this capacity to look at the Universe objectively from the outside, as if we were not only actors on the stage but have won some special dispensation that has allowed us to go behind the curtains and learn about the plot and the course of the play we are participating in. Now, back on the stage, we have the ability to alter the events in the play modifying its ultimate fate. Regarding your speculation about this pervading consciousness, I am afraid that the imaginative Denis Diderot, always concocting some great fantasies without straying too far from the truth, may have in fact anticipated your very same ideas, except that he spoke of sensations instead of consciousness. Nonetheless, he argued that all matter was capable of becoming sensitive. Sensations were the attributes of fibres that he likened to strings; they can be plucked and made to vibrate on their own and have the ability to stimulate other strings to resonate by sympathy. To him, in a rather vague manner, life was the result of fermentation and was inherent in all matter due to the intrinsic motions caused by heat. Thus, he postulated that complex organs and live beings were nothing but networks of these sensible fibres. He likened them to the spider communicating with its surroundings by feeling the pull and tear of its web; to stress that there is no limit to the complexity of these networks he said in jest that there might be a giant spider somewhere whose web is the entire Universe. To him these categories we create of inanimate matter, of sensible beings or conscious ones were artificial, like names we give from the top of a mountain to the peaks that rise above the low clouds and the fog. When this fog dissipates, we realize that all the peaks are linked together by other peaks and valleys, forming a continuous range where subdivisions and labels become meaningless.

Albertus: I do feel a true affinity toward these suggestions of Diderot, although I must confess that, unlike him, I would be reticent to put them in writing. I do not know whether it is due to some past experiences, but I have met individuals who believed that the stones had consciousness and that they had the means to communicate with them. There was no way to bring them to reason. I have also said that the stones have consciousness, but I am exaggerating, only because exaggerations are an expedient and concise way to stress a point of view. Nevertheless, let us grant that there is this pervading consciousness in all of matter. When we contemplate our own perfectibility or our capacity to transform the immediate world around us, we think that our awareness allows us to act purposefully, designedly.

On the other hand, we have witnessed the effects of much evolution, at least on the scale of the Earth's history, guided, for the most part, by random changes and natural selection. We have drawn an artificial distinction, as if an exception has been made with us, so that our awareness enables us to effect changes outside of the natural order. But we certainly would not pretend to place ourselves outside, nor at the end of the process of evolution. Why do we not state that all of evolution has taken place with some kind of awareness, that our awareness is Nature's eternal awareness, expressed in human form at this place and at this moment in its course? Why could we not admit that our doings, on our behalf or for the purpose of changing our surroundings, are also part of Nature's doings, that we are only one link in this chain of transformation and evolution? These questions may throw us into an unsolvable quandary, but we also recognize that man has always gained access to a more remarkable vantage point, offering the view of an even wider landscape, whenever he has abdicated any central position in the Universe, either in space, in time or in the general hierarchy of things. I speak of a quandary for this reason, that if Nature proceeds randomly and blindly in its process of evolution, bringing forth whatever chance seems to favour, then our tinkering with our own nature or with the world at large would cause little effect, being part of the large mixture of trials and rehearsals. On the other hand, if Nature proceeds advisedly and on purpose, then we could not fail to do its bidding, regardless of what we do, or how we decide to do what we do. In either case, I do not see much room to flatter our sense of responsibility or self-importance. At all events, we have now learnt that the Universe is not a frightful drama being staged for the first time. Its history is infinite and it is burdened with innumerable experiments. One of these produced, out of the condensation of a trace of cosmic ash, a minuscule solar system and the rise of mankind on this orbiting rock we call the Earth. This experiment will one day come to a close and, having run its course, it will return once again to cosmic ash. It is a sobering thought, but it is also comforting to think that we are what the world has always been. Amidst an infinity of such experiments, we must conclude that every particle we are made of must have already been everywhere it could have been. Every atom that I call myself has been in the core of a star and in the tail of a comet. We have been the tusk of an elephant and the wing of a butterfly. We have been the grass in the meadow and the sea-shell in the ocean. This is why I spoke of the consciousness of all matter. Is it not sensible to presume that each atom preserves its own identity by the memory of what it has been? Perhaps all these ideas are just pure fantasy. We have already seen in our long conversation that most theories about the Universe have been proven wrong and were eventually superseded by others. How could we claim an exception for our own? In any case, this should be no discouragement to continue searching.

Plebeius: Perchance one day we will realize that we are only an elaborate dream in the slumber of a sleeping god, in which case we will need again to think of some new and implausible fantasy to keep us busy and prevent him from waking up. Otherwise, our demise would be just as certain.

Albertus: In the mean time, Plebeius, let us forego considering any other of our fantasies until a future meeting. The hour is late, dusk has already fallen and we will barely be able to find our way out of this park.

Plebeius: You should not worry, Albertus, I often come for a stroll in the evening and know the path well. Have you noticed how all breezes cease shortly after sunset? It is always the most calm and placid moment of each day. Follow me, I will lead the way.

Index

red giants, 248

redshift, 276

    cosmological, 308

Regiomontanus, Johannes (1436-1476), 111

Relativity, 231–236

Remus, Johannes, 158, 165, 167

Rheticus, Georg (1514-1574), 122

Riccioli, Giovanni Battista (1598-1671), 169

Richard of Wallingford (*ca.* 1292-*d.* 1335), 96

Ritchey, George (1864-1945), 265

Roberts, Isaac (1829-1904), 263

Robertson, Howard Percy (1903-1961), 308

Röntgen, Wilhelm (1845-1923), 228

Rome's decline, 45

Römer, Ole (1644-1710), 171, 181

Royal Society, 167

Russell, Henry Norris (1877-1957), 249

Sandage, Allan (*b.* 1926), 284

Sarpi, Paolo (1552-1623), 152

save the phenomenon, 31

scholasticism, 88

Schwarzschild, Karl (1873-1916), 248

science & democracy, 330–333

science & responsibility, 326–328

Secchi, Angelo (1818-1878), 247

Seeliger, Hugo von (1849-1924), 243

Seleucus of Babylon (*fl. ca.* 150 B.C.), 30

Shapley, Harlow (1885-1972), 249, 252, 258, 273

al Shatir (*ca.* 1305-*ca.* 1375), 81, 121

de Sitter, Willem (1872-1934), 282, 308

Slipher, Vesto (1875-1969), 266

solar system, motion, 193, 240, 243, 255

spectroscopy

    nebulae, 262, 266
    stars, 247

spiral nebulae, 254, 261

stars

    distribution, 194, 238, 243, 253
    parallax, 192, 201
    variable, 205, 250

Strato of Lampsacus (*d. ca.* 268 B.C.), 28

Streete, Thomas (1622-1689), 175

Struve, Friedrich Wilhelm (1793-1864), 201, 239, 258

Struve, Otto Wilhelm (1819-1905), 240

al Sufi (903-986), 72

Sun's distance, 28, 43, 155, 157, 167, 170, 189

supergiants, 270

Sylvester II (*ca.* 930-*d.* 1003), 75, 82

Synesius of Ptolemais (*ca.* 370-*ca.* 410), 58